Biodiversity Conservation and Sustainable Development

Biodiversity Conservation and Sustainable Development

Dr. I.Sundar

Associate Professor of Economics
Directorate of Distance Education
Annamalai University
Tamil Nadu, India

SERIALS PUBLICATIONS
NEW DELHI (INDIA)

First Published - 2010

ISBN: 978-81-8387-410-6

Published by

SERIALS PUBLICATIONS
4830/24, 1st Floor, Ansari Road
Darya Ganj, New Delhi-110002 (India)
Phone : 23245225. Fax : 91-11-23272135
E-mail : serials@sify.com
serialspublications.india@gmail.com
Website : www.serialspublications.com

Contents

Preface

Biodiversity is the variety of plant and animal life found in an ecosystem and the variation in their genetic makeup. Biodiversity is a measure of the health of an ecosystem, with healthy ecosystems having greater variety and variation in plant and animal life than unhealthy ones. It is a large number and wide range of species of animals, plants, fungi, and microorganisms. Ecologically, wide biodiversity is conducive to the development of all species. Biodiversity, a contraction of the phrase "biological diversity," is a complex topic, covering many aspects of biological variation. In popular usage, the word biodiversity is often used to describe all the species living in a particular area. If we consider this area at its largest scale-the entire world - then biodiversity can be summarized as "life on earth." However, scientists use a broader definition of biodiversity, designed to include not only living organisms and their complex interactions, but also interactions with the abiotic aspects of their environment. Definitions emphasizing one aspect or another of this biological variation can be found throughout the scientific and lay literature. For the purposes of this module, biodiversity is defined as the variety of life on Earth at all its levels, from genes to ecosystems, and the ecological and evolutionary processes that sustain it.

There is also an important spatial component to biodiversity. The structure of communities and ecosystems for example the number of individuals and species present can vary in different parts of the world. Similarly, the function of these communities and ecosystems i.e. the interactions between the organisms present can vary from one place to another.

Different assemblages of ecosystems can characterize quite diverse landscapes, covering large areas. These spatial patterns of biodiversity are affected by climate, geology and physiography.

The structural, functional, and spatial aspects of biodiversity can vary over time; therefore there is a temporal component to the analysis of biodiversity. For example, there can be daily, seasonal, or annual changes in the species and number of organisms present in an ecosystem and how they interact. Some ecosystems change in size or structure over time (e.g. forest ecosystems may change in size and structure because of the effects of natural fires, wetlands gradually silt up and decrease in size). Biodiversity also changes over a longer-term, evolutionary, time-scale. Geological processes changes in sea-level.

The history of origin and the process of evolution have led to the diversity of life forms. The increase in environmental awareness over the last few decades has underlined the need to enhance our understanding of the ways in which human race and biodiversity interact. It is surprising that till 1992, the subject of biodiversity was never given the attention that it deserved, both at the national and the international level. It was only in June 1992, at the United Nations Conference on Environment and Development (UNCED) that the issue of biological diversity was taken up as a priority area for action for conservation and sustainable use. However, preparations for a Convention on Biological Diversity (CBD) were initiated by the Governing Council of the United Nations Environment Programme (UNEP) in 1987 through the establishment of an Ad Hoc Working Group of Experts on Biological Diversity, which met in 1988. The Ad Hoc Working Group was followed in 1991 by an Inter-Governmental Negotiating Committee for devising a Convention on Biological Diversity. The agreed text of the CBD was adopted by 101 governments in Nairobi in May 1992 and was signed by 159 governments and the European Union at the UNCED held in Rio de Janeiro in June 1992. The Convention finally entered in to force on December 29, 1993, which has since then been signed by 184 countries.

India became a signatory to the Convention by signing it on June 5, 1992.

The main objectives of the CBD, to be pursued in accordance with its relevant provisions, include conservation of biological diversity, sustainable use of its components and the fair and equitable sharing of benefits arising out of the utilisation of genetic resources by means of appropriate access to genetic resources, appropriate transfer of relevant technologies, and appropriate funding.

This book is organized in 15 chapters. The first chapter deals with biodiversity - concept meaning and Functions. This chapter discusses functions of bio diversity, species, functional types, and composition of biodiversity, ecosystem responses to changes in biodiversity. biodiversity and ecosystem stability, predictability and reliability, affiliated ecosystems, economics and biodiversity, importance of biodiversity, soils formation and protection, contribution to climate stability, maintenance of ecosystems and social benefits. The second chapter discusses agricultural biodiversity

It covers topics such as food crops and biodiversity, source regions of major crops, introduced crops, agriculture and genetic diversity, genetic engineering genetic diversity and livestock breeding genetic improvement of forest species risks of high-yielding crop varieties, the value of biodiversity and soil biodiversity. The third chapter is Indian Biodiversity. It narrates about Wetlands Forests Marine Environment Species Diversity Endemic Species, Threatened Species Protected Areas of the Western Ghats Coastal and Marine Biodiversity of India and Threats to coastal and marine biodiversity. The fourth chapter is biodiversity and food security. It analyses genetic erosion in agriculture, preserving options for the future, genetic inter-dependence, farmers as innovators and conservers of diversity, biodiversity and indigenous knowledge, seeds of survival dangers of genetic uniformity and importance of crop genetic diversity. The fifth chapter discusses about biodiversity loss. It covers topics such as massive extinctions from human activity, declining ocean biodiversity, people and forests, misuse of land and resources, long term costs, species at higher trophic levels and species with chronically small populations. The sixth chapter deals with climate change and biodiversity. It explains climate change affects all biodiversity, The links between climate change and biodiversity, Biodiversity, land use and climate change, Poverty, climate change and biodiversity, climate change, conservation and development, Biodiversity and climate change and research priorities. The eighth chapter deals with people's biodiversity register with basic behavioral issues, revealed preference and biodiversity conservation, objectives of the people biodiversity register and benefit sharing. The ninth chapter discusses about biodiversity economic valuation and this chapter discusses economic valuation techniques, travel cost method, valuing changes in environmental quality at a site, basic model of ITCM, application of the individual travel cost approach, advantages of the travel cost method, evaluation of travel cost method and random utility model. The tenth chapter is biodiversity measurement. It discusses species diversity: measures and scales, species richness, functional diversity and the niche preemption model or the geometric model. The eleventh chapter narrates about biodiversity survey techniques and this chapter explains biodiversity survey, standardized data collection methods, scat collection and point count method. The twelfth chapter is bioregional planning. It points out characteristics of bioregional planning, challenges to bioregional managers, integrated strategic planning, protected areas in a bioregional framework, bioregional planning frameworks and bioregional planning in India. The thirteenth chapter is gender and biodiversity. It deals with Gender Dimensions of Biodiversity Management, *FAO, gender and biodiversity*, Ecofeminism and Biodiversity and Global Action for Women towards Sustainable and Equitable Development. The fourteenth chapter is indigenous knowledge and biodiversity. The fifteenth chapter is biodiversity education. It deals with biodiversity education methods.

I express my sincere gratitude to my revered Vice Chancellor Dr. M.Ramanathan and my beloved Director

S.B. Nageshwara Rao, Directorate of Distance Education, Annamalai University who have been the source of constant inspiration. I express my sincere gratitude to Dr. N.Malathi, Professor and Head, Department of Economics, Annamalai University for her motivation in writing this book. I thank profusely Dr. D. Namasivayam, Professor and Economics Wing Head, Directorate of Distance Education, Annamalai University for for his constant encouragement in my academic career.

I thank and share my happiness and love with my beloved wife **Mrs. S. Chithra,** Secretary, Centre for Sustainable Development Trust, Chidimbaram Talik, Tamil Nadu and my beloved son **S. Naveen Nivas** for their encouragement and unbounded cooperation towards writing this book.

It is with heart and soul that I thank to Shri. S.K.Jha, Proprietor Serials Publications, New Delhi for evincing keen interest in publishing this book.

Dr. I. Sundar

1

Bio Diversity - Concept Meaning and Functions

Biodiversity is an all encompassing term to describe the variety of all life and natural processes on Earth. The Convention on Biological Diversity defines biodiversity as "the variability among living organisms from all sources this includes diversity within species, between species and of ecosystems". Biodiversity is synonymous with all things 'natural' and worth caring for in the environment such is the generality with which the term is used. According to the United Nations Convention on Biological Diversity, the "... variability among living organisms". Hence, the term relates specifically to variability rather than to the overall extent of the biological resource.

This variability can be considered at four levels:

- Genetic in terms of genes, nucleotides, chromosomes, individuals;
- Species in terms of kingdom, phyla, families, subspecies, species, populations;
- Ecosystem in terms of bioregions, landscapes, habitats; and, Functional in terms of ecosystem robustness, resilience, goods and services.
- Hence, the term can be applied to the variability of genetic material through to landscape diversity.

It is also important to recognise that variability and overall magnitude are intrinsically linked. It is conceivable that genetic variability could be protected in intensively managed small-scale facilities such as zoos and herbariums and that mini landscape preserves could be established to provide representations at the ecosystem level. But to ensure resilience and robustness at the functional level it is difficult to imagine a protection regime that does not involve scale as well as diversity. For instance, resilience implies an ability to recover from an externally imposed shock. Almost by definition, a small-scale attempt to protect biodiversity will be at greater risk from an external shock. Even at the genetic and species levels, the protection of variability in the biological resource is afforded by larger scale protection initiatives. Put simply, scale and variability in the biological resource are likely to be strongly complementary. It is, perhaps, because of this complementarity that biodiversity has taken on such an extensive mantle in the nature conservation debate.

Functions of Bio Diversity

Ecosystem functioning reflects the collective life activities of plants, animals, and microbes and the effects of these activities. Feeding, growing, moving, excreting waste, etc. have on the physical and chemical conditions of their environment. Note that functioning means showing activity and does not imply that organisms perform purposeful roles in ecosystem-level processes. A functioning ecosystem is one that exhibits biological and chemical activities characteristic for its type. A functioning forest ecosystem, for example, exhibits rates of plant production, carbon storage, and nutrient cycling that are characteristic of most forests. If the forest is converted to an agroecosystem, its functioning changes. Ecologists abstract the essential features of an ecosystem into two compartments, the biotic and the abiotic. The biotic compartment consists of the community of species, which can be divided functionally into plant producers, the consumers that feed on producers and on each other, and the decomposers. The abiotic compartment consists of organic and inorganic nutrient pools. Energy and materials move between these two compartments, as well as into and out of the system. Ecosystem processes are quantified by measuring rates of these movements e.g., plant production, decomposition, nutrient leaching or other measures of material production and transport or loss. Ecosystem functioning, in turn, is quantified by measuring the magnitudes and dynamics of ecosystem processes.

Ecologists describe Ecosystem functioning results from interactions among and within different levels of the biota as a nested hierarchy. For example, green plant production on land is the end product of interactions of individual plants nested within populations; interactions among populations nested within a single species; interactions among a variety of species nested within a group of functionally similar species; and so on up to the level of interactions between different types of ecosystems nested within landscapes.

Biodiversity: Species, Functional Types, and Composition

Every organism contributes to ecosystem processes, the nature and magnitude of individual contributions vary considerably. Research in biodiversity places much emphasis on the uniqueness of individual species and their singular contributions to ecosystem services. Yet most ecosystem processes are driven by the combined biological activities of many species, and it is often not possible to determine the relative contributions of individual species to ecosystem processes. Species within groups such as grazing mammals, large predators, perennial grasses, or nitrogen-fixing microbes may therefore be functionally similar despite their uniqueness in genes, life history, and other traits. Groups of species that perform similar roles in an ecosystem process are known as functional types or functional groups. Species may also be divided into functional types based on what they consume or by trophic status e.g., their place in the food web as producers, decomposers and predators. Within trophic groups, species may be further divided according to life history, climatic or nutrient needs, physiology or other biological traits. Researchers may place a species into several different functional categories depending on the ecosystem process they are studying.

Figure 1- Magnified portion of the biosphere showing its position within the three major biogeochemical spheres. Driven primarily by solar energy, producers, decomposers, and consumers annually move large quantities of materials containing many elements and compounds among the different spheres. The role of the tremendous diversity found within the biosphere is only recently beginning to be understood.

Figure 2- The Role of Earth's Biota in Biogeochemical Processes

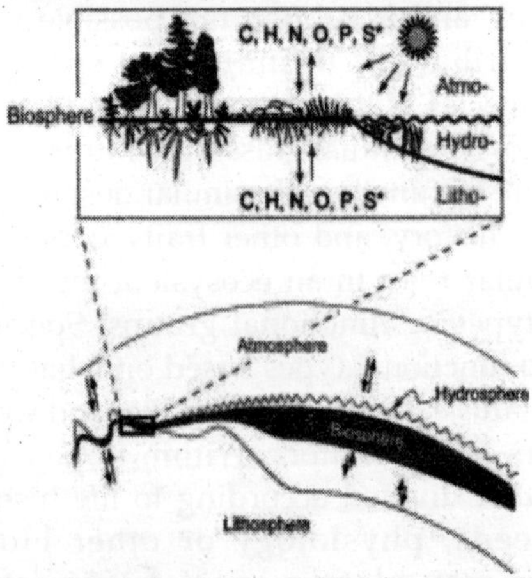

Figure 3- The major biogeochemical spheres consist of the lithosphere, hydrosphere, and atmosphere, where the biosphere is found.

Because species can vary dramatically in their contributions to ecosystem functioning, the specific composition or identity of species in a community is important. The fact that some species matter more than others becomes especially clear in the case of .keystone species. or .ecosystem engineers or organisms with high .community importance values.. These terms differ in usage, but all refer to species whose loss has a disproportionate impact on the community when compared to the loss of other species. For example, a species of nitrogen-fixing tree, Myrica faya, introduced to the Hawaiian Islands has had large-scale effects on nitrogen cycling, greatly increasing the amount of this essential plant nutrient in soils where the tree invades. The nitrogenfixing lupine Lupinus arboreus also enriches soils and, as a consequence, encourages invasions of weedy grasses. Among animals, moose i.e. Alces alces through their dietary preferences greatly reduce soil nitrogen levels and also influence the succession of trees in the forest. Beavers, too, through their feeding and dam-building not only alter soil fertility and forest succession but increase the diversity of ecosystems in a landscape. Even termites play critical roles in soil fertility and other ecological processes in many arid grasslands. On the other hand, there are some examples where additions or losses of particular species have had little effect on ecosystem processes.

Ecosystem Responses to Changes in Biodiversity

Since Darwin, prominent biologists have hypothesized about the relationship between biodiversity and ecosystem functioning. More recently, concerns about increasing loss of biodiversity and questions about resulting degradation of ecosystem services have stimulated unprecedented observational, theoretical, and experimental studies.

Observational Studies

It might seem that observational studies comparing one ecosystem type with another, or comparing similar ecosystems at different locations, could provide ready answers to questions about the impacts of species richness on ecosystem processes. But these studies have invariably proven problematic. For example, an ecosystem such as a tropical forest or a coastal wetland may vary from one site to another not only in species number and composition, but also in physical and chemical conditions such as soil type, slope, rainfall, or nutrient levels. Comparing different ecosystems is likely to yield an unclear result because the response to variations in biodiversity cannot easily be distinguished from responses caused by variations in environmental and other factors. It is possible, though difficult, to control statistically for such potentially confounding factors.

Experimental Studies

Experimental studies, if well-designed, can minimize the confounding factors that plague observational studies. Experiments can provide insights not only into the relationships between biodiversity and ecosystem functioning but also into the possible mechanisms behind the relationships. Studies to date have ranged from large outdoor experiments and trials in large controlled environment facilities to modest-sized pot experiments and tests in small laboratory microcosms (Figure). This research has attempted to address two different questions about the link between biodiversity and ecosystem functioning. First, how are levels of ecosystem functioning affected by changes in biodiversity, particularly species richness? Second, how are the dynamics of ecosystem functioning, particularly the resilience and stability of processes, affected by changes in biodiversity?

The following two sections review the experimental and theoretical results that shed light on these questions.

Biodiversity and Levels of Ecosystem Functioning

Results from many recent experimental studies conducted in North America and Europe demonstrate that ecosystem productivity increases with species richness. These studies range from large outdoor experiments to controlled laboratory experiments conducted in growth chambers, greenhouses, or small containers. Outdoor experiments such as those conducted in grasslands on nutrient-poor serpentine soils at Stanford, California and on prairie grasslands at Cedar Creek Natural History Area, Minnesota

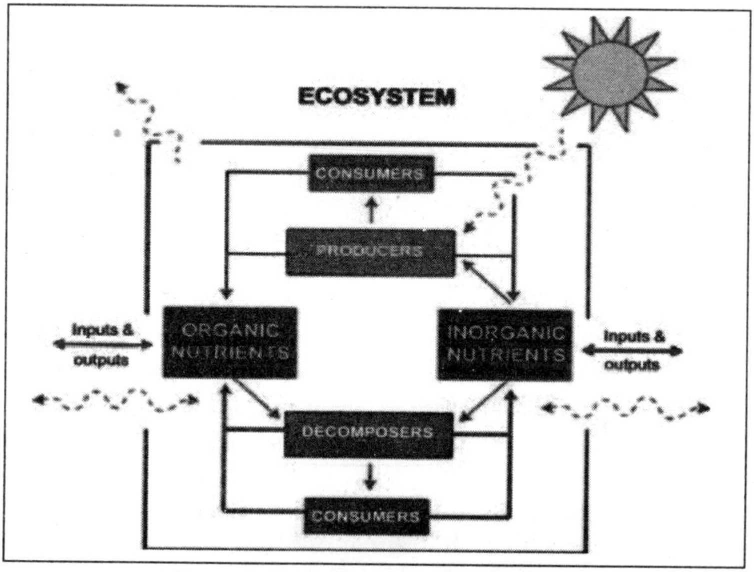

Figure 4- Basic Ecosystem Functioning

Producers acquire energy through photosynthesis and take up inorganic nutrients to produce living biomass, forming the food base for consumer species such as herbivores and their predators. Mortality leads to accumulation of organic nutrients which are transformed by decomposers into living biomass, forming the food base for consumers. Decomposers and consumers contribute to formation of inorganic nutrients by mineralization, completing the cycling of nutrients between organic and inorganic forms. Energy flows shown in the form of wavy, dashed lines begin with acquisition by producers and end in loss due to the respiration activities of all organisms. More recent laboratory experiments in Europe and North America have begun to examine the impact of other components of biodiversity, such as the diversity of soil microorganisms, on plant production and the role of bacteria, predators, and herbivores in freshwater microbial communities.

All of these studies show that ecosystem functioning is decreased as the number of species in a community decreases. Declines in functioning can be particularly acute when the number of species is low, such as in most managed ecosystems including croplands or timber plantations. In addition, recent experimental studies in grasslands indicate that the effects of biodiversity on production can depend on both the number of functional groups present and the identity of the plant species i.e., on community composition. Other studies have shown that loss of functional groups from a food web, or reductions in the number of species per trophic group producers, consumers and decomposers can also cause declines in ecosystem functioning. Finally, another study has shown that some species of plants may be more or less productive or show no response at all to changes in the diversity of their communities, even though total community productivity is, on average, lower at lower diversity. Studies

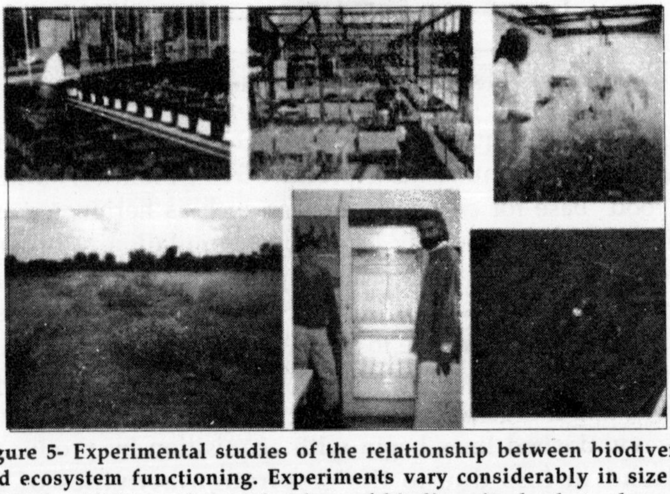

Figure 5- Experimental studies of the relationship between biodiversity and ecosystem functioning. Experiments vary considerably in size and methods. All have shown that loss of biodiversity leads to decreased ecosystem functioning. Clockwise from upper left, a researcher examining growth in potted assemblages of plants, transplanted Mediterranean vegetation in greenhouses, model ecosystems consisting of plants and small invertebrates in a growth chamber, field experiment in Switzerland, microbial microcosms in a growth chamber, and a field experiment in England.

on plants have been particularly revealing and support results from recent theoretical models which predict that decreasing plant diversity leads to lower plant productivity. These models predict that diversity and composition are approximately equal in importance as determinants of ecosystem functioning. Two possible mechanisms have been identified to explain why levels of ecosystem functioning increase with increasing biodiversity. First is the .sampling effect.: When the pool of species available in a region contains individual species that vary in productivity and other contributions to ecosystem functioning, then species-rich ecosystems have a higher probability of containing species with high levels of functioning. Second is the .Complementarily effect.: This occurs when increasing diversity results in increasing numbers of species that are complementary rather than competitive in their use of resources, exploiting different niches, such as rooting depths, and allowing more effective use of available resources.

Biodiversity and Ecosystem Stability, Predictability and Reliability

Few experimental studies of the impact of biodiversity on stability have been attempted, largely because stability is a long term attribute of a system and testing for it requires either long-running experiments or experiments with short-lived organisms. In the one available long-term ecological field study, however, reductions in plant species richness also lowered the resistance of grassland production to drought. Predictably lower, year-to-year fluctuations in community productivity was also significantly lower at lower diversity. In addition, studies of microbial communities in small experimental chambers have also shown that fluctuations in ecosystem functions such as productivity can be greater when species richness is reduced. Thus, the loss of diversity causes a loss of ecosystem stability.

Several mechanisms could account for these results. One mechanism comes from the ability of competing species to replace or compensate for one another and thus minimize, at higher diversity, the ups and downs in functioning. Another mechanism is the .portfolio effect,. a theory which suggests that cumulative properties such as ecosystem functioning show less severe fluctuations in systems with many species, much the way investment portfolios of varied stocks have lower long term variance than portfolios of one or a few kinds of stocks.

Three points emerge from this growing body of research. First, declining species richness can lead to declines in overall levels of ecosystem functioning. This is especially pronounced at lower levels of diversity. This finding is particularly relevant to current ecological change, since most ecosystems are being transformed into managed systems which typically contain only a few dominant species, whereas the natural ecosystems they replaced typically contained tens to hundreds of species. Second, at least one species per functional group is essential to ecosystem functioning. Having more than one species per functional group may or may not alter overall levels of ecosystem functioning, but it may nevertheless insure against loss of functioning in times of disturbance if species within functional groups are able to replace or compensate for one another. Third, the nature of an ecosystems response to declining biodiversity is dependent on community composition, that is, on which species are lost and which . Biodiversity: What is it, where is it, and why is it important?

Coral Reefs

Coral reefs and their associated communities cover an estimated 600,000 square kilometers, mostly between the Tropic of Capricorn and the Tropic of Cancer. Coral reefs represent less than 0.2 percent of the total area of oceans and cover an area equivalent to 4 percent of the world's cropland area. Coral reefs are found in shallow waters, extending to depths of 30 meters and cover 15 percent of the world's coastline.

Locally, coral reefs are even more important as a food source. Throughout southeast Asia, coral reef fisheries provide 10 to 25 percent of the protein available to people living along the coastlines. Coral reefs also protect coastal areas from erosion. In the case of coral atolls, coral provides the foundation of the island itself. In the Indian Ocean, 77 percent of isolated islands and island archipelagoes are built exclusively of reef depositions.

Coral reefs resemble tropical rainforests in two ways:

- Both thrive under nutrient-poor conditions where nutrients are largely tied up in living matter, yet support rich communities through incredibly efficient recycling processes.
- Both exhibit very high levels of species diversity.

Coral reefs and other marine ecosystems, however, contain more varied life forms than do land habitats. All but one of the world's 33 phyla (major kinds of organisms) are found in marine environments–15 exclusively so. Coral reefs are noted for some of the highest levels of total gross productivity on earth.

- The net primary productivity of coral reefs is approximately 2,500 grams of carbon per square meter per year, compared to 2,200 grams of carbon per square meter per year for tropical forests and only 125 grams of carbon per square meter per year in the open ocean.
- Coral polyps–the thin living layer covering reef structures–provide much of the energy that fuels these communities. These tiny animals contain algae, which convert sunlight to fuel, deriving nutrients from polyp wastes in the process.
- Reef-building corals and certain calcareous algae which may constitute more than half of a reef's stony substance lay down a foundation of calcium carbonate. Over generations this accumulation results in often massive structures, providing homes and hiding places for countless other creatures. Coral reefs, then, are the net result of thousands of years of growth. As such, many are among the planet's oldest living communities.

Affiliated Ecosystems

Coral reefs stand out from other marine environments because of their species diversity, but many coral reef species also depend on other affiliated ecosystems.

Often, coral reefs, mangroves, and sea grass beds are linked physically and biologically: reefs serve as breakwaters that allow coastal mangroves to develop; the calcium of the reef provides the sand and sediment in which mangroves and sea grasses grow; and the mangroves and sea grass communities provide energy input into the coastal ecosystem and serve as spawning, rearing, and foraging habitat for the many of the species associated with the reefs.

The link and dependency between plants, bees, and human agriculture is so crucial, the two scientists writing up years of research into the problem summarized with this warning: Humankind needs to act quickly to ensure that the ancient pact between flowers and pollinators stays intact, to safeguard our food supply and to protect our environment for generations to come. These efforts will ensure that bees continue to provide pollination and that our diets remain rich in the fruits and vegetables we now take for granted.

It noted how a few decades ago, some fishermen campaigned for killing whales because they were threatening the fish supply and thus jobs.

A chain of events eventually came full circle and led to a *loss* of jobs:

- The massive reduction in the local whale population meant killer whales in the region usually preying on younger whales moved to other animals such as seals;
- As seal numbers declined, the killer whales targeted otters;
- As otter numbers were decimated, the urchins and other targets of otters flourished;
- These decimated the kelp forests where many fish larvae grew in relative protection;
- The exposed fish larvae were easy pickings for a variety of sea life;
- Fishermen's livelihoods were destroyed.

Biodiversity providing Lessons for Scientists in engineering

For a number of years now, scientists have been looking more and more at nature to see how various species work, produce, consume resources, trying to mimic the amazing feats that millions of years of evolution have produced. As just one small example, *some spiders can produce their silk with a higher tensile strength than many alloys of steel* even though it is made of proteins. So biologists are looking at these processes in more depth to see if they can reproduce or enhance such capabilities.

Economics and Biodiversity

From the perspective of an economist, biological diversity or 'biodiversity' is of interest for two fundamental reasons. First, biodiversity is valuable to society. That is, the greater the biodiversity we have, the better off we are and if we lose some biodiversity, we consider ourselves to be worse off. Second, choices made by society have made and are continuing to have effects on biodiversity. That is, some of the resource use decisions made by society–albeit inadvertently – have left us with less biodiversity. Clearing land for agriculture, harvesting timber from forests, draining wetlands for housing estates for example, have caused depletions in biodiversity. Putting these two reasons together lead the economist to conclude that biodiversity is a scarce and valuable resource. And for an economist, that means their discipline has something to contribute to the biodiversity debate, simply because the focus of economics is on the analysis of the ways societies make choices about their scarce and valuable resources. The primary goal here is therefore to provide an understanding of what economics has to say about biodiversity. Because the overriding goal of economics is to deliver choice solutions that make society better off, a strong emphasis will be given to the value of biodiversity because it is the creation of value that makes society better off. Specifically, three questions will be addressed:

1. Why is biodiversity valuable?
2. How can the value of biodiversity be estimated? and,
3. How can the value of biodiversity be delivered to society?

The goal is not to advocate or promote any particular type of analysis. Rather, it is intended that we will give policy makers a background on what economics has to offer, with both the strengths and weaknesses of the economics approach being highlighted. This, it is hoped, will be a base for policy makers to determine what role economic analysis should play in the consideration of resource use choices that have biodiversity impacts. If a role is deemed appropriate, the secondary goal is to form a foundation for decisions regarding elements of the application and further development of economic analysis of biodiversity.

We begin with an attempt to define biodiversity. 'Attempt' is used advisedly here because it is immediately apparent that biodiversity is a complex concept that generates definitional conundrums. The importance to society and the policy significance of biodiversity are addressed to answer the question: 'Why is biodiversity valuable?' What economists have done to estimate the value of biodiversity. While the lists of valuation techniques developed and applications performed internationally are extensive, it is also apparent that the discipline does not find itself standing on 'safe' ground in this area. Some of the specific issues that have caused concerns both within and outside the economics profession are overviewed, especially in the light of observations made regarding the relevance of applications to policy making. Briefly, the third question of how the value of biodiversity can be delivered to society.

Importance of Biodiversity

Today, as ever, human beings are dependent for their sustenance, health, well-being and enjoyment of life on fundamental biological systems and processes. Humanity derives all of its food and many medicines and industrial products from the wild and domesticated components of biological diversity. Biotic resources also serve recreation and tourism, and underpin the ecosystems which provide us with many services. While the benefits of such resources are considerable, the value of biological diversity is not restricted to these. The enormous diversity of life in itself is of crucial value, probably giving greater resilience to ecosystems and organisms. Biodiversity also has important social and cultural values.

The Value of Biological Diversity's Components

Generally, benefits arising from the conservation of components of biological diversity can be considered in three groups: ecosystem services, biological resources and social benefits. Some examples of these benefits follow.

Ecosystem Services

Protection of Water Resources

Natural vegetation cover in water catchments helps to maintain hydrological cycles, regulating and stabilising water runoff, and acting as a buffer against extreme events such as flood and drought. Vegetation removal results in siltation of catchment waterways, loss of water yield and quality, and degradation of aquatic habitat, among other things. Vegetation also helps to regulate underground water tables, preventing dryland salinity which affects vast areas of agricultural lands, at great cost to the community. Wetlands and forests act as water purifying systems, while mangroves trap silt, reducing impacts on marine ecosystems.

These services translate into substantial financial benefits. Australian Government sponsored study, for example, calculated the financial benefit of water supplied to Melbourne from forested catchments at $250 million per year. This amount is based on a study which valued water collected in the Thomson Reservoir and supplied to Melbourne at $530 per megalitre, and the fact that the bulk of water supplied to Melbourne is harvested from 80 000 ha of catchment forested with ash-type eucalypts. Annual water yields from these forests vary from six to twelve megalitres per hectare, depending on whether the forest is 30 year old regrowth or old growth more than 200 years old. Presently most of the ash-type eucalypt forest in the catchments is 54 years old. Over the next 50 to 100 years, as the regrowth forests age, the value of water produced each year will increase by $150 million due to natural streamflow increases.

Soils Formation and Protection

Biological diversity helps in the formation and maintenance of soil structure and the retention of moisture and nutrient levels. The loss of biological diversity through clearing of vegetation has contributed to the salinisation of soils, leaching of nutrients, laterisation of minerals and accelerated erosion of topsoil, reducing the land's productivity. Trees, on the other hand, lower the water table and remove deposited salt from the upper soil horizons. Soil protection by maintenance of biological diversity can preserve the productive capacity of the soil, prevent landslides, safeguard coastlines and riverbanks, and prevent the degradation of coral reefs and riverine and coastal fisheries by siltation.

Trees and other vegetation also assist in soil formation. A significant contribution is the introduction of organic matter through litter formation and the decay and regeneration of tiny fibrous roots, both of which facilitate microbial activity. Another contribution is through the effects of root systems which break up soil and rock leading to, amongst other things, penetration of water. Root systems also bring mineral nutrients to the surface through root uptake. Organic matter formed by the decay of tiny fibrous roots can also bind with minerals, such as iron and aluminium, which can reduce the potential deleterious effects of these minerals on other vegetation.

Nutrient Storage and Cycling

Ecosystems perform the vital function of recycling nutrients. These nutrients include the elements of the atmosphere as well as those found in the soil, which are necessary for the maintenance of life. Biological diversity is essential in this process. Plants are able to take up nutrients from the soil as well as from the air, and these nutrients can then form the basis of food chains, to be used by a wide range of other life forms. The soil's nutrient status, in turn, is replenished by dead or waste matter which is transformed by microorganisms; this may then feed other species such as earthworms which also mix and aerate the soil and make nutrients more readily available.

Pollution Breakdown and Absorption

Ecosystems and ecological processes play an important role in the breakdown and absorption of many pollutants created by humans and their activities. These include wastes such as sewage, garbage and oil spills. Components of ecosystems from bacteria to higher life forms are involved in these breakdown and assimilative processes. Excessive quantities of any pollutant, however, can be detrimental to the integrity of ecosystems and their biota. Some ecosystems, especially wetlands, have qualities that are particularly well suited to breaking down and absorbing pollutants. Natural and artificial wetlands are being used to filter effluents to remove nutrients, heavy metals and suspended solids, reduce the biochemical oxygen demand and destroy potentially harmful microorganisms.

Contribution to Climate Stability

Vegetation influences climate at the macro and micro levels. Growing evidence suggests that undisturbed forest helps to maintain the rainfall in its immediate vicinity by recycling water vapour at a steady rate back into the atmosphere and through the canopy's effect in promoting atmospheric turbulence. At smaller scales, vegetation has a moderating influence on local climates and may create quite specific micro-climates. Some organisms are dependent on such micro-climates for their existence.

Maintenance of Ecosystems

Ecosystem relationships resemble a web of connections from one living thing to many other living and non-living things. They not only allow survival, but also maintain a balance between living things and the resources such as food and shelter they need to survive. Vegetation is integral to the maintenance of water and humidity levels and is essential for the maintenance of the oxygen and carbon dioxide balance of the atmosphere. Due to the complex nature of ecosystem relationships, the removal or disturbance of one part of the ecosystem could affect the functioning of many other components of the ecosystem. Our knowledge of these relationships is incomplete, and the results of disturbance are thus to some extent unpredictable.

Maintaining natural habitats helps ecosystem functions over a wider area. Natural habitats afford sanctuary to breeding populations of birds and other predators which help control insect pests in agricultural areas, thus reducing the need for, and cost of artifical control measures. birds and nectar loving insects roost and breed in natural habitats may range some distance and pollinate crops and native flora in surrounding areas.

Recovery from Unpredictable Events

Maintaining healthy ecosystems improves the chances of recovery of plant and animal populations from unpredictable natural catastrophic events such as fire, flood and cyclones and from disasters caused by humans. Inadequately conserved and isolated populations, and ecosystems which are degraded, are less likely to recover or to recover as quickly, to their former state. Populations of biota may end up with small, possibly non-viable, genetic bases, which can lead to extinctions.

Biological Resources Food

Human existence and that of most other organisms is heavily dependent on what biologists call primary producers, mainly plants. Five thousand plant species have been used as food by humans, but less than twenty now feed the majority of the world's population and just three or four carbohydrate crops are staples for a vast majority. One of the important benefits of conservation of biodiversity is the wild plant gene pool which is available to augment the narrow genetic base of these established food crops, providing disease resistance, improved productivity and different environmental tolerances.

India's native species are contributing to the global food capacity. India has important native fish and crustacean harvesting industries and is the reservoir of genetic diversity for the macadamia and quandong. Such reservoirs increase the opportunities for enhancing agricultural productivity. These may prove to be extremely valuable genetic stock in the future because, unlike current commercial varieties, many of these wild plants have genes that help them resist leaf rust diseases.

There is also great food potential in native Indian plants. The nutritional value of 'bush' foods is quite high, some having greater amounts of protein, fats, carbohydrates, minerals and vitamins than cultivated plant foods. For example, acacia seeds, some 50 types of which were used by Aboriginal people for food, are superior to rice and wheat in energy, protein and fats. The seeds of pigweed (*Portulaca oleracea*), which were commonly eaten by Aboriginal peoples, contain 20 per cent protein, 16 per cent fat, and high levels of iron. A native fig, *Ficus platypoda*, has very high levels of calcium (4000mg/100g), as well as higher protein and fat content than expected for fruits, while the wild Arnhem Land plum has spectacular amounts of Vitamin C - more than 50 times the level found in exotic citrus fruit. New chemical structures are being discovered all the time, and the conservation of biological diversity is essential for the continuation of this research.

The short and long-term values of these genetic resources are enormous and most improvements in agriculture and silviculture depend on their preservation. Moreover, the gene pool value of natural habitats will increase as remaining natural habitats become more scarce. These areas are therefore of great value as *in situ* gene banks, and need to be effectively managed.

Medicinal Resources

People have long used biological resources for medicinal purposes. Indian Aboriginal societies made use of many native plants as medicines. Examples cover many genera, and include acacias and eremophilas, as well as individual species such as *Isotoma petraea* and the parrot plant (*Crotolaria cunninghamii*). A few Aboriginal medicines have been widely used in western medicine, such as the ubiquitous eucalyptus oil for relief of respiratory tract infections, but many more are now being investigated.

A number of Indian species are the basis of medicinal products. Hyoscine or scopolamine, used to treat motion sickness, stomach disorders and the effects of cancer therapy, is a product of two species of corkwood (*Duboisia*). One of these, *D. leichhardtii*, is restricted to Queensland, and the hybrid between the two species produces more hyoscine than any other plant. The vine *Tylophora* is the source of the drug tylocrebrine, which has been effective in treating lymphoid leukemia, while the kangaroo apples *Solanum aviculare* and *S. laciniatum*, found in India and New Zealand and cultivated overseas, provide solasodine, which is easily converted to steroids. Wild plant, animal and microorganism resources are also of great importance in the search for new medically active compounds, and the potential of other

Studies of various chemicals produced by animals have led to discoveries of medicinally useful substances. A substance called Prostaglandin, which could be of importance in the treatment of gastric ulcers, was originally discovered in the two species of gastric brooding frogs (*Rheobatrachus*) found only in the rainforests of Queensland. Unfortunately, both species have not been sighted for some time, and it is conceivable that one at least (*R. silus*) is now extinct.

Wood Products

Wood is a basic commodity used worldwide, and is still largely harvested from the wild. It is a primary source of fuel, is used in construction, and forms the basis for paper production. Native plants have been of continued importance in the construction of buildings and furniture, and more recently for paper production. The timber industries form a significant part of our modern economy. The unique nature of species has been well recognised: huon pine (*Lagarostrobus franklinii*) does not decay; gidgee (*Acacia cambagei*) and mulga (*Acacia aneura sens. lat.*) are very hard woods currently being investigated for musical instruments; and the lignotubers of some eucalypts and banksias are favoured for highly decorative furniture.

Ornamental Plants

Native species are increasingly being used for ornamental and horticultural purposes, with new hybrids and strains being developed and marketed. One well known example is the *Grevillea* "Robyn Gordon". Natural areas provide support systems for commercially valuable environmental benefits and resources. Some habitats protect crucial life stages or elements of wildlife populations that are widely and profitably harvested outside these habitats, such as spawning areas in mangroves and wetlands. For example, when mangrove areas are cleared for resort and urban development, populations of commercial fish species which rely on mangroves for breeding habitats also diminish. Some of these crucial habitats have been declared protected areas, as their importance for maintaining stocks of fish, crustaceans such as prawns and mud-crabs, and other aquatic fauna has been recognised. Other habitats act as genetic reservoirs from which seed and other material can be assessed for enhancement of harvested species. The leaf oil content of several eucalyptus mallee species, for example, is being assessed from across their natural distributions to identify high yield populations for potential propagation.

Future Resources

As knowledge improves, new bio resources to increase human welfare will be discovered and developed. There is a clear relationship between the conservation of biological diversity and the discovery of new biological resources. The relatively few developed plant species currently cultivated have had a large amount of research and selective breeding applied to them. Many presently under-utilised food crops have the potential to become important in the future. The documentation of indigenous peoples' use of plants is often the source for ideas on developing plant species for wider use and economic benefit and there are a large number of as yet undiscovered plant species which could prove useful. Potential products which may be derived from biological resources include sunscreens from corals, light and high tensile fibres from spider silk, and instant adhesives from velvet worms or barnacles. Microorganisms are important in the production of extensive ranges of agrochemicals, protein for animal feed, enzymes and biopolymers.

There is also potential for further development of biotic resources for natural pesticides, similar to the insecticidal microorganism *Bacillus thuringiensis*, and other useful products such as fats, and oils. The conservation of diversity is also essential for finding effective biological control organisms and for breeding disease resistant species. Genetic engineering of microorganisms promises further advances in the production of new compounds and processes.

Social Benefits

There is still much to learn on how to get better use from biological resources, how to maintain the genetic base of harvested biological resources, and how to rehabilitate degraded ecosystems. Natural areas provide excellent living laboratories for such studies, for comparison with other areas under different systems of use, and for valuable research into ecology and evolution. Unaltered habitats are often essential for certain research approaches, providing controls against which the changes brought about by different management regimes may be measured and assessed.

Recreation

Biological diversity is an intrinsic part of many areas valued in India for tourism and recreational purposes. The aesthetic qualities of such areas are often strikingly different, in large part due to the range of biological diversity to be found on this continent. People value such areas for a variety of recreational pursuits: film, photographs or literature based on or using wildlife, natural habitats and natural features; birdwatching; and ecological field study and other scientific pursuits.

Cultural Values

The cultural value of biological diversity conservation for present and future generations is an important reason for conserving it today. Human cultures coevolve with their environment, and the conservation of biological diversity can be important for cultural identity throughout India. The natural environment provides for many of the inspirational, aesthetic, spiritual and educational needs of people, of all cultures, now and in the future.

The aesthetic values of our natural ecosystems and landscapes contribute to the emotional and spiritual well-being of a highly urbanised population. The conservation of biological diversity also has ethical benefits. The presence of a wide range of living organisms reminds people that they are but one interdependent part of Earth. Aboriginal relationships to the land and sea, and its animals and plants are complex. To these people the land and sea has a deep spiritual, economic, social, protective and recreational significance. By hunting and gathering, tribal Aboriginal people are not only supplementing their diet with food very high in nutritional value; they are also confirming their self-sufficiency and, more importantly, educating their children in relationships to the land and to other aspects of their culture. Biological diversity conservation can contribute to the conservation of Aboriginal cultural identity.

Benefits of Timely Action

Another benefit of conservation is the avoidance of the rising costs of inaction. Already India is suffering losses in production from environmental degradation and is spending considerable sums in environmental repair. Without remedial action degradation will inevitably increase and the costs of repair faced by the community in the future will be much greater. These costs can be reduced by strategic and timely conservation actions.

What does Biodiversity do?

To understand the economist's approach to the role of biodiversity, it is first necessary to understand the approach's underlying philosophy. It is: Anthropocentric, in that it is fundamentally centred on people; and it is Utilitarian, because it is focussed on improving the wellbeing of people. In other words, the economic approach involves the analysis of what makes people better off. Furthermore, because the anthropocentric nature of the approach implies that the individual matters, a democratic base is also implied. The individual is taken to know what is best for themselves. This is contrary to the paternalistic approach of governments and 'experts' knowing what is 'best' for society without reference to the preferences of individuals.

In addition, the economics approach is: Marginalist, in that it involves the consideration of the impact on people that will occur when a specified change is imposed. Hence, the concern of economists is strictly not for 'the value of biodiversity' but rather for 'the value of a change in biodiversity'. The approach used by Costanza et al (1998) to infer a value for the world's biodiversity was therefore a contradiction. It used value estimates that were based on the economic approach and were therefore marginalist in an attempt to value the entire stock of biodiversity. What these approach characteristics imply is that an economic analysis of biodiversity requires an understanding of the connections between the choices people make, the resultant changes in biodiversity and the subsequent changes in the wellbeing of people. Figure 6 provides a diagrammatic representation of this causal chain.

At the core of the interrelationships between people and biodiversity is the ecological system in which the state and scale of biological resources are intrinsically linked to biodiversity. This is particularly evident in a dynamic analysis of the interrelationship through the notion of resilience. This is because, in general, the greater the biodiversity, the greater is the resilience of the ecological system. Hence, with greater biodiversity, society has better 'insurance' against the impacts of a future adverse event.

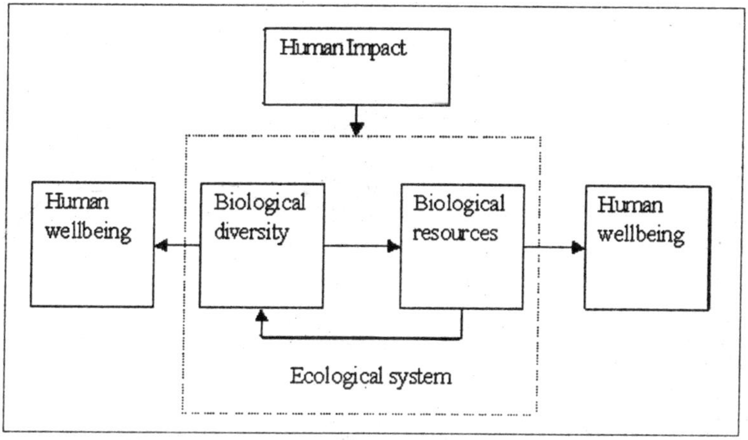

Figure 6: Human/biodiversity interrelationships

Human impacts on the ecological system may have impacts on the extent of biological resources and the their diversity. In turn, human wellbeing both now and in the future may be impacted. People may be affected both directly and indirectly. Hence biodiversity cannot be considered as an entity separate from the extent of the biological resource in the process of generating wellbeing. Nor can it be considered at just one point in time. Biodiversity's interaction with the extent and quality of the biological resource through time precludes it.

For the economist, the important thing is that there is a causal relationship between the resource use choices that people make, the status of the ecological system and the wellbeing of people. The nature of the relationship between human cause and human effect is clearly tied into the functioning of the ecosystem. Hence, to understand 'what biodiversity does' it is necessary to understand the way in which the ecological system works. Put simply, in order to be able to predict what will happen to human wellbeing as a result of a human impact, it is first necessary to be able to predict what happens to the extent and variability of the biological resource. This is the province of the biophysical scientists.

It is possible therefore to see a two-stage answer to the question of what biodiversity does. The first stage involves the prediction of the impacts on the ecological system of human actions. For instance, in the context of a choice regarding the clear felling of a forest, this stage would require forest ecologists to assess the likely impact on key parameters of biological diversity and extent through time. The second stage involves the consideration of what effect those impacts have on human wellbeing. So in the same forest-clearing context, this second stage would involve an assessment of the impact on wellbeing that results from the changes in biological diversity and extent. This is the domain of the economist.

The clear message from this two-stage characterisation is that a prerequisite to any economic analysis of changes in the wellbeing of people is a strong scientific analysis of the biophysical impacts of human choices.

To this point, the relationship between the extent and variability of the biological resource and human wellbeing has been left only vaguely specified. To pursue the question of what biodiversity does any further, it is important now to pay greater attention to this relationship. It is useful to consider biodiversity-induced changes in human wellbeing using the concept of 'total economic value'. This involves a reductionist approach. Specifically, wellbeing is defined in terms of the different types of value that individuals derive from biodiversity. The first category of value is known as 'direct use' value. It includes: The benefits arising from marketed goods such as pharmaceuticals and agricultural products that are impacted by the diversity and extent of biological resource.

- The benefits generated by tourism and recreation activities that are dependent on biological resources

  The second category is known as 'passive use' value and includes:

- Life support services such as nutrient removal, flood control, climate stabilisation etc.

  Finally, there are the so-called 'non use' values of biodiversity that derive from:

- Human ethical considerations relating to matters such as the extinction of species and ecosystems i.e. existence value
- Philanthropic and bequest motives whereby individuals enjoy the pleasure of others both in the current and future generations in the continuing availability of the biological resource
- The 'insurance' benefit that is provided to society through the protection that a resilient ecological system provided option value.
- In short then, biodiversity provides a range of contributions to human wellbeing. It makes these contributions through the interaction between biodiversity and the extent of the biological resource. For the economist, the interest is in being able to understand the impact on wellbeing that arises because resource-use choices made by people impact on biodiversity.

The importance of risk and time in these concepts of value must be stressed. First, the protection of biodiversity provides value streams through time. The values generated are not only available immediately. Biodiversity is natural capital, and therefore supplies potentially a stream of value to current and future generations. Second, biodiversity provides resilience to ecosystems and hence provides a greater level of security for the values that those ecosystems generate. In other words, biodiversity lowers the risk of adverse outcomes. This too involves a future perspective in that the risks being considered confront both current and future generations. This is the option value of biodiversity protection.

An example is instructive to illustrate the point. Consider a choice made to drain a wetland for irrigated cropping. Wetland ecologists and hydrologists may be able to predict some consequences of this choice on the extent and variability of the biological resource. For instance, a species of bird may face an increased threat of extinction, some duck shooting opportunities may be lost and the extent of flooding downstream may be increased. In turn, these impacts may result in a range of impacts on human wellbeing. People may be worse off because of the concerns they hold for the fate of the species. Without that species, the ecosystem's capacity to withstand an adverse turn of events–say an explosion in the population of insects that damage crops–may be diminished. Hunters may be disadvantaged by not being able to enjoy the preferred location for their recreation. Downstream residents may be harmed by the increased losses from flooding.

From this example, another key point can be drawn. None of the values described are certain. Uncertainty is first introduced because we cannot be sure about the ecological consequences of peoples' actions. That is we cannot predict exactly the fate of species impacted by the draining of the wetland. Furthermore, the fate of the species is unlikely to be a black or white matter. That is, the draining of the wetland will have an impact on the probability of the species being made extinct. In other words, the economics of biodiversity protection is a matter of probabilities.

2

Agricultural Biodiversity

The variety and variability of animals, plants and micro-organisms used directly or indirectly for food and agriculture including, in the FAO definition, crops, livestock, forestry and fisheries. It comprises the diversity of genetic resources viz., varieties, breeds, etc. and species used for food, fodder, fibre, fuel and pharmaceuticals. It also includes the diversity of non-harvested species that support production e.g. soil micro-organisms, predators, pollinators and so on and those in the wider environment that support agro-ecosystems viz., agricultural, pastoral, forest and aquatic, as well as the diversity of the agro-ecosystems themselves.

It has also been defined as:

Agricultural biodiversity encompasses the variety and variability of animals, plants and micro-organisms which are necessary to sustain key functions of the agro-ecosystem, its structure and processes for, and in support of, food production and food security. Agricultural biodiversity of all food species is a vital sub-set of general biodiversity, highly threatened by globalization of food markets and tastes, intellectual property systems and the spread of unsustainable industrial food production, but it provides the basis of the food security and livelihood security of billions of people and the development of all food production, including for industrial agriculture and for the biotechnology industries. It is the first link in the food chain, developed and safeguarded by farmers, herders and fishers throughout the world.

Although the term "agricultural biodiversity" is relatively new-it has come into wide use in recent years as evidenced by bibliographic references-the concept itself is quite old. It is the result of the careful selection and inventive developments of farmers, herders and fishers over millennia. Agricultural biodiversity is a vital sub-set of biodiversity. It is a creation of humankind whose food and livelihood security depend on the sustained management of those diverse biological resources that are important for food and agriculture. Agricultural biodiversity, also known as agrobiodiversity or the genetic resources for food and agriculture, includes:

- Harvested crop varieties, livestock breeds, fish species and non-domesticated wild resources within field, forest, rangeland and in aquatic ecosystems;
- Non-harvested species within production ecosystems that support food provision, including soil micro-biota, pollinators and so on; and
- Non-harvested species in the wider environment that support food production ecosystems such as agricultural, pastoral, forest and aquatic ecosystems.

FAO, (1999) reported that agricultural biodiversity results from the interaction between the environment, genetic resources and the management systems and practices used by culturally diverse peoples resulting in the different ways land and water resources are used for production. It thus encompasses the variety and variability of animals, plants and micro-organisms which are necessary to sustain key functions of the agro-ecosystem, its structure and processes for, and in support of, food production and food security.

Agricultural biodiversity has spatial, temporal and scale dimensions especially at agro-ecosystem levels. These agro-ecosystems-ecosystems that are used for agriculture-are determined by three sets of factors: the genetic resources, the physical environment and the human management practices. There are virtually no ecosystems in the world that are "natural" in the sense of having escaped human influence. Most ecosystems have been to some extent modified or cultivated by human activity for the production of food and income and for livelihood security.

What is happening to agricultural biodiversity?

These locally diverse food production systems are under threat and, with them, the accompanying local knowledge, culture and skills of the food producers. With this decline, agricultural biodiversity is disappearing and the scale of loss is extensive and with the disappearance of harvested species, varieties and breeds goes a wide range of unharvested species.

- More than 90 per cent of crop varieties have disappeared from farmers' fields;
- Half of the breeds of many domestic animals have been lost.
- In fisheries, all the world's 17 main fishing grounds are now being fished at or above their sustainable limits, with many fish populations effectively becoming extinct.

The genetic erosion of agricultural biodiversity is also exacerbated by the loss of forest cover, coastal wetlands and other 'wild' uncultivated areas, and the destruction of the aquatic environment. This leads to losses of 'wild' relatives, important for the development of biodiversity, and losses of 'wild' foods essential for food provision, particularly in times of crisis.

Underlying Causes of the Losses

There are many causes of this decline, which has been accelerating throughout the 20th century in parallel with the demands of an increasing population and greater competition for natural resources. The principal underlying causes include:

- The rapid expansion of industrial and Green Revolution agriculture, intensive livestock production, industrial fisheries and aquaculture some production systems using genetically modified varieties and breeds that cultivate relatively few crop varieties in monocultures, rear a limited number of domestic animal breeds, or fish for, or cultivate, few aquatic species.

- Globalization of the food system and marketing, and the extension of industrial patenting and other intellectual property systems to living organisms, which have led to the widespread cultivation and rearing of fewer varieties and breeds for a more uniform, less diverse but more competitive global market.

As a consequence there has been:

- Marginalization of small-scale, diverse food production systems that conserve farmers' varieties of crops and breeds of domestic animals, which form the genetic pool for food and agriculture in the future.
- Reduced integration of livestock in arable production, which reduces the diversity of uses for which livestock are needed.
- Reduced use of 'nurture' fisheries techniques that conserve and develop aquatic biodiversity.

Genetic erosion is the loss of genetic diversity, including the loss of individual genes, and the loss of particular combinations of genes i.e. of gene-complexes such as those manifested in locally adapted landraces. The term "genetic erosion" is sometimes used in a narrow sense, i.e. the loss of genes or alleles, as well as more broadly, referring to the loss of varieties. The main cause of genetic erosion in crops, as reported by almost all countries, is the replacement of local varieties by improved or exotic varieties and species. As old varieties in farmers' fields are replaced by newer ones, genetic erosion frequently occurs because the genes and gene complexes found in the diverse farmers' varieties are not contained in the modern variety. In addition, the sheer number of varieties is often reduced when commercial varieties are introduced into traditional farming systems. While some indicators of genetic erosion have been developed, according to FAO (1998) there have been few systematic studies of the genetic erosion of crop genetic diversity which have provided quantifiable estimates of the actual rates of genotypic extinction. Nearly all countries say, in Country Reports to FAO in 1996, that genetic erosion is taking place and that it is a serious problem.

Variety replacement is the main cause of losses. The replacement of local varieties or landraces by improved and/or exotic varieties and species is reported to be the major cause of genetic erosion around the world. It is also cited as the major cause of genetic erosion in all regions except Africa.

Agro-ecosystems may be identified at different levels or scales, for instance, a field/crop/ herd/pond, a farming system, a land-use system or a watershed. These can be aggregated to form a hierarchy of agro-ecosystems. Ecological processes can also be identified at different levels and scales. Valuable ecological processes that result from the interactions between species and between species and the environment include, inter alia, biochemical recycling, the maintenance of soil fertility and water quality and climate regulation in terms of micro-climates caused by different types and density of vegetation. Moreover, the interaction between the environment, genetic resources and management practices determines the evolutionary process, which may involve, for instance, introgression from wild relatives, hybridization between cultivars, mutations, and natural and human selections. These result in genetic material viz., farmers' crop varieties or animal breeds that is well adapted to local abiotic and biotic environmental variation.

So, agricultural biodiversity is not only the result of human activity but human life is dependent on it not just for the immediate provision of food and other goods, but for the maintenance of areas of land that will sustain production and for the maintenance of the wider environment.

Agro-ecosystems comprise polycultures, monocultures, and mixed systems, including crop-livestock systems like rice and fish, agroforestry, agro-silvo-pastoral systems, aquaculture as well as rangelands, pastures and fallow lands. Their interactions with human activities, including socio-economic activity and sociocultural socio-cultural diversity, are determinant. Some of the key functions for maintaining stable, robust, productive and sustainable agro-ecosystems may include the following:

- breakdown of organic matter and recycling of nutrients to maintain soil fertility and sustain plant and consequently animal growth;
- breakdown of pollutants and maintenance of a clean and healthy atmosphere;
- moderation of climatic effects such as maintaining rainfall patterns and modulation of the water cycle and the absorption of solar energy by the land and its subsequent release;
- maintenance and stability of productive vegetative, fish and animal populations and the limitation of invasion by harmful or less useful species;
- protection and conservation of soil and water resources, for example through a vegetative cover and appropriate management practices, and the consequent maintenance of the integrity of landscapes and habitats;
- sequestration of CO2 by plants.

This was discussed at an international Agricultural Biodiversity workshop organised by the Food and Agriculture Organization of the United Nations (FAO) and the Convention on Biological Diversity (CBD) where the multiple dimensions of agricultural biodiversity were also summarised as providing for:

- Sustainable production of food and other agricultural products emphasising both strengthening sustainability in production systems at all levels of intensity and improving the conservation, sustainable use and enhancement of the diversity of all genetic resources for food and agriculture, especially plant and animal genetic resources, in all types of production systems.

- Biological or life support to production emphasising conservation, sustainable use and enhancement of the biological resources that support sustainable production systems, particularly soil biota, pollinators and predators.
- Ecological and social services provided by agro-ecosystems such as landscape and wildlife protection, soil protection and health in terms of fertility, structure and function, water cycle and water quality, air quality, CO2 sequestration, etc.

Food Crops and Biodiversity

Human beings have used about five thousand species of plants as food, but only 150 or so have entered world commerce and less than twenty provide most of the world's food. Just three crops–wheat, rice, and maize–account for roughly 60 percent of the calories and 56 percent of the protein that humans consume directly from plants. Many of the most important food crops belong to just a few plant families. The grass family–including such crops as wheat, rice, maize, barley, sorghum, millet, oats, and rye–provides some 80 percent of calories consumed by humans, and the legume family has yielded soybeans, peanuts, common beans, peas, chickpeas, cowpeas, and other protein-rich crops. Forty percent of an estimated 2,300 species of cultivated plants belong to just four families: Graminae (grasses), Leguminosae (legumes), Rosaceae (apples, pear, etc.), and Solanaceae (potatoes). The remaining species belong to a diverse array of more than 160 families.

Source Regions of Major Crops

Like economically valuable species in general, many of the major food crops originated in regions that are not particularly species rich. Crops were domesticated in warm temperate and subtropical zones and in tropical mountainous regions. Wheat and barley were first grown in the steppes and woodlands of southwestern Asia, and the origin of maize has been traced to the seasonally dry central highlands of

Mexico. The highlands of Peru contributed the tomato and potato, though the tomato was probably first cultivated in Mexico.

Most important food crops appear to have originated where seasons are pronounced, so it makes sense to look there–and not in rain forests–for promising new crops. The cereals of both the Old and New Worlds come from regions with well-marked wet and dry seasons, while root and tuber agriculture, a mainstay of tropical regions, seems to have developed in tropical lowlands with distinct dry periods. As for why, scientists point in part to the tendency of plants in seasonal environments to store nutrients during the growing season; often, these reserves are what human beings eat. Several regions–known as Vavilov Centers of Diversity after N.I. Vavilov, the Russian botanist who first described the pattern–have been identified as locations of highly diverse crop genetic resources. The centers of crop genetic diversity–including the Mediterranean, the Mexican highlands, Central China, and the Northern Andes–are characterized by a long agricultural history, ecological diversity, mountainous terrain, cultural diversity, and a lack of heavy forest cover. These centers may or may not be located where the crop was first domesticated; wheat and barley were domesticated in southwest Asia, but a current center of their varietal diversity is in Ethiopia; the tomato originated in northwest Peru, but the greatest domestic varietal diversity is in Mexico.

Introduced Crops

Much of the world's agriculture is based on introduced crops. In developing countries in the Americas, only 32 percent of production, by value, is of crops of American origin. The comparable figure for African developing countries is 30 percent. Only in Asian developing countries is most production–70 percent–by native species. Dependence on introduced species reaches its extreme in Australia, the Mediterranean, northern Europe, northern Asia, and the United States and Canada. In these regions, more than 90 percent of production is derived from introduced species.

None of the world's twenty most important food crops are native to Australia or to North America north of Mexico.

Whether introduced or native, the most important crops in any region of the world originated or diversified in places with climates similar to those where they are now grown. The main crops grown in temperate zones are thus not the same as those of the tropics where most developing countries lie. Rice, which has origins in either India or China, is the eighth most important crop in the developed world by weight but by far–a factor of two–the leading crop in the developing world, and is the most important source of calories in tropical developing countries.

Similarly, cassava also known as manioc–native to tropical America–is not grown in developed countries but is the fourth most important crop in the developing world by weight and provides more than half of the caloric requirements for over 420 million people in 26 tropical countries. In Africa, cassava is a fundamental subsistence crop, and in tropical developing countries it is the fourth most important dietary source of calories after rice, maize, and sugarcane.

Agriculture and Genetic Diversity

In agriculture and forestry, genetic diversity can enhance production. Several varieties can be planted in the same field to minimize crop failure, and new varieties can be bred to maximize production or adapt to adverse or changing conditions. In the United States from 1930 to 1980, plant breeders' use of genetic diversity accounted for at least one-half of a doubling in yields of rice, barley, soybeans, wheat, cotton, and sugarcane; a threefold increase in tomato yields; and a fourfold increase in yields of corn, sorghum, and potato.

As important a genetic diversity is to increasing yields, it is at least as important in maintaining existing productivity. For example, crop yields can be increased by introducing genetic resistance to certain insect pests, but since natural selection often helps insects quickly overcome this resistance, new genetic resistance has to be periodically introduced into the crop just to sustain the higher productivity.

Pesticides are also overcome by evolution, so another important agricultural use of genetic diversity has been to offset productivity losses from pesticide resistance. Indeed, the record shows that pesticides only temporarily conquer pests. Over 400 species of pests now resist one or more pesticides, and the proportion of U.S. crops lost to insects has approximately doubled–to 13 percent–since the 1940s, even though pesticide use has increased. Even more significantly, as crop yields increase, so must efforts to sustain the gains. In general the easy gains come first.

Genetic Engineering

Until the advent of genetic engineering, the spectrum of genetic resources available for plant breeding ranged from other varieties of the crop to wild relative so the species. Wild relatives of crops have contributed significantly to agriculture, particularly in disease resistance. Thanks to wild wheats, domesticated wheat now resists fungal diseases, drought, winter cold, and heat. Rice gets its resistance to two of Asia's four main rice diseases from a single sample of rice from central India.

The spectrum of genetic resources available to breeders is now expanding to still more distantly related species, thanks to genetic engineering techniques. However, much more work remains to be done before the potentials of these techniques are realized. Moreover, most new genetic engineering techniques will at least initially involve single-gene modifications of species, and in many cases such modifications are less useful than the multiple-gene changes that result from traditional breeding programs. A transfer of a single gene to a crop plant may confer resistance to a certain pest, but that resistance is often relatively quickly overcome by the pest. In only one species–tobacco–has single-gene resistance been useful for extended periods of time.

Genetic Diversity and Livestock Breeding

Genetic diversity has also been of significance in breeding programs for species other than edible plants. It is becoming increasingly important in forestry and fisheries, and the use of genetic resources in livestock breeding has markedly increased yields. The average milk yield of cows in the United States has doubled over the past 30 years, and genetic improvement accounts for more than 25 percent of this gain in at least one breed.

Still, for several reasons, genetic diversity has been less useful in livestock breeding than in crop breeding. First, whereas one major use of the genetic diversity of crops has been in the development of strains resistant to specific pests and diseases, livestock husbandry has relied largely on vaccines since animals unlike plants can develop immunity to disease. Second, maintaining livestock germplasm is tougher logistically than maintaining the genetic material of plants: since domesticated animals do not go through dormant stages comparable to the seed stage of plants, long-term storage is a problem. Finally, many of the closest relatives of domesticated animals are extinct, endangered, or rare, and thus unavailable for breeding.

Genetic Improvement of Forest Species

For different reasons, genetic improvement of forest species has also received less attention than crop improvement. Until recently, most timber harvested has been wild, so breeding programs seemed unnecessary. In addition, because trees are so long-lived, the rate of genetic improvement of tree species is quite slow. Tests and measurements of growth characteristics have been made for some 500 species primarily conifers over the years, but less than 40 tree species are being bred. Yet, impressive gains have been made with these species. In intensive breeding programs, a 15 to 25 percent gain in productivity per generation has been attained for trees growing on high-quality sites without inputs of fertilizer, water, or pesticides.

Aquaculture

Most of the fish that humanity eats or converts to livestock feed are wild, so breeding has not been widely utilized in fisheries to enhance yields either. An exception is aquaculture. In one case, the domestic carp like *Cyprinus carpio* was bred with a wild carp in the Soviet Union to enhance the cold resistance of the domestic species and allow a range extension to the north.

Risks of High-Yielding Crop Varieties

The development of high-yielding varieties of crops has greatly benefitted some segments of society, but some risks attend use of these new varieties. With the advent of modern plant breeding techniques, a trend toward genetically more uniform agriculture has developed in areas suited to the high-yielding, high-input modern varieties. Whereas traditional mixed farming systems produce modest but reliable yields, planting a single modern crop variety over a large area can result in high yields but the crop may be extremely vulnerable to pests, disease and severe weather. In 1970, for instance, the U.S. corn crop suffered a 15 percent reduction in yield and losses worth roughly $1 billion when a leaf fungus *Helminthosporim maydis* spread rapidly through the genetically uniform crop. Similarly, the Irish potato famine in 1846, the loss of a large portion of the Soviet wheat crop to cold weather in 1972, and the citrus cancer outbreak in Florida in 1984 all stemmed from reductions in genetic diversity.

To stabilize production, breeders use one of several tactics to maintain a genetically diverse crop array. Typically, varieties are replaced with higher-yielding relatives after four to ten years even if they still resist disease or pests. In effect, the spatial diversity of traditional agriculture is replaced with a temporal diversity created by a continuous supply of new cultivars. In the United States, the average life-time of a cultivar of cotton, soybean, wheat, maize, oats, or sorghum is between 5 and 9 years.

Newer strategies for stabilizing production involve the use of varietal blends i.e. a mix of strains sharing similar traits but based on different parents or multilines i.e. varieties containing several different sources of resistance. In each case, the crop represents a genetically diverse array that can better withstand disease and pests. Despite these efforts, genetic uniformity still places some crops at risk of disease outbreaks and in some regions that risk is considerable. Some 62 percent of rice varieties in Bangladesh, 74 percent in Indonesia, and 75 percent in Sri Lanka are derived from one maternal parent.

The External Costs of Biodiversity Loss in Agroecosystem

In the case of endemics the local deletion of some species may well imply global extinction, but even in such cases there is a need to consider the costs to local resource users. This is rather different from much recent work on biodiversity loss which tends to focus on the global value of local conservation, and the scope for local capture of global values. One concern, however, is with the local efficiency of biodiversity loss, and the scope for developing local incentives for biodiversity conservation. A local focus draws attention to the role of biodiversity in the provision of locally valuable ecological services. Biodiversity supports a range of ecological services including, at the global level, the maintenance of the gaseous quality of the atmosphere and amelioration of climate. But it also supports a number of much more localised services: the operation of the hydrological cycle including flood control and water supply, waste assimilation, recycling of nutrients, conservation and regeneration of soils, pollination of crops and so on. The local value of biodiversity derives from the value of these services. In agro ecosystems, for example, the most important ecological services are those influencing the productivity of the system, and its capacity to maintain productivity over a range of environmental conditions. These comprise both on and off-farm services. Watershed protection, for example, offers a range of off-farm services to agriculture including regulation of surface run-off, groundwater recharge, erosion control and localised climatic effects. The conversion of watersheds as a by-product of extensive agricultural growth often means the loss of these services.

Habitat Conversion and Degradation: The Extent of the Problem

The main proximate cause of biodiversity loss is the habitat loss associated with the processes of deforestation and desertification. Both processes are associated with areas where a high proportion of output and employment derives from agriculture. Specifically, biodiversity loss due to agricultural growth at the extensive margin is associated with regions of low population density but high population growth. Biodiversity loss due to agricultural growth at the intensive margin is associated with regions with high rural population density and growth. This said, as the productivity gains of agricultural intensification have faltered in Asia, so pressure in that region has gone back on to remaining forested areas, and recent rates of deforestation are higher in South and East Asia than elsewhere.

The Value of Biodiversity

Heywood, (1995) noted that the costs of the biodiversity loss associated with these two processes include the loss of globally important services such as carbon fixation and sequestration, together with the loss of genetic information. There are few estimates of value of these costs but all indicate that the sums involved are not trivial. The point has already been made, however, that these processes also leads to changes in ecological services that are of primarily local importance. These include watershed protection and the derivative services of flood control, water supply and storage already mentioned. But they also include amelioration of microclimate, soil conservation, nutrient cycling, as well as timber and non timber forest products. These services have both current use value, and option value-the potential value of such services in the future.

It has been recognized for some time that the local value of such services is relevant to the analysis of investments in natural resource-based sectors. Anderson (1987), for example, argued that inclusion of indirect environmental benefits substantially improved the economic rate of return on forestry investments. Illustratively, an estimate of the indirect benefits of forest conservation in Korup National Park, Cameroon, Ruitenbeek, (1989) found the net benefits of watershed protection, flood control and soil fertility maintenance to be roughly comparable to the forgone benefits from timber production.

Soil Biodiversity

Soil biodiversity refers to all organisms living in the soil. Depending on the size class organisms mav be divided into macro, meso and microfauna. Beyond that, bacteria, fungi, protozoa and algae are grouped as microorganisms. Regarding the preferred living environment, aboveground (e.g. foraging on top of the ground, or inside the litter/fine woody debris layer) and belowground specialists can be distinguished. Soil is one of the most diverse habitats on earth and contains the most diverse assemblages of living organisms. Biological activity in soils is largely concentrated in the topsoil. The biological components occupy a tiny fraction (<0.5%) of the total soil volume and make less than 10% of the total soil organic matter. This living component consists of plant roots and soil organisms. Soil microorganisms are responsible for a large part of biological activity (60- 80%) which is associated with processes regulating nutrient cycles and decomposition of organic residues.

Earthworms often form a major part of the soil fauna biomass and can represent up to 50% of the soil fauna biomass in some temperate grasslands, and up to 60% in some temperate forests. High levels of biodiversity, particularly at the microbial scale accompany the huge numbers of soil organisms. 104 bacterial species g-1 soil is not an extreme number. Next to this, hundreds of other species are present. The organisms interact in the soil in food webs, resulting in a flux of matter, nutrients and energy through the biological component of the soil.

Human Practices that Threaten Soil Biodiversity

Soil management strongly influences soil biodiversity, e.g. in agricultural ecosystems. Different practices cause shifts in habitat quality and in substrate availability, resulting in changes in abundance of individual species. A question is, which management practices threaten the functioning of soil ecosystems. Many processes carried out by soil organisms persist in native ecosystems as well as in intensively cultivated soil. There is only a limited insight to what extent these changes in management intensities are accompanied by changes in spectrum of soil microorganisms responsible for the processes involved.

The most important factors affecting the soil biodiversity comprise: Habitat fragmentation;-Resource availability (amount and quality of nutrients and energy sources);- Temporal heterogeneity (seasonal effects);-Spatial heterogeneity (spatial differences in the soil); Climate variability; Interactions within the biotic community; Reduction in biological diversity of soil macrofauna is one of the most profound ecological consequence of modern agriculture, as an example, the number of earthworm species is largely decreased in agricultural soils.

The biodiversity of the soil organisms lead to the control (natural biological suppression) of plant root diseases. The management practices used in many agro-ecosystems (e.g. monocultures, extensive use of tillage, chemical inputs) degrade the fragile web of community interactions between pests and their natural enemies and lead to increased pest and disease problems.

Decline in soil biodiversity is expected to affect soil turnover, decrease natural soil aggregation, increase crusting, reduce infiltration rates, and thus exacerbate soil erosion. The main driving forces that influence biodiversity in agricultural soils are:-*Intensification of land-use.* The introduction of agriculture and its expansion have changed the diversity of habitats, and thus the number of species occurring in the environment at the landscape scale. The increasing intensity of land use on the other hand has destroyed habitat and thus has substantially decreased biodiversity. E.g. a consequence of agricultural practices is the loss of trees and surface litter and consequently of the groups of macrofauna dependant on trees and surface litter (e.g. termites, ants, soil-dwelling insect larvae). Increased use of heavy machines in agriculture leads to soil compaction, and thus to degradation of habitat for soil organisms.

- *Influences of Crops.* Systems that increase belowground inputs of C and N through inclusion of legumes or fibrous rooted crops in rotations may increase microbial populations and activities in comparison to application of commercial fertilisers. The chemical composition of crop residues may have a significant effect on the structure of decomposer communities. E.g., the application of animal manure leads generally to increased abundance and activity of a specific part of the soil biota.

- *Influences of Plants.* Plants have an impact on soil microbial communities through C flow and competition for nutrients. It has been shown that there are distinct differences in bacterial community structure between the bulk, non-rhizosphere and rhizosphere soil. Numbers of bacteria in the rhizosphere are greater than numbers in non-rhizosphere soil. Bacterial activities are stimulated in this area because of the nutrients provided by roots. The variability in chemical composition of root exudates may also influences the composition of soil microbial communities. Also, crop rotation is a key component, which influences the composition of the soil microbial community. The fact that crop rotation can change aggressiveness of pathogens approves changes in soil biodiversity and function because of management.

- *Influence of Fertilisers and pH.* Application of fertilisers and the soil pH both influence the structure of the soil biota. Low pH favours fungi over bacteria, and high nitrogen concentrations result in increased bacterial concentrations. pH influences on soil fauna are also clear. E.g. a low pH in the soil leads to a decrease in abundance of earthworms.

- *Influence of Tillage Crop Residues.* Periodic tillage reverts soil to an earlier stage of ecosystem succession.

Physical disturbance caused by tillage is a crucial factor in determining soil species diversity in the agro-ecosystem. Tillage causes the loss of stratified soil microhabitat, which results in a decreased abundance of species that inhabit such agro-ecosystems. Tillage aerates the soil and therewith causes rapid mineralisation of organic matter and an often substantial loss of nutrients. Activity and diversity of soil microbial communities are influenced by distribution of crop residues. Soil tillage can indirectly impact physical processes in soils through changes in the diversity and activity of soil communities. Reduced tillage with surface placement of residues creates relatively stable environments, which results in more diverse decomposer communities and slower nutrient turnover. No-till system favours fungi over bacteria, as decomposition of plant residues occurs on top of the soil.

- *Pesticides Application*. Pesticides have both targeted and non-targeted effects that may cause a shift in the composition of the soil biota. When organisms are suppressed others can proliferate in the vacant ecological niches. The effect of pesticides strongly depends on soil physical and chemical properties, which affect their availability.
- *Influence of Pollution on Soil Biodiversity and Functioning (ecotoxicological studies)*. Pollutants in general influence the organisms living in the soil. Exposure of organisms to sublethal doses of stress (e.g. a pollutant chemical) over a long time period results in different effects:
- Initially these interactions may occur at the level of biochemical and cellular processes and lead to physiological effects;

Subsequently the structure of the DNA in organisms may be affected in the organism, leading to modification and eventual evolution of organisms; - Consequently such patterns of evolution of resistance or tolerance to the stress factors also occurs in entire communities (e.g. shifts in the composition of plant communities in the vicinity of polluted sites, which might result in the evolution of plant species that accumulate metals).

In soils contaminated by heavy metals the ratio of the resistant and sensitive bacteria increases in the contaminated soil, and the metal-resistant bacteria are much less effective in the decomposition of a number of organic pollutants than the trace elements sensitive bacteria. Interference of different soil stresses complicates the assessment of effects of single stresses and pollutants.

Role and Functions of Soil Organisms

Soil organisms interact in a soil food web, where each trophic layer is food for the next trophic layer. In general a soil food web is based on the degradation of roots and dead organic material (detritus). The stability of the performance of an ecological function is dependent on the stability of the soil food web. It can be said, that the soil food web stability increases with an increasing number of interactions between the organisms. When predators have an increasing choice in their food, the chance that they will predate on a specific species until its extinction decreases. This is an important interest of biodiversity of soil organisms.

Different classes of soil organisms have a different function. In this paragraph a short overview is given of the functions of different groups of soil organisms.

The role of macrofauna includes: - Direct processing of organic matter (e.g. snails, earthworms, enchytraeids, millipedes, ants, and termites). Earthworms can improve the incorporation of organic matter below the soil surface, increase the numbers of water stable soil aggregates, improve water infiltration, aeration and root penetration and increase microbial activity; - Predation (e.g. spiders, ants); - Creation of soil structure (e.g. earthworms, termites);

The micro- and mesofauna (protozoa, nematodes, collembola, and mites) are important regulators of the bacterial and fungal activities. They predate on these organisms (and on each other) and are involved in the decomposition of organic matter. Soil nematodes occupy an important position in the soil detritus food web and are thus significant regulators of decomposition and nutrient mineralisation.

Soil particles may be moved into various positions by earthworms, roots, wet-dry and freezethaw cycles and other forces thus forming soil structure. Production of faeces by mesofauna, fungal hyphae and polysaccharides of microbial origin play an important role in soil. development, nutrient cycling and the maintenance of soil structure. One of the most essential soil functions controlled by soil biota is the recycling of organic matter. Soil organic matter (SOM) is a direct product of the combined biological activity of plants, microorganisms and animals plus the myriad of abiotic factors. It is responsible for crucial aspects of soil function such as aeration and fertility. Production of SOM, including extracellular polysaccharides and other cellular debris, increases the capacity of the soil to maintain structure once it is formed. This means directly, that exogenous organic matter such as compost and sewage sludge is very different from soil organic matter and only after processing by soil organisms may contribute to its content. Such wastes contain high concentrations of organisms. These organisms interfere with the local soil organisms and may cause shifts in soil communities. The survival of such "exogenous" organisms is not always clear. Many will not survive in soil and on the long run will contribute to its organic content. However, some may survive and these may be unwanted organisms, e.g. pathogenic fungi.

Determination of Soil Biodiversity

Species diversity Soil biodiversity comprises many different types of organisms: bacteria, fungi, protozoa, nematodes, enchytraeids, earthworms, mites, springtails, etc. Many techniques exist to characterise the microbial community. Also for the higher organisms there are several welldescribed techniques. All these techniques are widely applied, especially in the field of microbiological research, with modifications and recent new method developments. Using them in a harmonised and comparative way requires method comparisons and standardisation. However the most substantial research gap currently relates to the lack of knowledge regarding the link between species abundance and diversity parameters and soil functioning. Soil biological activity The functional role of soil biodiversity is difficult to quantify directly. Due to the high variation in environmental conditions in the soil in space and time (daily and seasonal courses, water content, temperature, vegetation), the actual activity is highly variable and not easy to measure in comparable ways between different (national or local) inventory systems and research projects. First solutions can be found in the determination of soil respiration, potential nitrogen or carbon mineralisation. However, based on ecological insights there is also the possibility to assess fluxes of material and energy by the use of ecosystem models and food web models, filled with monitoring data.

Organic matter decomposition, nutrient mineralisation etc. are processes driven by the organisms, and when organism activity is restricted (e.g. by water-logging) the processes slow interpret the monitoring data; All these frame conditions of course initially depend on the monitoring objectives and targets.

Indicators for Soil Biodiversity in Monitoring Programmes

In the preceding paragraph some major problems on the set up of monitoring programmes are given. However, when a set up has been achieved, the monitoring data need interpretation. Therefore a framework is needed to evaluate the data, based on soil quality and health standards. Therewith problems in production areas may be identified, to make realistic estimates of sustainable food production, to monitor changes in environmental quality in relation with management and to assist government agencies in formulating and evaluating sustainable agriculture ant other land-use policies.

A huge number of methods to investigate soil organisms activity, biomass, biodiversity, toxic effects or bioaccumulation exists, but for reasons of cost efficiency, inventory quality, and inventory "repeatability", often only a few of them can be selected as indicator parameters. In many cases the selection of indicator or flagship species helps to come to an efficient operational system.

3

Indian Biodiversity

India is the seventh largest country in the world and Asia's second largest nation with an area of 3,287,263 square km. The Indian mainland stretches from 8 4' to 37 6' N latitude and from 68 7' to 97 25' E longitude. It has a land frontier of some 15,200 kms and a coastline of 7,516 km. India's northern frontiers are with Xizang (Tibet) in the Peoples Republic of China, Nepal and Bhutan. In the north-west, India borders on Pakistan; in the north-east, China and Burma; and in the east, Burma. The southern peninsula extends into the tropical waters of the Indian Ocean with the Bay of Bengal lying to the south-east and the Arabian Sea to the south-west. For administrative purposes India is divided into 24 states and 7 union territories. The country is home to around 101 million people, about 16% of the World's population. Summary data for India are given in.

Physically the massive country is divided into four relatively well defined regions - the Himalayan mountains, the Gangetic river plains, the southern Deccan plateau, and the islands of Lakshadweep, Andaman and Nicobar. The Himalayas in the far north include some of the highest peaks in the world. The highest mountain in the Indian Himalayas is Khangchenjunga (8586 m) which is located in Sikkim on the border with Nepal. To the south of the main Himalayan massif lie the Lesser Himalaya, rising to 3,600- 4,600 m, and represented by the Pir Panjal in Kashmir and Dhaula dhar in

Himachal Pradesh. Further south, flanking the Indo-Gangetic Plain, are the Siwaliks which rise to 900-1,500 m.

The northern plains of India stretch from Assam in the east to the Punjab in the west at a distance of 2,400 km, extending south to terminate in the saline swamplands of the Rann of Kachchh, in the state of Gujarat. Some of the largest rivers in India including the Ganga (Ganges), Ghaghara, Brahmaputra, and the Yamuna flow across this region. The delta area of these rivers is located at the head of the Bay of Bengal, partly in the Indian state of west Bengal but mostly in Bangladesh. The plains are remarkably homogenous topographically: for hundreds of kilometres the only perceptible relief is formed by floodplain bluffs, minor natural levees and hollows known as 'spill patterns', and the belts of ravines formed by gully erosion along some of the larger rivers. The agriculturally productive alluvial silts and clays of the Ganga-Brahmaputra delta in north-eastern India, for example, contrast strongly with the comparatively sterile sands of the Thar Desert which is located at the western extremity of the Indian part of the plains in the state of Rajasthan.

The climate of India is dominated by the Asiatic monsoon, most importantly by rains from the south-west between June and October, and drier winds from the north between December and February. From March to May the climate is dry and hot.

Wetlands

India has a rich variety of wetland habitats. The total area of wetlands excluding rivers in India is 58,286,000ha, or 18.4% of the country, 70% of which comprises areas under paddy cultivation. A total of 1,193 wetlands, covering an area of about 3,904,543 ha, were recorded in a preliminary inventory coordinated by the Department of Science and Technology, of which 572 were natural. India's most important wetland areas. Two sites - Chilka Lake in Orissa and Keoladeo National Park in Bharatpur-have been designated under the Convention of Wetlands of International Importance in terms of Ramsar

Convention as being especially significant waterfowl habitats. The country's wetlands are generally differentiated by region into eight categories: the reservoirs of the Deccan Plateau in the south, together with the lagoons and the other wetlands of the southern west coast; the vast saline expanses of Rajasthan, Gujarat and the gulf of Kachchh; freshwater lakes and reservoirs from Gujarat eastwards through Rajasthan called Kaeoladeo Ghana National park and Madhya Pradesh; the delta wetlands and lagoons of India's east coast viz., Chilka Lake; the freshwater marshes of the Gangetic Plain; the floodplain of the Brahmaputra; the marshes and swamps in the hills of north-east India and the Himalayan foothills; the lakes and rivers of the montane region of Kashmir and Ladakh; and the mangroves and other wetlands of the island arcs of the Andamans and Nicobars.

Forests

India possesses a distinct identity, not only because of its geography, history and culture but also because of the great diversity of its natural ecosystems.The panorama of Indian forests ranges from evergreen tropical rain forests in the Andaman and Nicobar Islands, the Western Ghats, and the north-eastern states, to dry alpine scrub high in the Himalaya to the north. Between the two extremes, the country has semi-evergreen rain forests, deciduous monsoon forests, thorn forests, subtropical pine forests in the lower montane zone and temperate montane forests.

One of the most important tropical forests classifications was developed for Greater India and later republished for present-day India. This approach has proved to have wide application outside India. In it 16 major forests types are recognised, subdivided into 221 minor types. Structure, physiognomy and floristics are all used as characters to define the types.

The main areas of tropical forest are found in the Andaman and Nicobar Islands; the Western Ghats, which fringe the Arabian Sea coastline of peninsular India; and the greater Assam region in the north-east. Small remnants of rain forest are found in Orissa state. Semi-evergreen rain forest is more extensive than the evergreen formation partly because evergreen forests tend to degrade to semi-evergreen with human interference. There are substantial differences in both the flora and fauna between the three major rain forest regions (IUCN, 1986; Rodges and Panwar, 1988).

The Western Ghats Monsoon forests occur both on the western coastal margins of the ghats and on the eastern side where there is less rainfall. These forests contain several tree species of great commercial significance such as Indian rosewood Dalbergia latifolia, Malabar Kino Pterocarpus marsupium, teak and Terminalia crenulata, but they have now been cleared from many areas. In the rain forests there is an enormous number of tree species. At least 60 percent of the trees of the upper canopy are of species which individually contribute not more than one percent of the total number. Clumps of bamboo occur along streams or in poorly drained hollows throughout the evergreen and semi-evergreen forests of south-west India, probably in areas once cleared for shifting agriculture.

The tropical vegetation of north-east India which includes the states of Assam, Nagaland, Manipur, Mizoram, Tripura and Meghalaya as well as the plain regions of Arunachal Pradesh typically occurs at elevations up to 900 m. It embraces evergreen and semi-evergreen rain forests, moist deciduous monsoon forests, riparian forests, swamps and grasslands. Evergreen rain forests are found in the Assam Valley, the foothills of the eastern Himalayas and the lower parts of the Naga Hills, Meghalaya, Mizoram, and Manipur where the rain fall exceeds 2300 mm per annum. In the Assam Valley the giant Dipterocarpus macrocarpus and Shorea assamica occur singly, occasionally attaining a girth of up to 7 m and a height of up to 50 m. The monsoon forests are mainly moist sal Shorea robusta forests, which occur widely in this region (IUCN, 1991).

The Andamans and Nicobar islands have tropical evergreen rain forests and tropical semi-evergreen rainforests as well as tropical monsoon moist monsoon forests (IUCN,

1986).The tropical evergreen rain forest is only slightly less grand in stature and rich in species than on the mainland. The dominant species is Dipterocarpus grandiflorus in hilly areas, while Dipterocarpus kerrii is dominant on some islands in the southern parts of the archipelago. The monsoon forests of the Andamans are dominated by Pterocarpus dalbergioides and Terminalia spp.

Marine Environment

The nearshore coastal waters of India are extremely rich fishing grounds. The total commercial marine catch for India has stabilised over the last ten years at between 1.4 and 1.6 million tonnes, with fishes from the clupeoid group viz., sardines *Sardinella* sp., Indian shad *Hilsa* sp. and whitebait *Stolephorus* sp. accounting for approximately 30% of all landings. In 1981 it was estimated that there were approximately 180,000 non-mechanised boats about 90% of India's fishing fleet carrying out small-scale, subsistence fishing activities in these waters. At the same time there were about 20,000 mechanised boats and 75 deep-sea fishing vessels operating mainly out of ports in the states of Maharashtra, Kerala, Gujarat, Tamil Nadu and Karnataka.

Coral reefs occur along only a few sections of the mainland, principally the Gulf of Kutch, off the southern mainland coast, and around a number of islands opposite Sri Lanka. This general absence is due largely to the presence of major river systems and the sedimentary regime on the continental shelf. Elsewhere, corals are also found in Andaman, Nicobar, and Lakshadweep island groups although their diversity is reported to be lower than in south-east India.

Indian coral reefs have a wide range of resources which are of commercial value. Exploitation of corals, coral debris and coral sands is widespread on the Gulf of Mannar and Gulf of Kutch reefs, while ornamental shells, chanks and pearl oysters are the basis of an important reef industry in the south of India. Sea fans and seaweeds are exported for decorative purposes, and there is a spiny lobster fishing industry along the south-east coast, notably at Tuticorin, Madras and

Mandapam Commercial exploitation of aquarium fishes from Indian coral reefs has gained importance only recently and as yet no organised effort has been made to exploit these resources. Reef fisheries are generally at the subsistence level and yields are unrecorded.

Other notable marine areas are seagrass beds, which although not directly exploited are valuable as habitats for commercially harvested species, particularly prawns, and mangrove stands. In the Gulf of Mannar the green tiger prawn *Penaeus semisulcatus* is extensively harvested for the export market. Seagrass beds are also important feeding areas for the dugong *Dugong dugon*, plus several species of marine turtle.

Five species of marine turtle occur in Indian waters: Green turtle Chelonia mydas, Loggerhead Caretta caretta, Olive Ridley Lepidochelys olivacea, Hawksbill Eretmochelys imbricata and Leatherback Dermochelys coriacea. Most of the marine turtle populations found in the Indian region are in decline. The principal reason for the decrease in numbers is deliberate human predation. Turtles are netted and speared along the entire Indian coast. In south-east India the annual catch is estimated at 4,000-5,000 animals, with C. mydas accounting for about 70% of the harvest. C. caretta and L. olivacea are the most widely consumed species (Salm, 1981). E. imbricata is occasionally eaten but it has caused deaths and so is usually caught for its shell alone. D. coriacea is boiled for its oil which is used for caulking boats and as protection from marine borers. Incidental netting is widespread. In the Gulf of Mannar turtles are still reasonably common near seagrass beds where shrimp trawlers operate, but off the coast of Bengal the growing number of mechanized fishing boats has had the effect of increasing incidental catch rates.

Biodiversity

Species Diversity

India contains a great wealth of biological diversity in its forests, its wetlands and in its marine areas. This richness is shown in absolute numbers of species and the proportion they represent of the world total.

Table 1
Comparison Between the Number of Species in India and the World

Group	*Number of Species in India (SI)*	*Number of Species in the World (SW)*	*SI/SW (%)*
Mammals	350(1)	4,629(7)	7.6
Birds	1224(2)	9,702(8)	12.6
Reptiles	408(3)	6,550(9)	6.2
Amphibians	197(4)	4,522(10)	4.4
Fishes	2546(5)	21,730(11)	11.7
Flowering Plants	5,000(6)	250,000(12)	6.0

India has a great many scientific institutes and university departments interested in various aspects of biodiversity. A large number of scientists and technicians have been engaged in inventory, research, and monitoring. The general state of knowledge about the distribution and richness of the country's biological resources is therefore fairly good.

Inventories of birds, mammals, trees, fish and reptiles are moderately complete. Knowledge of special interest groups such as primates, pheasants, bovids, endemic birds, orchids, and so on,is steadily improving through collaboration of domestic scientists with those from overseas. The importance of these biological resources cannot be overestimated for the continued welfare of India's population.

Endemic Species

It is observed from Botanical Survey of India, (1983) that India has many endemic plant and vertebrate species. Among plants, species endemism is estimated at 33% with c. 140 endemic genera but no endemic families. Areas rich in endemism are north-east India, the Western Ghats and the north-western and eastern Himalayas. A small pocket of local endemism also occurs in the Eastern Ghats (MacKinnon & MacKinnon, 1986). The Gangetic plains are generally poor in endemics, while the Andaman and Nicobar Islands contribute at least 220 species to the endemic flora of India (Botanical Survey of India, 1983).

WCMC's Threatened Plants Unit (TPU) is in the preliminary stages of cataloguing the world's centres of plant diversity; approximately 150 botanical sites worldwide are so far recognised as important for conservation action, but others are constantly being identified. Five locations have so far been issued for India: the Agastyamalai Hills, Silent Valley and New Amarambalam Reserve and Periyar National Park all in the Western Ghats, and the Eastern and Western Himalaya.

The 396 known endemic higher vertebrate species identified by WCMC from source references are listed in. Endemism among mammals and birds is relatively low. Only 44 species of Indian mammal have a range that is confined entirely to within Indian territorial limits. Four endemic species of conservation significance occur in the Western Ghats. They are the Lion-tailed macaque *Macaca silenus*, Nilgiri leaf monkey *Trachypithecus johni* locally better known as Nilgiri langur *Presbytis johnii*, Brown palm civet *Paradoxurus jerdoni* and Nilgiri tahr *Hemitragus hylocrius*.

Only 55 bird species are endemic to India, with distributions concentrated in areas of high rainfall. These areas, mapped by BirdLife International formerly the International Council for Bird Preservation are shown in. They are located mainly in eastern India along the mountain chains where the monsoon shadow occurs, south-west India the Western Ghats, and the Nicobar and Andaman Islands.

In contrast, endemism in the Indian reptilian and amphibian fauna is high. There are around 187 endemic reptiles, and 110 endemic amphibian species. Eight amphibian genera are not found outside India. They include, among the caecilians, *Indotyphlus*, *Gegeneophis* and *Uraeotyphlus*; and among the anurans, the toad *Bufoides*, the microhylid *Melanobatrachus*, and the frogs *Ranixalus*, *Nannobatrachus* and *Nyctibatrachus*. Perhaps most notable among the endemic amphibian genera is the monotypic *Melanobatrachus* which has a single species known only from a few specimens collected in the Anaimalai Hills in the 1870s (Groombridge, 1983). It is possibly most closely related to two relict genera found in the mountains of eastern Tanzania.

Threatened Species

Groombridge, (1993) observed that India contains 172 species of animal considered globally threatened by IUCN, or 2.9% of the world's total number of threatened species. These include 53 species of mammal, 69 birds, 23 reptiles and 3 amphibians. India contains globally important populations of some of Asia's rarest animals, such as the Bengal Fox, Asiatic Cheetah, Marbled Cat, Asiatic Lion, Indian Elephant, Asiatic Wild Ass, Indian Rhinoceros, Markhor, Gaur, Wild Asiatic Water Buffalo etc. The number of species in various taxa that are listed under the different categories of endangerment is shown below in Table 2.

Table 2
Globally Threatened Animals Occurring in India by Status Category

	1994 IUCN Red List Threat Category					
Group	*Endangered*	*Vulnerable*	*Rare*	*Indeterminate*	*Insufficiently*	*TOTAL*
Mammals	13	20	2	5	13	53
Birds	6	20	25	13	5	69
Reptiles	6	6	4	5	2	23
Amphibians	0	0	0	3	0	3
Fishes	0	0	2	0	0	2
Invertebrates	1	3	12	2	4	22
Total	26	49	45	28	24	172

Source: Groombridge, B. (ed). 1993. The 1994 IUCN Red List of Threatened Animals. IUCN, Gland, Switzerland and Cambridge, UK. lvi+286 pp.

A workshop held in 1982 indicated that as many as 3,000-4,000 higher plants may be under a degree of threat in India. Since then, the Project on Study, Survey and Conservation of Endangered species of Flora (POSSCEP) has partially documented these plants, and published its findings in Red Data Books (Nayar and Sastry, 1987). *Appendix 7* provides a conservation status listing of many of these plants, based on information maintained at WCMC, whilst Table 3 provides summary statistics for this information.

Table 3
Summary of Plant Conservation Status Information at WCMC

IUCN Threat Category	*Number of Species*
Extinct	19
Extinct/Endangered	43
Endangered	149
Endangered/Vulnerable	2
Vulnerable	108
Rare	256
Indeterminate	719
Insufficiently Known	9
No information	1441
Not threatened	374
TOTAL	3120

Source: WCMC Species Unit.

Protected Areas Network

Development and History

The protection of wildlife has a long tradition in Indian history. Wise use of natural resources was a prerequisite for many hunter-gatherer societies which date back to at least 6000 BC. Extensive clearance of forests accompanied the advance of agricultural and pastoral societies in subsequent millennia, but an awareness of the need for ecological prudence emerged and many so-called pagan nature conservation practices were retained. As more and more land became settled or cultivated, so these hunting reserves increasingly became refuges for wildlife. Many of these reserves were subsequently declared as national parks or sanctuaries, mostly after Independence in 1947. Examples include Gir in Gujarat, Dachigam in Jammu & Kashmir, Bandipur in Karnataka, Eravikulum in Kerala, Madhav (now Shivpuri) in Madhya Pradesh, Simlipal in Orissa, and Keoladeo, Ranthambore and Sariska in Rajasthan.

Wildlife, together with forestry, has traditionally been managed under a single administrative organisation within the forest departments of each state or union territory, with the role of central government being mainly advisory. There have been two recent developments. First, the Wildlife (Protection) Act has provided for the creation of posts of chief wildlife wardens and wildlife wardens in the states to exercise statutory powers under the Act. Under this Act, it is also mandatory for the states to set up state wildlife advisory boards. Secondly the inclusion of protection of wild animals and birds in the concurrent list of the constitution, has proved the union with some legislative control over the states in the conservation of wildlife (Pillai, 1982). The situation has since improved, all states and union territories with national parks or sanctuaries having set up wildlife wings.

The adoption of a National Policy for Wildlife Conservation in 1970 and the enactment of the Wildlife (Protection) Act in 1972 lead to a significant growth in the protected areas network, from 5 national parks and 60 sanctuaries to 69 and 410 respectively, in 1990 (Panwar, 1990). The complete United Nations List of National Parks and Protected Areas for India (1993) is given in. These protected areas, shown in are distributed throughout mainland India and its islands.

The network was further strengthened by a number of national conservation projects, notably Project Tiger, initiated in April 1973 by the Government of India with support from WWF (IBWL, 1972; Panwar, 1982), and the crocodile Breeding and Management Project, launched on 1 April, 1975 with technical assistance from UNDP/FAO (Bustard, 1982).

Protected Areas of the Western Ghats

The Western Ghats are a chain of highlands running along the western edge of the Indian subcontinent, from Bombay south to the southern tip of the peninsula, through the states of Maharashtra, Karnataka, Kerala and Tamil Nadu. Covering an estimated area of 159,000 sq. km, the Western Ghats are an area of exceptional biological diversity and conservation interest, and are "one of the major Tropical Evergreen Forest regions in India" (Rodgers and Panwar, 1988). As the zone has already lost a large part of its original forest cover (although timber extraction from the evergreen reserve forests in Kerala and Karnataka has now been halted) it must rank as a region of great conservation concern. The small remaining extent of natural forest, coupled with exceptional biological richness and ever increasing levels of threat (agriculture, reservoir flooding plantations, logging and over exploitation), are factors which necessitate major conservation inputs."

There are currently seven national parks in the Western Ghats with a total area of 2,073 sq. km (equivalent to 1.3% of the region) and 39 wildlife sanctuaries covering an area of about 13,862 sq. km (8.1%). The protected areas of Kerala State are shown in .

The management status of the wildlife sanctuaries in this part of India varies enormously. Tamil Nadu's Nilgiri wildlife sanctuary, for example, has no human inhabitants, small abandoned plantation areas and no produce exploitation, while the Parambikulam wildlife sanctuary in Kerala includes considerable areas of commercial plantations and privately owned estates with heavy resource exploitation. Summary sheets describing some of the protected areas in Kerala State are given in

Coastal and Marine Biodiversity of India

India is one among 12 mega-biodiversity countries and 25 hotspots of the richest and highly endangered eco-regions of the world. Among the Asian countries, India is perhaps the only one that has a long record of inventories of coastal and marine biodiversity dating back to at least two centuries. However, these are so diverse in space, time and taxon. India has a coastline of about 8000 km, an Exclusive Economic Zone of 2.02 million km^2 adjoining the continental regions and the offshore islands and a very wide range of coastal ecosystems such as estuaries, lagoons, mangroves, backwaters, salt marshes, rocky coasts, sandy stretches and coral reefs, which are characterized by unique biotic and abiotic properties an processes. A network of 14 major, 44 medium and numerous minor rivers together with their tributaries cover practically the entire country except for the western arid region of

Rajasthan Desert. The total length of the rivers is estimated at over 40,000 km.

Diatoms

A total of 102 species of diatoms belonging to 17 families are know from the east coast, with the largest diversity pertaining to 21 species of Naviculaceae and 11 species of Chaetoceraceae and Coscinodiscaceae. Several other families like Biddulphiaceae, Lithodesmiaceae, Nitzchiaceae, Thalassionemataceae and Thalassiosiraceae are represented by fewer species. The diatom diversity along the west coast is relatively higher, with 148 species under 22 families. Naviculaceae among them is the dominant with 22 species, followed by 16 species of Biddulphiaceae, 15 species of Lithodesmiaceae and 12 species of Thalassiosiraceae. It could be noted that the number of pinnate diatoms in the world oceans could range from 500 to 784 and that of centric diatoms, from 865 to 999. Compared with these, not more than 25% of diatoms species are recorded in Indian waters.

Dinoflagellates

The dinoflagellate species diversity in the east coast estuaries is relatively small comprising 15 species in 7 families compared to the west coast estuaries 76 species from 10 families. The family Dinophyceae is he dominant with 18 species followed by Peridiniaceae and Ceratiaceae with 13 and 10 species each, respectively. Unlike the diatoms, the number of estimated dinoflagellate species in the marine environment varies from 1000 to 2000. Compared to these, the current inventory of dinoflagellates in the Indian waters appears too small. Such inventories, however, do not distinguish between truly tropical species and others that are cosmopolitan.

Marine Algae

Marine algae from Indian coasts have been fairly well surveyed since several decades. The latest systematic account lists 844 species including forma and varieties distributed among 217 genera. The most abundant among them are

Rhodophyta with 434 species, followed by 216 species of Chlorophya, 191 species of Phaeophyta and 3 species of Xanthophyta.

The seaweeds are harvested mainly for use as raw materials for the production of agar, alginates and seaweed liquid fertilizer. The red algae Gelidiella acerosa, Gracilaria edulis, G. crassa, G.foliifera and G.verrucosa are used for manufacture of agar and the brown algae Sargassum spp., Turbinaria spp. And Cystoseria trinodis, for alginates and seaweed liquid fertilizes. Bulk of the harvest is from the natural seaweed beds of Gulf of Mannar Islands. Recently, Euchema cottonii has been introduced in the Gulf of Mannar for commercial farming. Its effect on native species, not known so far, remains a matter of great concern.

Seagrasses

It could be noted that fourteen species of searasses under 6 genera are known from Indian seas. They are often found in association with coral reef areas. Eleven species are known from the Palk Bay, which include Cymodocea serrulata, Halophila ovalis sub sp. ovalis, Halodule pinifolia and Syringodium isoetifolium. Thirteen species occur in the Gulf of Mannar Biosphere Reserve, with Halophila, Halodule, Enhalus and Cymodocea being common among them. Thalassia and syringodium are dominant in coral reef areas and coral rubbles whereas others are distributed in muddy and fine sandy soils. Along the west coast, only Halophila and Halodule species are cosmopolitan in distribution and Cymodocea species and Syringodium isoetifolium occur as very small patches at the southern most end of Thiruvananthapuram. Nine species, among which Thalassia hemprichiii and Cymodicea rotundata are dominant are found in the Andaman and Nicobar Islands. From lakshadweep Islands, 7 species are known, among which Thalassia hemprichi is dominant.

Mangroves

The Indian mangroves cover about 4827km^2, with about 57% of them along the east coast, 23% along the west coast and the remaining 20% in Andaman and Nicobar Islands. The mangrove formations are of three types-deltaic, back water-estuarine and insular. The deltaic mangroves occur mainly along the east coast, the backwater-estuarine type along the west coast and the insular in Andaman and Nicobar islands. A total of 39 mangrove species are known from India. While several of them are cosmopolitan in distribution, five of them-Aegialitis rotundifolia, Heritiera fomes, H. kanikensis, Rhizophora annamalayana and R.stylosa-are restricted to the east coast and one, Lumnitzera littorea, is present only in the Andaman and Nicobar Islands.

Protozoa

The known number of protozoan species from Indian seas is 2577, equivalent to about 8% of the total world protozoan fauna. Among them, 52% are free-living and the remaining, parasitic species. Out of seven protozoan phyla only one viz., Labyrinthomorpha has not yet been reported from India.

Foraminifera

Bhatia and Bhalla identified 14 benthic species form Puri Beach, Satyavathy (Waltair Coast), Sarojini (Waltair Coast), Subba Rao & Vedantam (distribution of 32 species on the continental shelf off Visakhapatnam at depths of 20-200 m).

Tintinnid

The order Tintinnida comprises of more than 1000 species of marine ciliates that from an important component of the micro zooplankton. A total of 32 species belonging to 12 genera are known from Indian waters. The degree of abundance of tintinnid populations. Seems to coincide with diatom and dinoflagellate blooms, however, the persistence of such "swarms" appear to be controlled by the larger zooplankton grazers, such as copepods, chaetognaths, bivalve and gastropod veligers.

Sponges

This group has an evolutionary history of about 570 million years and so far, 486 species have been described in India. The sponge fauna of India is dominated by species of Desmospongia followed by those of Hyalospongiae and Calciospongiae. Also 34 species of coral boring sponges of which 20 from Gulf of Mannar and Palk Bay, 5 from Andaman and Nicobar Islands and 18 from Lakshadweep reefs have been recorded. The Gulf of Mannar and Palk Bay region has the highest diversity with 319 species followed by Andaman and Nicobar islands with 95 species, Lakshadweep with 82 species and Gulf of Kachchh with 25 species.

Cnidaria

The global estimates of cnidarian diversity vary between 9000 and 12,000 species. In India 212 species of Hydrozoa, 25 species of Scyphozoa, 5 species of Cubozoa and 600 species of Anthozoa have been found out till now.

Hydrozoa

The first description of hydrozoans in India was by Annandale from Chilka Lagoon and subsequently by Menon reporting 35 species under 28 genera and of Mammen who has recorded 116 species belonging to 13 families.

Siphonophora

Siphonophora are abundant in the Indian seas and constitute an important part of the marine plankton. The siphonophores from the Indian Ocean have been studied by several workers. A comprehensive account of Siphonophora of India shows 116 valid, one variety and 3 doubtful species known from the Indian Ocean, of which 89 occur in the Indian seas.

Anthozoa-Scleractinia

Brook found out 8 species of Acropora from Rameswaram, South East coast of India, out of which A. Multicaulia, A. thurstoni and A. indica were described as new. Subsequent contributions to inventory of coral species were made by Alcock, Gardiner, Matthai, Gravely and Sewell. Contemporary studies on corals are those of Pillai and Venkataraman et al., which lists a total of 218 species under 60 genera and 15 families. Among the four major reef areas of India, Andaman and Nicobar Islands are rich in coral species diversity whereas those of Gulf of Kachchh are poorer. Lakshadweep Islands have more number of species than the Gulf of Mannar. Among the deepwater ahermatypic corals, so far 686 species belonging to 110 genera and 12 families have been reported from the world of which 227 species belonging to 71 genera and 12 families have been reported from the Indian Ocean region. However, meager attention has been paid so far to inventories the deepwater corals and as a result, only 44 species are known until now from Indian seas.

Scyphozoan-Scyphomedusae

In the Indian seas several cruises of the R.I.M.S. 'Investigator' and coastal surveys by the Officers of the Zoological Survey of India have yielded a collection of 24 species, which form the Indian National Collections in the Zoological Survey of India, Calcutta. In addition, several cruises of the R.V. 'Chota Investigator' along the Madras coast from 1972 to 1983 revealed the occurrence of 19 species of which 11 were already known from the Indian seas. Thus, out of the 200 species of Scyphomedusae known from the World Oceans; 34 are know from the Indian seas.

Ctenophore

Only 12 species of Ctenophores, among the 100-150 species known from the world ocean, occur in the Indian seas.

Annelida-Archiannelida

Rao and Ganapati recorded 15 species of archiannelids from the beach sands along Waltair coast. Thus, compared to the vast stretch of Indian coast, the investigations hitherto carried out on Archiannelida are quite limited and any further intensive surveys of the fauna in other areas are quite likely to yield interesting results. The world records of Archiannelida hitherto made fall under 5 families, 18 genera and over 90 species, of which about 20 species are reported from Indian coasts.

Polychaeta

In the phylum Annelida, the Polychaeta have received considerable attention from 1909. Survey of this group actually started with Southern's work on "Polychaeta of Chilka Like" followed by the littoral fauna of Krusadai Island in the Gulf of Mannar by nearly 36 species under 11 families of Gravely and by 119 species under 22 families of Fauvel. Perusal of literature shows that most of the records pertaining to this group are either from the Madras coast or the Gulf of Mannar. Central Marine Fisheries Research Institute has listed 200 species under 46 families in the catalogue of types and reference collections. From the collections of Zoological Survey of India and the Indian Museum, Fauvel described 300 species under 30 families and in his later monograph raised this to 450 species.

Oligochaeta

Marine Oligochaete fauna is not well known in India, and most of the species are recorded from littoral zones of small freshwater bodies like ponds, tanks, pools, ditches, etc., all over the country. The enchytraeidae viz., pot-worms occur in terrestrial, littoral and marine habitats, being abundant in acidic soils with high organic matter. As compared to the world fauna, only 3% of enchytraeid species have so far been reported from this region, mainly from Orissa.

Sipuncula

Of the 145 species under 17 genera known from the world oceans, 35 species under 10 genera and 5 families occur in the Indian coasts. So far as the distributional pattern of the sipunculan fauna is concerned the major areas of species concentration are Andaman and Nicobar Islands, Lakshadweep Islands, Gulf of Mannar and Gulf of Kachchh.

Echiura

The phylum Echiura has 127 species under 32 genera and 5 families. It is observed that 33 species under 11 genera is fairly rich in comparison with what is known 43 species under 14 genera from the Indian Ocean. Maximum abundance of echiurans is in Gulf of Kachchh, Gulf of Kambath, Lakshadweep, Andaman and Nicobar Islands and Gulf of Mannar. Mud-dwelling form are few in numbers and are found in Kerala, West Bengal and Orissa.

Chaetognatha

Chaetognaths are called arrow worms or glass worms rank second in terms of abundance after copepods in marine zooplankton and are cosmopolitan in distribution. They are mostly marine, but a few species are estuarine. Among the 120 species known from the world oceans, about 30 are reported from the Indian Seas.

Tardigrada

Tardigrades occur as meiofauna in the sandy beaches up to 2 or 3 m from water's edge of seacoast. Among the three orders of the Phylum Tardigrada, the Heterotardigrada is found in marine, freshwater and high altitude mountains. So far 214 species are reported from the world under five families and 20 genera. However in India only 10 species under two families and three genera have been reported as meiofauna of marine region.

Arthropoda: Crustacea

Global estimate of Crustacean species diversity is 150,000 of which 40,000 have been well known so far. Of the 2934 species of Crustacea that have been reported so far. In India as many as 139 species of stomatopods with 4 families and 26 genera, 26 species of lobsters with 4 families and 11 genera, 162 species of hermit crabs with 3 families and 40 genera, 705 brachyuran crabs with 28 families and 270 genera, 84 species of shrimps and prawns with 7 families and 19 genera and 159 species of

Caridea with 15 families and 56 genera have been recorded so far. Other than these, 540 species of copepods, 104 species of cirripeds, 120 species of ostrocods have also been recorded.

Copepoda

A total of 106 species belonging to 23 families are known from the east coast estuaries. Among them the calanoids are the dominant, distributed in 16 families, followed by harpacticoids with 5 families and cyclopoids with 2 families. The diversity in the west coast estuaries is relatively higher, with 179 species in 31 families, Calanoids are dominant with 20 families. Though the number of families of harpacticoids and cyclopoids are the same with 6 families, the latter is more diverse, with 22 species compared with 7 species of harpacticoids.

Ostracoda

The Ostracoda are one of the most successful aquatic crustacean groups with approximately 8000 living species. The six extant and extinct orders are ubiquitous and diverse, with over 50,000 named species and genera. It could be noted that only 60+ species of ostracods are known from the Indian coast of which 38 species are known from the east coast and 28 species from the west coast.

Branchiura

Indian species of arguulids are found parasitic on fishes, that the group received adequate attention. There are five species of the genus Argulus of which three were described as new to science.

Girripedes

In India only 36 species having been observed so far. Even these records are rather sketchy and have low geographical coverage.

Malacostraca-Mysidacea

Mysidaceans, with a total number of about 75 species, are known.

Cumacea

There are 23 species of bodotriidae, 3 species of Disstylidae, 4 species of Nannastaeidae and one species of Camylaspididae found in Indian region.

Tanaidacea

It has a species of the group from the Chilka Lake.

Isopoda

The Bopyrid isopods of Indian Macrura has 33 species pertaining to 13 genera are found in Andaman Islands, Delta of Ganges and Madras and other areas. Chopra contributed another paper on the Bopyrid isopods on Indian Macrura.

Amphipoda

Sivaprakasam in a series of contributions enriched our knowledge on the amphipods of east coast of Indian and listed 61 species. Nayar dealt with the amphipods of the Madras coast and Gulf of Mannar. In his monographs on the Gammaridean amphipods of the Gulf of Mannar he dealt with 78 species, of 26 families. Surya Rao enumerated a detailed account of the intertidal Gammarid amphipods from the Indian coasts and listed 132 species pertaining to 54 genera.

Euphausiacea

Among the Indian coasts 23 species of euphasids from the Laccadive and Maldive as well as from adjoining regions and two species from south west coast of India have been found out till date.

Stomatopoda

India has 139 species of stomatopods.

Decapoda-Macrura

It could be noted that large number of species of prawns and lobsters are known to occur in and along the Indian coast, work on this group of species is very limited. World over 17 families, 67 genera and 383 species have been recorded as commercially important. A total of 55 species of commercial shrimps and prawns have been recorded in India. The east coast of India contributes to about 24.5% and the west coast contributes to 75.3% of country's shrimp production.

Brachyura

Kemp identified 38 species under six families collected from the Chilka Lake and in 1923 accounted for crabs collected from the mouth of Hoogly River. Many other Indian authors added to the earlier works raising the total carcionological fauna to above 250 species. There are about 254 species of crabs belonging to 120 genera under 24 families recorded along the west coast of India. Among these, the names of 100 species have been revised. 22 families and 37 subfamilies represents brachyuran crabs. Family Leucosiidae represented the higher number with 20 species followed by subfamily Thalamitae of family Portunidae with 19 species. Family Xanthidae alone is represented by 10 subfamilies of which the subfamily Zosiminae is represented by 14 species.

Anomura

Reddy and Ramakrishna listed twenty species pertaining to the families Paguridae and Coenobitidae. There is a need of more studies on Anomuran crabs.

Mollusca

In India, till today, 5070 species of Mollusca have been recorded of which, 3370 species are from marine habitats. From the available data, it is possible to identify certain areas having rich molluscan diversity. Andaman and Nicobar Islands have a rich molluscan diversity, which includes over 1000 species from the marine region. Gulf of Mannar and Lakshadweep have 428 and 424 species respectively. Eight species of oysters, two species of mussels, 17 species of clams, six species of pearl oysters, four species of giant clams, one species of window-pane oyster and other gastropods such as Sacred chank, Trochus, Turbo as well as 15 species of cephalopods are exploited from the Indian marine region.

Bryozoans

Bryozoans have as many as 20,000 species actually occupying an intermediate position in the hierarchy of animal phylum in respect of species representation. Of these, approximately 4,000 species are living. At least 200 valid species occur in India. The bryozoa are grouped under three classes i.e. Phylactolaemata the freshwater species, Stenolaemata and Gynolaemata. A total of 126 families are recognized -100 from Gymnolaemata 15 from the order ctenostomata and 85 Cheilostomata, 21 from Stenolaemata and five from Phylactlaemata. In India, as in other parts of the world, only a few species of bryozoans inhabit freshwater lakes and rivers viz., Phylactolaemata and most others are marine or estuarine.

Entoprocta

The Entoprocta are predominantly marine having about 60 species known from the world with the exception of one genus in freshwater. The diversity in Entoprocta is limited and restricted basically to the following three families viz., Loxosomatidae, Pedicellinidae is commonly represented in India by two genera viz., Loxosoma and Loxocalyx. The family Pedicellinidae is represented by the genera Pedicellina, Myosoma, Chitaspis, Loxosomatoides, Pedicellinopsis, Barentsia, Gonypodaria and Arthropodaris.

Echinodermata

India has 765 species including Crinoidea: 13 families, 43 genera 95 species; Astereroida with 20 families, 81 genera and 180 species; Ophiuroidea with 15 families, 67 genera 150 species; Echioida with 28 families, 79 genera 150 species; Holothuroidea with 14 families, 62 genera 160 species recorded until today and about 257 species are known from Andaman and Nicobar Islands. Lakshadweep has 77 species and the Gulf of Mannar, 112 species. Economically, only Holothuroidea are exploited on a commercial scale for export. Twelve species of Holothurians belonging to the genera Actinopyga, Bohadschia, Holothuria, Stichopus and Thelenota are known to be of commercial importance in India. However, only three species Bohadschia marmorata, Holothuria scabra and H. spinifera are being exploited to a large extent in the Gulf of Mannar.

Hemichordata

Phylum Hemichordata is divided into three classes, i.e. Enteropneusta, Pterobranchia and Plancto-sphaeroidea. Of the four families known from the world only three families are observed from India. So far, 102 species are recorded from the world of which 12 are known from India.

Protochordata

This phylum consists of two subphylum cephalochordate and Urochordata. Worldwide the diversity of cephalochordates includes 2 families, 2 genera and 24 species and in India 6 species are reported under 2 families and 2 genera. The subphylum Urochordata is divided into class Ascidiacea that are sessile or benthos attached to substratum to the coral reef. About 2000 species of ascidians are reported from all over the world of which 47 are reported in India with 9 families, 21 genera. Out of 57 species of Thaliacea reported from the world, 48 species with four families and 19 genera occur in India and out of 25 species of Larvacea reported from the world 18 with two families and 14 genera are reported from India.

Fishes

The number of fishes in Indian Ocean is 1367 species. The Lakshadweep islands have a total of 603 species of fishes. Over 1000 species are found in the Andaman and Nicobar Islands and about 538 in the Gulf of Mannar Biosphere Reserve. The categories of fishes occurring in coral reef ecosystem of India includes groups such as the damselfishes with 52 species, butterfly fishes with 32 species, sweet lips with 16 species, angelfishes with 16 species, parrot fishes with 14 species, snappers with 42 species and most of the wrasses with 53 species, groupers with 43 species and surgeonfish with 18 species.

Reptiles

About 26 species of sea snakes belonging to the family hydrophiiidae and five species of sea turtles have been reported from seas around India. All the sea snakes and four species of turtles in their marine environment are known from islands of Andaman and Nicobar. Seven species of sea turtles are found in the world's warm oceans of which five species are reported in India.

Marine Mammals

Marine mammals belong to three orders Cetacea, Carnivore and Sirenia. A little over and of these 40 are reported from Indian Ocean and 25 species of marine mammals belonging to the order Cetacea and Sirenia arc reported from Indian waters. However, a majority of these are oceanic forms and occasionally a few individuals may get stranded on the shore waters of Gulf of Mannar, Gulf of Kachchh and Andaman and Nicobar Islands. Dolphins and some of the whales that live or breed in tropical waters, such as humpbacks, are occasionally seen near shore areas.

Threats to Coastal and Marine Biodiversity

Marine resources have traditionally been a major source of food for local inhabitants and of major economic value in terms of commercial exploitation. The human exploitation of marine resources has increased dramatically in the last few decades for reasons, both commerce and subsistence living. Marine ecosystems of India have been exploited since long time but it is only in the last century that the rate of exploitation has increased dramatically, due mostly to the increase in the human population. Except for some of the Andaman-Nicobar Islands, no pristine area exists today. Noticeable decline of marine resources of India go back at least as far as Gardiner who in 1936 expressed concern about the degree to which the degradation takes place where the habitation occurs near marine areas. At the end of the last century or in the beginning of this century, very few marine areas of India remained unaffected, whereas most were partially deteriorated and a few were severely affected.

Natural Threats

Cyclonic disturbances develop during October-November along the coast. These cyclones have sustained winds with speed ranging from 65 to 120 km per hour. High-speed winds cause extreme wave action that kills many fauna and flora, also break coral into rubbles and sometimes large amounts of sand and other materials may be dumped onto the coral reef. Also freshwater runoff kills many fauna and flora in semi-enclosed bays and lagoons by lowering salinity and depositing large amounts of sediments and nutrients.

Human Impacts

Varied human activities which are, a cause for concern over and above the natural disturbances, included runoff and sedimentation from development activities, eutrophication from sewage and agriculture, physical impact of maritime activities, dredging, collecting and destructive fishing practices, pollution from industrial sources and oil refineries and the synergistic impacts of anthropogenic disturbance.

International Programmes and Conventions

India participates with many international agreements and programmes concerned with aspects of nature conservation and sustainable development. These range from legal instruments such as the Convention on Biological Diversity, which place obligations on those nations which become contracting parties, to scientific programmes such as the UNESCO Man and the Biosphere Programme, a global programme of international scientific cooperation. Examples of agreements and programmes with which India is collaborating include:

Convention on International Trade in Endangered Species (CITES)

Since India became a party to CITES on 18th October 1976 it has provided data annually to the CITES secretariat on the trade of endangered species through its CITES *Management Authority*. The text of the *CITES convention* along with the *CITES appendices* are provided.

World Heritage Convention

India ratified the *World Heritage Convention* in 1977 and since then five natural sites have been inscribed as areas of 'outstanding universal value'. These sites are:

- *Kaziranga* National Park
- *Keoladeo* National Park
- *Manas* National Park
- *Sundarbans* National Park
- *Nanda Devi* National Park

Convention on Biological Diversity

India signed the *Convention on Biological Diversity* on 5th June 1992, ratified it on 18th February 1994 and brought it into force on 19th May 1994. This convention will provide a framework for the sustainable management and conservation of India's natural resources.

Ramsar (Wetlands) Convention

India has been a contracted party to the *Ramsar Convention* since 1st February 1982. India has now six sites covering some 192,973 hectares of important wetlands.

4

Biodiversity and Food Security

Biodiversity is threatened at all levels; we are losing genes within species, species within ecosystems, and ecosystems within .regions. Human destruction of natural habitats, whether exploited for commercial of subsistence reasons, is the greatest threat to biodiversity. The clearing of land for agriculture, over-grazing of grasslands, cutting and burning of forests, logging, fuel wood collection, indiscriminate use of pesticides, over-exploitation of fisheries, draining and filing of wetlands, urbanization, pollution of air and water; the devastating impacts of war and natural disasters-these are jus a few examples. Neither terrestrial nor aquatic habitats are immune to the ravages of human-induced destruction.

Loss of biodiversity is frequently characterized as an environmental problem, but the underlying causes are social, economic and political. The excessive and unsustainable consumption of resources by a small but rich minority of the world's population, coupled with the destructive impacts of the world's poor in a desperate bid for survival, have altered, over-exploited or destroyed natural habitats worldwide. While the pressures of human population growth are real and undeniable, it is by no means clear whether over-consumption or poverty ultimately leads to greater loss of biodiversity. Growing disparity between rich and poor promotes unsustainable human activities that deplete the earth's biotic wealth. Over the past 30 years, the poorest 20% of the world's people saw their share of global income shrink from 2.3% to 1.4% meanwhile, the share of the richest 20% of the world's population rose from 70.2% to 82.7%. In more graphic terms, the assets of the world's 358 billionaires exceed the combined annual incomes of countries with 45% of the world's people. Inequities within nations drive the destruction of biological resources no less than those among nations. UNEP's Global Biodiversity Assessment observes, "In most countries, ownership and control of land and biotic resources, and all the benefits they confer, are distributed in ways that work against biodiversity conservation and sustainable living".

Genetic Erosion in Agriculture

Genetic erosion, the reduction of diversity within and between species, is a global threat to agriculture. The concern is not the loss of a single species like wheat or rice, but the loss of diversity within species of the same population.

The greatest factor contributing to the loss of crop and livestock genetic diversity is the spread of high-input industrial agriculture and the displacement of more diverse, traditional agricultural systems. Beginning in the 1960s and 1970s. The Green Revolution introduced high-yielding varieties of rice and wheat to the developing world, replacing farmer's traditional crop varieties and their wild relatives on a massive scale. The same process continues today. New, uniform plant varieties are replacing farmer's traditional varieties and the traditional ones are becoming extinct.

- In the United States, more than 7000 apple varieties were grown in the last century. Today over 85 percent of those varieties-more than 6000-are extinct. Just two apple varieties account for more than 50% of the entire US crop.
- In the Philippines, where small farmers once cultivated thousands of traditional rice varieties, just two Green Revolution varieties occupied 98% of the entire rice growing area in the mid-1980s.

The same is true with animal genetic resources. The introduction of modern breeds that are better suited for high production demands of industrial agriculture has displaced indigenous livestock breeds worldwide.

- FAO's 1995 World Watch List for Domestic Animal Diversity predicts that of the 4,000-5,000 breeds thought to exist, some 1,200-1,500 breeds worldwide are currently under threat of extinction. If only five percent of these breeds are being lost per year, then average rate of breed loss could be about three breeds every two weeks.
- In India, just 3 decades after the introduction of so-called "modern" livestock breeds, an estimated 50% of indigenous goat breeds, 20% of indigenous cattle breeds, and 30% of indigenous sheep breeds are in danger of disappearing.

Why are Agricultural Genetic Resources Important?

Whether they are used in traditional farming systems, conventional breeding, or new biotechnologies, plant and animal genetic resources are the foundation for sustainable agriculture and global food security-now and in the future. Genetic diversity in agriculture enables plants and animals to adapt to new pests and diseases, changing environments and climates. The ability of a certain variety to withstand drought, grow in poor soil, resist an insect or disease, give higher protein yields, or produce a better-tasting food are traits passed on naturally by the variety's genes. This genetic material constitutes the raw material that plant breeders use to breed new crop varieties. Without genetic diversity, options for long-term sustainability and agricultural self-reliance are lost.

Preserving Options for the Future

As genetic diversity erodes, our capacity to maintain and enhance agricultural productivity decreases along with the ability to respond to ever-changing needs and conditions. Scientists predict that the build-up of greenhouse gases in the atmosphere will cause global temperatures to rise 1 to 3 degrees Centigrade during the next century; melting glaciers and thermal expansion of the ocean will bring an associated rise in sea level of 1-2 meters. Each 1 degree rise in temperature will displace the tolerance of terrestrial species some 125 km. towards the poles, or 150 meters in altitude. In other words, global warming will wreak havoc on the world's living organisms. Approximately 30% of the Earth's vegetation will experience a shift as a result of climate change. But since climate will be changing faster than the migration rate of most species, experts predict a "drastic reduction" in global species diversity.

If we are to adapt food production systems to radically changing conditions in the coming decades, plant and animal genetic diversity will be the single most critical resources for doing so. Agricultural genetic resources thus hold the key to increasing food security, environmental stability, and improving the human condition.

Where is Diversity Found?

Genetic diversity is important wherever it is found, but not all nations are equally endowed. In general, a small number of countries lying within the tropics and subtropics account for a very high percentage of the world's biodiversity. It is well known that all major food crops, the staple crops grown and consumed by the vast majority of the world's population, have their origins and centers of diversity in the tropics and subtropics. By and large, other forms of animal and plant genetic diversity are also heavily concentrated in the South. The following are just a few examples:

- There may be more plant species in the area covered by Botswana, Lesotho, Nambia, south Africa and Swziland, than in any other region of the world of comparable size.
- Most domesticated animals have their centers of origin in the south; North America and Ocenia have no indigenous mammalian live stock species. Domestic animal diversity is perhaps greatest in the South, although data on non-European indigenous breeds is shockingly incomplete.

- Species richness of the deep ocean rivals that of tropical rain forests. New research reveals that the diversity of species at the bottom of the ocean is richest in the tropics and diminishes as one moves toward either pole.
- Covering over 600,000 sq. km, coral reefs contain more species per unit than any other marine ecosystem; over half this area is in the Indo-Western Pacific and about 15% are in the tropical Atlantic.
- Tropical Lake Malawi has 245 species of fish, while Lake Windermere (UK) has only nine. Brazil claims more than 3,000 freshwater fish species, three times more than any other country.
- The Indo-West Pacific Ocean contains an estimated 1,500 species of fishers and over 6,000 mollusc species, compared to only 280 fish and 500 mollusc species in the Eastern Atlantic.
- Indonesia ranks first, second, sixth and eighth in the number of species of endemic birds, mammals, reptiles and amphibians respectively. In contrast, Britain has few or no endemic species in any of these four groups.
- Tropical forests contain at least one-half of al known plant and animal species. In Malaysia, an area of forest covering just 50 hectares was found to contain 830 tree species. By contrast, there are 50 indigenous tree species in Europe north of the Alps.

Genetic Inter-dependence

All nations are inter-dependent in terms of access to biological resources. The richest nations on earth are home to the smallest pockets of genetic diversity, while some of the poorest nations are stewards of the very richest reservoirs. Although agricultural genetic diversity is concentrated in the South, no country or region is "self-sufficient" in plant or animal genetic resources. Even the most genetically abundant regions of the world look beyond their own borders for at least half of the germplasm required for their staple foods. In Brazil, for example, a "megacentre" of biodiversity, half of the population's energy from plants comes from rice, wheat and maize-all of which originated elsewhere. Sugar, which provides one-fifth of energy intake in Brazil, originated in Southeast Asia. Cassava (supplier of 7% of energy intake) is the only major food source originating in Brazil. For industrialized regions dependency on "imported" germplasm often exceeds 95 percent.

Given the importance of agricultural genetic diversity, it's not surprising that there has been enormous debate in the last 15 years over the control and ownership of biological resources. Fortunately, there is growing appreciation for the fact that all nations on earth are genetically interdependent. This reality underscores the need for international cooperation in conserving and using these materials, and in recognizing the vital role of rural communities who nurture and manage genetic resources for food and agriculture.

"The major factor driving genetic erosion is that traditional farmers who develop and conserve agrobiodiversity, are generating a "public good", without adequate incentives. They are producing global values for which they obtain no return. Without appropriate and urgent solutions to this paradox, the loss of biodiversity will accelerate, and the consequences will be serous, irreversible and global".

Farmers as Innovators and Conservers of Diversity

Farmers have managed genetic resources for as long as they have cultivated crops. For over 10,000 years they have selected an improved traditional varieties and livestock breeds to meet specific conditions of their agro-ecological environments and diverse nutritional and social needs. The immense genetic diversity of traditional farming systems is the product of human innovation and experimentation-both historic and ongoing. Indigenous farmers of the Andes, for instance, maintain a gene poor of some 3000 varieties of sweet potatoes representing eight species. In Papua New Guinea, as many as

5000 varieties of sweet potatoes are under cultivation, with as many as 20 varieties being planted in a single garden.

Traditional Maori weavers of Aotearoa/New Zealand recognize over 80 genetically and geographically distinct forms of local flax (Phormium cookianum and P. tenax). In Java, small farmers cultivate 607 crop species in their home gardens, with an overall species diversity comparable to a deciduous tropical forest. Traditional pastoralists have selected and maintained hardy livestock breeds that are capable of thriving in extreme environments such as deserts; at least a dozen breeds of camel are known in the Sudan alone.

The "Hindu" Harvest: Food Security and Wild Genetic Resources

It is only in recent years that Western observers have begun to understand the value and importance of wild resources to people's livelihoods. The extent to which local people use food harvested from wild species is typically hidden to outsiders, and thus the role of farmers and rural people (especially women) in maintaining and conserving wild plants and animals is seldom recognized. In their groundbreaking work on this subject, researchers Ian Sconnes, Mary Melnyk and Jules Pretty refer to wild resources-all non-domesticated plant and animal species which are used by people-as "the hidden harvest".

In many parts of the world, wild species and natural habitats are managed and used on a daily basis for household food security. Wild resources are collected to improve diets, to tide people over in times of famine, to supplement income, to provide genetic material for experimentation, as a source of medicine, food, utensils, craft and building materials. Many wild areas also have significant cultural or spiritual importance. Depending on the context and season, between a fifth and half of all foods consumed by poor is not cultivated but harvested from forests, un-tilled fields, or steams. In two districts of West Bengal, 124 "weed" species collected from rice fields were found to possess local economic importance to farmers and their families. In the Uxpanapa region of

Veracruz, Mexico, peasant farmers utilize 435 wild plant and animal species, of which 229 are eaten. In one Tanzanian village over 80% of the vegetable side dishes were composed of wild plants. The agro pastoral people of Tswana use 126 plant species and 100 animal species as sources of food.

New research on the value of "wild" species reveals what millions of rural poor people have known and practiced for millennia: There is no clear boundary between "domesticated" and "wild" species. The distinction between the fallow field and the cultivated may be illusory. Many species long considered to be "wild" are actively managed and improved by people. Food and livelihood security for the poor depends not just on cultivated crops and domestic livestock, but on non-cultivated foods and wild species. Ultimately, conservation of wild genetic resources depends on policies that acknowledge and support local people's rights to use, access and benefit from the genetic resources they have nurtured and developed.

Biodiversity and Indigenous Knowledge

Traditional medicines, although based on biological resources, are products of human knowledge that are of utmost importance to human health worldwide. Over 80% of the world's population relies on local health practitioners and traditional medicines for their primary medical needs.

Between 35,000 and 70,000 species of higher plants are used directly as medicines. Over 60 species of plants are used to treat skin infections in the Amazon region alone. Worldwide, at least 3,000 plants are used by indigenous peoples to control fertility. In Ghana, where the ratio of dentists to total population is only 1:150,000, some 27 woody species with anti-bacterial properties are used as chewing sticks, a popular and effective form of dental hygiene. A study of ethno veterinary practices in northwest Cameroon reveals that local people use nearly 400 medicinal plants to treat animal health problems.

Traditional medicines and indigenous knowledge also have immense commercial value. An estimated three-quarters of all plant-derived prescription drugs were discovered because of their prior use in indigenous medicine. Globally, an estimated US$32,000 million per annum in pharmaceutical industry profits are drawn directly from traditional remedies. Between the late 1950s and 1980, drugs derived from medicinal plants consistently accounted for not less than a quarter of all prescription drug sales in the United States. When the overall benefits to society are taken into account, the value of plant derived pharmaceuticals is astounding. It is estimated that the total economic value of plant-derived pharmaceuticals is astounding. It is estimated that the total economic value of plant-derived pharmaceuticals exceeds US$68,000 million annually in the US alone.

A growing number of pharmaceutical corporations and biotechnology companies are stalking the forests, fields, and waters of the South in search of biological riches and indigenous knowledge. One bio-prospecting company has had extraordinary success in searching for valuable medicinal compounds in tropical countries because they use indigenous knowledge as the basis for their plant collecting. More than half of the company's collected plant samples show promise as new drugs, as compared to less than 1 percent for conventional mass screening techniques. A North American dental institute is studying anti-infective medicinal plants used by curanderos (healers) of the Kekchi Indians of Guatemala, the Paya of Honduras and other indigenous peoples of Central America. It was a woman healer in Samoa who led a Western botanist to a tree, Homolanthus nutans, that she uses to treat viral illnesses. In laboratory tests, a chemical derived from the bark of the tree, known as prostratin, appears to protect immune cells from being destroyed by the HIV virus. In 1995, as US-based biotechnology company announced that it had identified 32 extracts from traditional Chinese herbal medicines that hold significant potential for preventing the sexual transmission of AIDS.

Traditional knowledge is increasingly important for the development of sustainable agricultural practices, and as a source for environmentally sound innovations in many areas of science and industry. The following are just a few examples:

Endod, Phytolacca dodecandra, commonly known as the African soapberry, is a perennial plant that has been cultivated for centuries in many parts of African where its berries are traditionally used as a laundry soap and shampoo. Ethiopian biologist Aklilu Lemma, observed that downstream from where women were washing clothes with endod barriers, dead snails were found floating in the water. Further research revealed that sun-dried and crushed endod snails were found floating in the water. Further research revealed that sun-dried and crushed endod berries are lethal to all major species of snails-but do not harm animals or people, and are completely biodegradable. Endod is now being developed as a low-cost, natural molluscicide for controlling schistosomiasis, a deadly disease transmitted by fresh-water snails that infects more than 200 million people per annum. In addition to the use of endod to control the spread of schistosomiasis in tropical countries, research is now under-way in the United States on the potential use of endod to control the infestation of zebra mussels in North American waters.

"We are witnessing a transformation of farmers as breeders and producers of their own seed supply to farmers as consumers of proprietary seed from the seed industry. It is also a shift from a food economy based on millions of farmers as autonomous producers to a food system controlled by a handful of multinationals that control both output and input. This is a recipe for food insecurity, biodiversity, erosion and uprooting of farmers from the land".

Products of plant origin make up an estimated 93% of human foodstuffs. How many plants feed the world? The answer depends on whom we ask, and where we look for evidence. Widely cited statistics, based on global production data, suggest that just a handful of major crop species especially rice, wheat, maize, barley, sorghum/millet, potato, sweet potato/yam, sugar cane and soybean supply most of the energy humans derive from plants. There's no doubt about the global economic importance of these major crops, but the tendency to focus on a small number of species masks the importance of plant species diversity to the world food supply.

Seeds of Survival

Some 12,000 yeas ago, agriculture began when farmers started to gather seeds from wild plants and began sowing them to grow food. Though frequently overlooked, it was largely women cultivators who first domesticated plants and invented grain milling. All major food crops, the staple crops grown and consumed by the vast majority of the world's population, have their origins and centres of diversity in the tropics and subtropics of Asia, Africa and Latin America. Over the past 12,000 years, farmers in these areas selected and domesticated all major food crops on which humankind survives today.

Wheat and barley originated in the Near East, for example. Soybeans and rice came from China. Sorghum, yams and coffee come from Africa. The genetic homeland of maize, tomato and cacao is Central America. The only major crop originating in North America is the sunflower.

By and large, crop genetic diversity is still concentrated in regions knows as "centres of diversity", located in the developing world. Farmers in these areas who practice traditional agriculture cultivated community-bred varieties known as "landraces" selected over many generations. Closely related species that survive in the wild are known as "wild relatives" of crops. Both farmer's crop varieties and their wild relatives serve as the world's richest repositories of crop genetic diversity.

Thousands of different and genetically distinct varieties of our major food crops owe their existence to thousands of yeas of evolution and to careful selection and improvement by our farmer ancestors. This diversity protects the crop and helps it adapt to different environments and human needs. The potato, for instance, originated in the Andes, but can be found today growing below sea level behind Dutch dikes, or high in the Himalayan Mountains. One variety of rice grows in 7 and one-half metes of water, while another survives on just 60 centimeters of annual rainfall.

Dangers of Genetic Uniformity

Industrialized agriculture favours genetic uniformity. Vast areas are typically planted to a single, high-yielding variety or a handful of genetically similar cultivars using capital intensive inputs like irrigation, fertilizer and pesticides to maximize production. A uniform crop is a breeding ground for disaster because it is more vulnerable to epidemics of pests and diseases. A pest or disease that strikes one plant spreads quickly throughout the crop.

The Irish Potato Famine of the 1840s is a dramatic example of the dangers of genetic uniformity. Potatoes originated in the Andes mountains of South America. In the 1500s, when New World potatoes were introduced in Europe, none of the introduced varieties were resistant to a fungus that struck Ireland's potato crop in the 1840s. when the disease struck, the potato crop was wiped-out. Over 1.5 million people died in the famine. The potato blight is not merely an historical footnote in a long list of crop epidemics. The same fungus, in new and more virulent forms, today poses a grave threat to food security.

In 1970, genetic uniformity in the United States maize crop was responsible for destroying almost $1 billion worth of US maize, and reducing yields by as much as 50%. The problem was that over 80% of the commercial maize varieties being grown in the United States at that time carried a gene that made them genetically susceptible to a virulent disease known as southern leaf blight. Further catastrophe in maize was averted due to intensive breeding programmes. The epidemic and its consequences for food security drew worldwide attention to the problem of genetic vulnerability in major food crops.

Are crops more or less vulnerable today than in 1970 In 1993, plant breeder Garrison Wilkes observed that, "Clearly our priorities with regard to genetic vulnerability and food stability strategies are maize has steadily eroded due to the dominance of a handful of high-yielding Green Revolution varieties. In Bangladesh, for example, Green Revolution wheat varieties covered about 96% of the wheat area in 1984 with 67% of the wheat land planted to a single variety. In the Philippines, two rice varieties developed by the International Rice Research Institute (IRRI) occupied about 90% of the entire rice-growing area during the 1984 dry season. With intensive cultivation of fewer rice varieties throughout the developing world, rice diseases and pests are reportedly growing in number, intensity and geographic distribution. In 1993, the National Academy of Sciences' Committee on Managing Global Genetic Resources made this somber prediction about the state of genetic vulnerability in the South:

"Lack of support for public plant breeding efforts in many developing countries makes it unlikely that they will be able to mobilize new varieties in sufficient time to prevent disaster".

Importance of Crop Genetic Diversity

The high-yielding, elite cultivators of industrial agriculture depend on a steady stream of new, exotic germplasm. Plant breeders call this "the varietal relay race"-they are constantly trying to develop and release new varieties to stay one step ahead of thousands of pests and diseases. Without access to exotic germplasm, industrial agriculture would literally grind to a halt. The Untied States government estimates that for just two major crops, access to exotic germplasm adds a value of US$3,200 million to the nation's US$11,000 million annual soybean production and about $7,000 million to its $18,000 million annual maize crop. Italian scientists estimate that the benefits of exotic germplasm for a singe crop, durum wheat, amount to US$300 million per year in Italy alone. Using exotic maize germplasm from Mexico, the Caribbean and Brazil, US breeders recently developed a commercial maize variety with genetic resistance to armyworm leaf damage-a pest that causes up to US$30 million in damage per annum in the southeastern United States. Rust-resistant genes from an ancient sunflower variety cultivated by the Havasupaid Indians of the southwestern United States are now being incorporated into sunflower hybrids in Australia, China, South Africa, India and the Untied States where new races of rust have threatened the commercial sunflower industry.

Not only cultivated species found in the farmer's fields, but also the genes from wild relatives are enormously valuable. Canadian researchers estimate that between 1976 and 1980, wild species contributed $340 million per year in yield and disease resistance to the US farm economy. Genes from a single wild tomato species gathered in the Peruvian Andes contributes $8 million per annum to US tomato processors. Genetic engineers at Germany's Hoechst corporation achieved genetic tolerance of glufosinate, the company's best-selling herbicide in crops through the introduction of two resistant genes-one of which is derived from a Cameroonian soil sample. AgrEvo is one of the industry's leading developers of transgenic herbicide tolerant plants, and glufosinate is the company's flagship, with sales of over 2500 tonnes per year.

Farmer-Led Food Security

Crop genetic diversity is not just a raw material for industrial agriculture; it is the key to food security and sustainable agriculture because it enables farmers to adapt crops suited to their own ecological needs and cultural traditions. Without this diversity, options for long-term sustainability and agricultural self-reliance are lost. The type of seed sown to a large extent determines the farmer's need for fertilizers, pesticides and irrigation. Communities that lose community-bred varieties and indigenous knowledge about them, risk losing control of their farming systems and becoming dependent on outside sources of seeds and the inputs needed to grow and protect them.

The South's poor farmers in marginal areas were largely by-passed or forgotten by the Green Revolution because high-yielding seeds perform best in rainfed and irrigated regions, and their success depends on capital intensive inputs. In spite of success in raising yields and food production in some high potential areas, the Green Revolution's universalistic approach to high-input, high-yielding plant breeding has been largely unsuccessful in less hospitable, site-specific farming environments. For the majority of the world's farmers, therefore, self-reliance in food production depends on adapting technologies and germplasm to a wide range of poor production environments.

Ultimately, farming communities hold the key to conservation and use of agricultural biodiversity, and to food security for millions of the world's poor. They are the innovators best suited to develop new technologies, germplasm, and management to their diverse ecosystems. As plant collector David Wood observes. "There are about 3 billion farming people in the world. They have almost infinite capacity, experience and application to select and maintain crop germplas. In the long run, the conservation of plant genetic diversity depends not so much on a small number of institutional plant breeders in the formal sector, but on the vast number of poor farmers who select, improve and use crop diversity, especially in marginal farming environments. But neither institutional breeders nor farmer breeders can succeed alone. Success depends on integrated approaches that combine the best of traditional knowledge and institutional technologies.

The challenge for the world community is to link conservation and development by enabling farm communities to assume a major role in managing and benefiting from the genetic resources on which their livelihoods depend. To succeed in these efforts, farmers must have greater control over their genetic resources, access to technologies, research information and a wider range of genetic resources and enhanced germplasm. This requires that the formal sector (governments, scientists and institutional plant breeders) build upon the knowledge and experience of farmers, involve farmers in setting the research agenda, enable them to select and assess technologies, and work with them as partners in the maintenance and further development of their own seeds and livestock breeds.

Over the past 30 years, plant breeding in the industrialized countries would has become increasingly commercialized. In the marketplace today plant breeding, agricultural biotechnology and commercial seed sales are now dominated by transnational seed and agrichemical corporations. Privatization of plant breeding in the industrialized world led to the development of "Plant Breeders Rights", a system of patent-like production that gives formal breeders private monopoly rights over the production, marketing and sale of their varieties for a period of up to 25 years. Many governments in the industrialized world adopted Plant Breeders' Rights as a mechanism to promote innovation in plant breeding and to allow seed companies to recoup their investment by collecting royalties on proprietary plant varieties. In recent years, intellectual property systems have been expanded and strengthened to afford the biotechnology industry greater control over seeds and germplasm. But intellectual property systems have evolved with little consideration for the impacts on farmers, food security and plant genetic resources. Intellectual property regimes increasingly deny farmers the right to save and propagate their seed, prohibit researchers from using proprietary germplasm even for non-commercial purposes, and thus profoundly restrict access to and exchange of germplasm.

Beginning in the early 1980s, representative from the South, together with NGOs, began to question the inequitable and contradictory nature of free access to plant genetic resources of the South in the face of monopoly rights for new varieties developed by industrial plant breeders. As the United Nations, South diplomats began to ask: Why are patented seeds, based on genes of Third World origin, bringing profits to transnational seed corporations without corresponding compensation for the original donors/ innovators of the genetic material? Who is responsible for conserving plant genetic resources? Who controls access to genetic material, and what mechanisms are needed to ensure reciprocal benefits between the "technology rich" countries of the industrialized world and the "gene rich" countries of the South? It could be observed that "On-farm conservation is not a new concept to Ethiopian farmers. They have been conserving and utilizing indigenous landraces from the dawn of agriculture until the present time. All that is required is to encourage them to continue to use something they already do, planting landraces for both subsistence and commercial purposes.

Conserving Domestic Animal Diversity

Like plants, animal genetic resources can be conserved both in situ and ex situ. Ex situ involves the preservation of animals in a setting removed from their normal habitat. It includes "cryogenic preservation techniques"-the collection and freezing in liquid nitrogen of animal genetic resources in the form of living semen, ova or embryos, or the preservation of DNA segments in frozen blood or other tissues. Ex situ conservation also includes the captive breeding of wild or domesticated species in zoos or other situations removed from their indigenous environment. Ex situ conservation complements conservation of live populations and provides a safeguard when population numbers are dangerously low. Despite the potential of new molecular technologies, however, scientists are not yet able to artificially re-create extinct animal breeds from bits and pieces of frozen DNA.

The genetic diversity of livestock can only evolve in use-and only in use can it retain its value for future generations. In situ conservation enables animal populations to continue to adapt, evolve and be selected for use in their natural environments. Unlike cryogenic techniques which require technology, equipment, knowledge and training for collection and storage, in situ conservation can be carried out at any level, in any country, with the skills and resources already available. In situ livestock conservation programmes are currently administered by national governments, by non-governmental organizations, by cooperative groups of farmers and individuals.

While ex situ conservation will always play a vital complementary role in preserving animal genetic resources, it will never be an adequate substitute for rural communities who conserve and use livestock genetic resource on a daily basis. Throughout history, agriculture has been shaped by the genius and innovation of millions of livestock breeders dispersed far and wide. In recent decades the spread of industrial livestock production has not only eroded livestock diversity-it has also reduced the number of breeders, conservers and users of animal genetic resources. These trends do not bode well for conservation of co-evolution and interdependence and this partnership is key to the future conservation and use of animal genetic resources, particularly in the South. Both in situ and on-farm conservation and use of animal breeds must play an increasingly important role in the future of genetic resource conservation. Ultimately, conservation of domestic animal diversity depends on the diversity of human cultures, environments and production systems that helped to shape them over millennia.

"In the end, the fate of the earth's wooded land is tied to the fate of their inhabitants. They will rise of fall together. Either rights to ancestral lands will be defended with the full force of the law, or the forest will fall. Either they will be accorded a share of the economic worth of the ecological services their forests provide, or the forests will fall. Either they will b allowed into the corridors of power where policies are made, or the forests will fall. Other things can help save forests, but these things are fundamental. Tenure, price, and power".

Forests: Food and Livelihood Security

Nobody knows exactly how many people live in or depend on forests in the South. According to FAO, forests are home for an estimated 300 million people-shifting cultivators and hunter-gatherers-around the world. But country specific estimates complied by NGOs suggest that the global population of peoples living in or dependent on forest resources has been drastically under-counted in the past. In six Southeast Asian countries alone (India, Indonesia, Nepal, Philippines, Sri Lanka, Thailand) the forest-dependent population exceeds 600 million according to NGO estimates.

Food security, income, nutrition, employment, energy sources and overall well-being of rural people are linked to the forests. Rural people living in and around forests depend on a large variety of forest products for subsistence and income, including foods and species, building materials, medicines, fibres, fodder, resins, oils, latex and industrial materials. A new study by scientists from Cornell University conservatively estimates that the value of non-timber products harvested from tropical forests is $90,000 million per annum. When forest ecosystems are degraded, and when local people lose access to forest resources, the livelihood and survival of millions of forest-dependent people is at stake.

Benefits and Use of Forests

Forest benefits and services go far beyond timber. Food and fodder from forests is a major contributor to household food security in the South. Forest foods-from both "wild" and domesticated species-include leaves, seeds and nuts, fruit, honey, roots and tubers, saps and gums, fungi and animals. In wooded areas of Thailand, for example, 60 percent of all food comes directly from the forests. In the Upper Shaba area of Zaire, local people gather and consume more than 20 tonnes of forest mushrooms per annum, an important source of protein and minerals. In Nigeria, people living near forests consume over 84% of their animal protein from forest game.

Worldwide, forests and forest industries provide wood products valued at over US$400,000 million per annum including timber, furniture, pulp and paper and fuel wood. Following timber, rattans are the second most important source of export earnings from tropical forests, account for more than US$1,000 million annually. Approximately 90% of the rattan used commercially comes from the "wild". Of the 104 species of rattan found in the Malay Peninsula alone, 98 are threatened or endangered.

Three-quarters or more of the South's population depend on woody species as their primary energy source. Because so many people depend on fuelwood to prepare, process and preserve food, there is an implicit relationship between fuelwood and food security. Fuelwood accounts for 58 percent of energy use in Africa, 15 percent in Latin America and 11 percent in Asia. An estimated 100 million people in the South cannot get sufficient fuel to meet energy needs and almost 1.3 billion are consuming fuelwood faster than it is being replenished. He demand for firewood is not just in rural areas; urban and industrial consumption of fuelwood and charcoal are also major factors degradation and deforestation.

Forests provide vital ecological functions. Their absorption of carbon dioxide and release of oxygen through photosynthesis help to control the level of greenhouse gases. This process, in turn, helps moderate fluctuations in global temperature and provides the atmospheric elements essential for all living things. The widespread conversion of forested ecosystems in the South to grassland and pasture contributes to the increase in atmospheric carbon dioxide and the build-up of greenhouse gases. During the 1980s, conversion of tropical forest to grassland contributed approximately 1.6 gigatons of carbon per year to the atmosphere. This is equivalent to about 16% of the carbon emissions released by the global consumption of fossil fuels for industry and transport, currently estimated at 5.5 gigatons of carbon per year.

Forest vegetation helps support the resource base by nutrient cycling. Forests keep soil from eroding into rivers and aid in flood control and the prevention of silting of reservoirs. An estimated 40 percent of the South's farmers depend on forested watershed as a source of water for irrigating crops or watering livestock. Many forest species, their potential use to society, and their ecological importance have yet to be discovered. Untapped treasure contained in the genes of forest-dwelling plants, animals and microorganisms include undeveloped medicines, crops, animals, pharmaceuticals, timbers, fibers, pulp, soil-restoring vegetation, petroleum substitutes and countless other products and amenities. The bark of the Pacific northwestern United States, was recently found to be the source of taxol, a potent anticancer chemical. The US National Cancer Institute recently discovered a promising anti-AIDS drug, Michellamine

B, that comes from a rainforest vine collected in Southwestern Cameroon. If forest feeling continues at its present rate, and if reservoirs of traditional knowledge continue to be lost, new and existing sources of scientific information will be forfeited and vast potential biological wealth will be destroyed.

World Forest Decline

The world's forests are being destroyed at unprecedented rates. Major threats to forest genetic resources are deforestation and atmospheric pollution. A third threat is the narrowing of the genetic base of tree species as a result of commercial forestry operations. According to FAO data from 1985, more than 400 temperate and tropical tree species are endangered in whole or in significant parts of their gene pools. One cause of genetic erosion of forest tree species is the intensive breeding of a few economically important species in the absence of conservation programs. In the face of rapid environmental change, the future productivity of forests depends on conserving and using the genetic resources of trees.

Tropical Forests

FAO reports that between 1980 and 1990, tropical forest areas have been shrinking by an average of 15.4 million hectares per year. This is an annual loss of about 0.8%; and a total loss over the decade of an area the size of Peru and Ecuador combined. Six countries Brazil, Indonesia, Zaire, Mexico, Bolivia and Venezuela, accounted for about half of all tropical deforestation.

The causes of tropical deforestation vary from region to region. The primary activities associated with deforestation include: the permanent conversion of forest land to agricultural use; harvesting of fuelwood and charcoal; commercial logging; dams; oil and mining projects; shifting cultivation; expansion of urban and industrial areas; overgrazing and fodder collection.

Although small farmers are frequently blamed as the leading cause of tropical deforestation, it is increasingly recognized that shifting cultivation systems as traditionally practiced by forest dwelling people in the South are not only sustainable, they also actively encourage diversity. A study of shifting cultivators in Southeast Asia, Africa and the Amazon prepared by FAO in 1991 concludes:

"Shifting cultivation is a complex agricultural system that is well-adapted, under certain conditions, to the environmental limitations of the tropics. It is not primitive, nor necessarily destructive. In requires in-depth knowledge of the tropical environment and a high degree of managerial skill to succeed.

Today, poverty, land use patterns and population pressures are reducing the land available for shifting cultivation. Shorter fallow periods and overuse of available land are turning traditionally sustainable methods into destructive ones. It is more often the newly arrived migrants, poor and landless people without generations of traditional knowledge, who are driven to over-exploit forest resources.

Conversion of forests to agriculture-primarily by small-scale, subsistence farmer is frequently cited as the single greatest cause of forest destruction in the South. The Consultative Group on International Agricultural Research (CGIAR), in a 1996 press release, reported that "Poor farmers could destroy half of the remaining tropical forest. Whether they are victims of unjust land tenure systems, refugees from political or social unrest, or settlers from poorly-conceived development programmes, poverty and the lack of access to land and jobs are the underlying causes of deforestation linked to poor farmers. Government polices in both the South and North are often at the root of these problems. Examples include subsidies to cattle-ranching and timber industries; agricultural development and colonization policies that encourage clearing of forests; tax incentives for land development that lead to concentration of land ownership; under-valuation of forest resources and the people who inhabit them; lack of legal recognition for indigenous peoples land rights.

From 1980-1990, commercial logging for international and domestic consumption increased in Africa, Asia and Latin America-with a rise in both area and volume harvested. According to FAO, 5.9 million hectares were logged annually in the tropics from 1986-1990, with 83% of the cutting in primary forests. While many believe that commercial timber can be harvested "sustainably", the evidence is hard to find. When the International Tropical Timber Organization conducted a study of "sustainable timber production" in 1988, it succeeded in identifying only 1 million hectares-or 0.12% of all timber producing tropical forests-that met their definition. FAO claims that "little progress has been made in the sustainable management of natural forests. Logging is only one of the causes of forest destruction, but its impact is magnified because it opens up the forest to further encroachment, and because most logging takes place in primary, species-rich forests. Logging roads allow both displaced people and extractive industries to enter previously inaccessible forest areas.

Evidence from the Brazilian Amazon and West Africa suggest that the loss of biodiversity in forest ecosystems far exceeds the deforested area. Degradation and fragmentation of forests magnifies the destructive impacts of deforestation. "Fragmentation" refers to patches of forest that are not deforested, but are tool small to support remaining populations of plants and animals. Forest edges are particularly vulnerable to degradation. Because of the so-called "edge effect" the negative effects of deforestation extend approximately 1k. into adjacent habitats. A 1993 study based on satellite imagery of the Brazilian Amazon found that the rate of deforestation averaged about 15,000 sq. km. per year from 1978-1988, while the rate of habit fragmentation and degradation was about 38,000 sq. km per annum.

Temperate and Boreal Forests

The loss of forest genetic resources is not confined to the tropics. Temperate and boreal forests account for almost half of global forest cover, but because they contain less biological diversity than tropical forests, area in Europe and the former USSR from 1980-1990. Although temperate and boreal forests are generally considered "stable", this description masks the rapid disappearance of old-growth stands, and the forests degradation occurring in some regions. Old growth forests are among the richest habitats found in temperate areas. In some cases, the rate at which they are disappearing exceeds the rate of forest deforestation in many tropical countries. In Western Europe, old growth forests now account for less than 1% of total forest area. Among temperate countries only Canada and New Zealand hold more than 20% of their total forest area as old growth forests. In Europe and North America, pollution from industrial and transportation activities and wildfires have threatened forests and the genetic resources of a range of species. Forests in Germany and Czechoslovakia, in particular, have been severely affected.

The effects of global warming could be catastrophic for forests, especially in higher latitudes. According to WWF International, even a 1 degree C absolute change in temperature could eradicate 25% of the world's boreal forests. Tree species found in both temperate and boreal forests have slow migration rates and thus have limited ability to adapt to new climatic zones. Scientists predict that higher temperatures and drought caused by global warming will trigger forest fires and invasions of pests and diseases, thus accelerating the loss of forest biodiversity. A 2 degree C increase in temperature, for instance, would quadruple the are of forest in British Columbia that is susceptible to attack by spruce weevil.

Tree Plantations

Worldwide, an estimated 100-135 million hectares are now devoted to tree plantations. The FAO estimates that at the end of 1990 there were 43.8 million hectares of industrial and nonindustrial Forest plantations in the South. The largest share, 73%, is found in the regions of tropical Asia and he Pacific. Just five countries-India, Indonesia, Brazil, Vietnam, and Thailand account for 85% of all tropical plantations. About 6 hectares of tropical forest are cut per year for every hectare put into forest plantations. About 2.6 million hectares of new plantations are planted yearly in the South, but only about 1.8 million survive. Eucalyptus, pine teak and acacia are the main species planted in tropical plantations.

Plantations play an important role in satisfying future demands for wood and fiber. However, when tree monocultures replace native forests, they transform diverse forest ecosystems into high-yielding, genetically uniform tree farms. When industrial tree plantations are based on uniform, introduced species, the native biodiversity is inevitably lost. According to WWF forestry consultant, Nigel Dudley, "Replacing old-growth, natural forests with plantations gives the impression that forests are being restored. But a plantations is about as similar to a natural forest as a football pitch is to a flower-rich meadow.

Fish genetic resources must be conserved and utilized because they are the key to maintaining the viability of cultured and natural fish populations. They enable species to adapt to environmental change and they provide the opportunity for genetic improvement programs in aquaculture.

Tropical waters are the richest in terms of species diversity. The Indo-West Pacific Ocean, for example, contains as estimated 1,500 species of fish and over 6,000 mollusc species, compared to only 280 fish and 500 mollusc species in the Eastern Atlantic. Brazil claims more than 3,000 freshwater fish species, three times more than any other country. An estimated 40 percent of freshwater fishes in South America have not yet been classified. Thailand may have as many as 100 species of freshwater fish, but only 475 have actually been documented.

Aquatic Harvest

Over 75% of the fish consumed by people still comes from the hunting of wild species in natural environments. Aquaculture, which accounts for the remainder of the catch, refers to all types of farming in enclosures such as ponds, tanks, pens, etc., as well as inland culture-based fisheries. Over the past decade, the fastest growing portion of the world fish supply has come from aquaculture. Approximately 85 percent of the global catch is finfish, with shellfish (molluscs and crustacean) accounting for most of the remainder. Numerous other resources such as aquatic plants (seaweeds) and turtles, though important for local economies, account for a small percentage of harvested organisms.

The contribution of aquaculture to world fish production is increasing dramatically, and has offset losses from capture fisheries over the past decade. Fisheries experts believe that future expansion of aquaculture offers the best hope for maintaining per capita fish consumption levels in the future. In 1993, aquaculture accounted for 16 percent of total world production of fish, and 23 percent of food fish supplies. Aquaculture production of freshwater fishes already exceeds that of production from freshwater capture fisheries. Globally, half of all salmon is no longer caught in the wild, but farmed; the same is true of shrimp culture. By contrast, aquaculture supplies only 5% of the total marine fish production. Aquaculture production is heavily concentrated in the South Asia accounts for 84% of world aquaculture and China accounts for about half of the total world production.

Importance of Fisheries

Fishing, fish processing and fish trading have provided the basis for food security, employment, income and cultural traditions in coastal and inland communities for centuries.

Fish contribute substantially to the world food supply, either directly for human consumption, or as feedstuff for livestock about 28% of the total world catch is used as animal feed. Worldwide, fish provides about 17% of the animal protein in the human diet. Over 200 million people around the world depend on fishing and related industries fro their livelihoods. For millions in the South, fish is often the main source of animal protein. According to FAO, fish provides 29% of the total animal protein consumed by Asians, 19% by Africans and 8% by Latin Americans. Fisheries also provide a significant source of employment.

Threats to Marine Biological Diversity

The loss and degradation of biodiversity in marine ecosystems has profound implications for food and livelihood security-particularly in the South. In 1990, eminent marine biologists solemnly concluded that "the entire marine realm, from estuaries and coastal waters to the open ocean and the deep sea, is at risk. Tragically, no place in the ocean is so remote that it has not been marred by human activities-pollutants have even been found on the deep-sea floor. Though fish stocks are a naturally-renewable resource they are being depleted at a non-sustainable rate and aquatic ecosystems are being destroyed.

The marine species and ecosystems suffering most are in coastal waters closest to humankind. Already 66% of the Earth's population lives within 60 km. of coasts, and migration towards these areas is increasing. Inland pollution and other environmentally-degrading activities ultimately affect marine biodiversity because oceans function as a sink for carbon dioxide, eroded soils, contaminants, fertilizers, human and industrial wastes. More than 90% of the global fish cash comes from the 10% of the oceans closest to land. An estimated 44% of all pollution entering the ocean is from runoff and discharges from land mainly through rivers, 33% from atmospheric pollution, 12% from maritime transportation such as oil spills and discharges, 10% from the deliberate dumping of wastes and 1 percent from offshore mining.

Over-exploitation and over-capacity of fisheries, pollution and habitat destruction, and the introduction of exotic species are the primary activities that imperil fish genetic resources worldwide. Both marine and freshwater ecosystems are imperiled; but the following section focuses primarily on marine genetic resources.

Fished Beyond Limits-Overexploitation and Overcapacity

FAO's 1995 world fisheries report estimates that 70% of the world's marine fish stocks are either fully exploited, over fished, depleted or recovering from over fishing. Since 1989, the global fish catch has stagnated and its quality has declined. Virtually all commercially valuable marine species are overexploited. Of the world's 15 major marine fishing regions, productivity in all but two has fallen. In four of the Atlantic fisheries and one of the Pacific fisheries total output has dropped over 30%. Only the Indian Ocean fisheries are still increasing total output-though these are likely pushing the limit. Declining catches have translated directly into job losses for over 100,000 people in recent years-and now threaten the livelihoods of millions more. With the collapse of the Canadian cod fishery, for example, some 80,000 fishermen and women have gone on welfare.

Fueled by huge capital investments, abundant fossil fuels and modern technology, the global fish catch has increased more than four-fold in the past 40 years. The world's industrial fishing catch has increased more than four-fold in the past 40 years. The world's industrial fishing fleet has doubled in size since 1970-from 585,000 to 1,200,000 vessels and now has twice the capacity needed to bring in the maximum sustainable catch. Experts predict that if Iceland and the European Union cut their fleets by 40%, and Norway by two-thirds, these countries would still catch as much fish as they do today.

The over-capacity of industrial fleets is a direct result of government subsidies that underwrite the growth of national fleets. But bigger, high-technology fleets have proved a sound investment. The FAO estimates that in 1989 government subsidies world-wide amounted to a staggering US$54,000 million-resulting in a catch valued at only $70,000 million. An estimated 46% of the value of all fish landed is required as return on capital invested in industrial fleets. Government subsidies in both the North and the South favor commercial fishers over small-scale, traditional ones. These subsidies not only consolidate fish resources in the hands of the rich and powerful, but they also threaten the future livelihoods of 15 to 21 million small-scale and traditional fishers (90% of all fishers), who are the mainstay of coastal communities worldwide.

The threat to marine biodiversity from over-exploitation cannot be measured in extinct species. While no commercially fished marine species is known to have become extinct in modern times, non-sustainable fishing practices have devastated fish stocks, genetic diversity and marine ecosystems. Experts conclude that "overexploitation not only diminishes species populations and reduces economic return, but also causes genetic changes in the exploited populations and alters ecological relationships with the species predators, symbionts, competitors, and prey.

Using costly and sophisticated technologies such as depth-sounding equipment, satellite data and spotter aircraft, high-intensity lamps, 100-metre factory-freezer trawlers, non-selective drift nets, and bottom trawls, industrial fleets are driving some species to the brink of extinction and destroying natural ecosystems in the process. The following are just two examples:

- Industrial fleets in the North Pacific have employed massive driftnets to capture squid and tuna, using up to 3.5 million kilometers of synthetic netting per annum-enough to circle the globe 88 times. Not surprisingly, some 40% of the catch netted in these "wall of death" driftnets is reportedly discarded as bycatch, including up to 200 nontarget species.
- Industrial shrimp trawlers, capable of scraping 1sq. kilometer of seabed in 10hrs., profoundly disturb the seabed and its species Trawling destroys the burrows of bottom dwelling species, mangles huge numbers of nontarget species, and increases suspended sediments. Shrimp fisheries in the northern Gulf of Mexico (USA) and the Sea of Cortez (Mexico) sweep the entire trawling grounds several times per year.

Inefficient and wasteful practices capture fish that are too small, and leave too few capable of reproducing. As a result, stocks are being depleted to the point where they cannot recover. Shrimp trawl fisheries have one of the highest levels of by catch and discard rates: 70 to 100% of the catch is thrown back into the sea either dead or drying. The FAO estimates that some 27 million tons of fish, or about 25% of all reported marine landings are discarded annually as by catch.

Soil Biodiversity

Soil biodiversity influences a huge range of ecosystem processes that contribute to the sustainability of life on earth. For example, soil organisms maintain critical processes such as carbon storage, nutrient cycling and plant species diversity. Soil biodiversity plays a role in soil fertility, soil erosion, nutrient uptake by plants, formation of soil organic matter, nitrogen fixation, the biodegradation of dead plant and animal material, reducing hazardous waste, the production of organic acids that weather rocks, and control of plant and insect populations through natural bio-control.

Through production of food, fibre and renewable forms of energy, soil-based plant productivity supports the livelihood of every person on earth. Soil biota enhance crop productivity because they recycle the basic nutrients required for all ecosystems, including nitrogen, phosphorous, potassium and calcium. Soil organisms enhance the productivity of the soil by increasing water infiltration, thereby reducing surface water runoff and decreasing soil erosion.

Termites, earthworms and other .burrow-building soil organisms enhance soil productivity by churning and mixing the upper soil, which redistributes nutrients, aerates the soil and increases surface water infiltration. Earthworms and other invertebrates can bring to the surface from 10 to 500 tonnes/ per hectare/per year of soil, and thus play a critical role in the formation of topsoil. Nitrogen from natural and commercial sources is vital to plants and animals. It is the main nutrient required for growth in plants and for building proteins in animals. Biologically fixed nitrogen primarily nitrogen-fixing microorganisms that live symbiotically on the roots of leguminous plants and trees makes an enormous contribution to global agricultural productivity. In poor soil, where alternative sources of fertilizer are either unavailable or unaffordable, biological nitrogen-fixation is vital to crop production. Worldwide, an estimated 140 to 170 million tones of nitrogen, valued at approximately US$90,000 million are fixed by microorganisms in agricultural and natural systems each year.

Soil biota plays a major role in stabilizing and regulating the earth's climate. Global warming is the result of increasing levels of carbon dioxide and other greenhouse gases in the Earth's atmosphere-primarily caused by the burning of fossil fuels by humans. The rate of exchange of carbon between the earth's surface, the oceans and the atmosphere, known as "the carbon cycle", is the primary mediating force with regard to climate change. Through the process of photosynthesis, green plants absorb carbon dioxide from the atmosphere. It is well known that trees and forests store the absorbed carbon in woody biomass. But it is actually soil organic matter that is the major global storage reservoir for carbon. The living microbes, fungi and invertebrates found in the soil are responsible for decomposing carbon and nitrogen and making them available for plant growth, while at the same time contributing to the rate of production and consumption of carbon dioxide, methane and nitrogen.

Despite the importance of soil biodiversity to life-sustaining ecosystem processes, soils are one of the most neglected habitats on earth. In most cases, soil biologists simply don't know which organisms or groups or organisms play the most important roles in ecological processes, they don't know which soil taxa are being lost, or what impact these losses will have in the future.

Soil ecologists believe that it is essential and urgent to establish the cause and effect relationships between the loss of soil biodiversity and the impact on terrestrial and global ecosystem processes. Only by knowing and understanding life in the soil can we begin to conserve and better utilize its life-sustaining services.

Industrial agriculture has contributed to the neglect of soil biodiversity because conventional soil science has generally relied on the use of purchased farm inputs to overcome constraints and modify the soil environment. For example, if the soil is dry, irrigate; if soil fertility is low, buy synthetic fertilizer; if pests and weed invade, spray chemicals. With growing awareness and need for low-input and sustainable agriculture, knowledge of soil biodiversity is increasingly important to future farming systems. A better understanding of soil biota will enable farmers to depend less on modification of the natural environment and place greater emphasis on using biological processes to optimize nutrient cycling, minimize the use of purchased inputs, and maximize the efficiency of their use.

The Value of Microbial Genetic Resources

Microorganisms are tiny living things that are not visible except with a microscope. They include algae, bacteria, fungi including yeasts, certain protests one celled animals that are not bacteria and viruses. Microbial biodiversity is a vast frontier and a potential goldmine for the biotechnology industry because it offers countless new genes and biochemical pathways to probe for enzymes, antibiotics and other useful molecules.

Worldwide, the economic value of microorganisms is estimated to be "at least many tens of billions of US dollars". Pharmaceuticals of microbial origin account for sales of approximately $35-50 billion per annum in the North. It is the invisible world of microbes that has given us more than 3,222 antibiotics, for example, many derived from soil samples. In 1993, five of the pharmaceutical industry's top-selling drugs were derived from microbes; accounting for more than $4,500 million in annual sales.

The commercial value of microbials extends beyond pharmaceuticals. The total world market for industrial enzymes, all produced by microorganisms, is $1,300 million. Enzymes are natural catalysts that can speed up a chemical reaction. Because the process is biological, they are biodegradable and can be used instead of synthetic chemicals. For example, industrial enzymes are used to enhance detergents, as biological pesticides, to clean up toxic wastes, to replace chemicals in paper and pulp processing, and for oil extraction.

With the use of modern biotechnology, the potential applications of microorganisms is vast, scientists are experimenting with genetically engineered bacteria that are capable of producing products such as biodegradable plastics, artificial skin, and fibres that are as strong as spider silk. Maize, rice, potato and cotton are among the crops that have been genetically engineered to produce insecticidal genes from a common soil bacterium. Bacillus thuringiensis (Bt). The Bt genes enable the crops to produce a toxic protein that kill insects which feed on the plant. Microbial diversity can play an important role in the decomposition of hazardous wastes. Molecular biologists are attempting to harness specific organisms, or groups of organisms, to clean-up toxic wastes in the environment, or reduce hazardous waste production in industrial processes.

Today, transnational microbe hunters are especially interested in exotic and hostile environments-including boiling hot springs, undersea hydrothermal vents, alkali lakes and the frozen tundra of Antarctica-as a source of unexplored microbial diversity. Bioprospecting for microbes goes, quite literally, to the ends of the earth. The following are just a few examples:

A subsidiary of German agrochemical giant agrevo is conducting intensive soil sampling in India. According to German researcher Michael Flitner, the company has already screened over 90,000 Indian soil samples and is building a new, high-efficiency system in Frankfurt for screening plants and soil collected in India.

5

Biodiversity Loss

The loss of biological diversity may take many forms but at its most fundamental and irreversible it involves the extinction of species. Over geological time, all species have a finite span of existence. Species extinction is therefore a natural process which occurs without the intervention of man. However, it is beyond question that extinctions caused directly or indirectly by man are occurring at a rate which far exceeds any reasonable estimates of background extinction rates, and which, to the extent that it is correlated with habitat peturbation, must be increasing. Unfortunately, quantifying rates of species extinction, both at present and historically, is difficult and predicting future rates with precision is impossible. Documenting definite species extinctions is only realistic under a relatively limited set of circumstances, where a described species is readily visible and has a well-defined range which can be surveyed repeatedly. Unsurprisingly, most documented extinctions are of species that are easy to record e.g. land snails, birds) and inhabit sites which can be relatively easily inventoried (e.g. oceanic islands.

The large number of extinct species on oceanic islands is not solely an artifact of recording, because island species are generally more prone to extinction as a result of human actions. Rather than being derived from observed extinctions, therefore, quoted global extinction rates are derived from extrapolations of measured and predicted rates of habitat loss, and estimates of species richness in different habitats. These two estimates are interpreted in the light of a principle derived from island biogeography which states that the size of an area and of its species complement tend to have a predictable relationship; fewer species are able to persist in a number of small habitat fragments than in the original unfragmented habitat, and this can result in the extinction of species.

Even on best available present knowledge, these estimates involve large degrees of uncertainty, and predictions of current and future extinction rates should be interpreted with very considerable caution. Pursuit of increased accuracy in the estimation of global extinction rates, however, whilst of great concern, is not a crucial activity; it is more important to recognise in general terms the extent to which populations and species which are not monitored are likely to be subject to fragmentation and extinction. Loss of biodiversity in the form of crop varieties and livestock breeds is of near zero significance in terms of overall global diversity, but genetic erosion in these populations is of particular human concern in so far as it has implications for food supply and the sustainability of locally-adapted agricultural practices. For domesticated populations, loss of wild relatives of crop or timber plants is of special concern for the same reason. These genetic resources may not only underlie the productivity of local agricultural systems but also, when incorporated in breeding programmes, provide the foundation of traits in terms of disease resistance, nutritional value, hardiness, etc. of global importance in intensive systems and which will assume even greater importance in the context of future climate change. Erosion of diversity in crop gene pools is difficult to demonstrate quantitatively, but tends to be indirectly assessed in terms of the increasing proportion of world cropland planted to high yielding, but genetically uniform, varieties.

Massive Extinctions From Human Activity

As well as the need for biodiversity for continued ecosystem survival, from a human perspective, from common drugs to possible cures for cancers, most of our medicine comes from plants, many of which are now endangered. However, it has long been feared that human activity is causing massive extinctions. It could be noted that the current extinction rate is now approaching 1,000 times the background rate and may climb to 10,000 times the background rate during the next century, if present trends continue. At this rate, one-third to two-thirds of all species of plants, animals, and other organisms would be lost during the second half of the next century. A huge report known as the Millennium Ecosystem Assessment, started in 2000, was released in March 2005. Amongst many warnings for humankind, it noted that there has been (as summarized from the *BBC*) a substantial and largely irreversible loss in the diversity of life on Earth, with some 10-30% of the mammal, bird and amphibian species currently threatened with extinction, all due to human actions.

A report from the World Wide Fund for Nature (WWF) in 2006 confirmed concerns from the previous year, noting that Already resources are depleting, with the report showing that vertebrate species populations have declined by about one-third in the 33 years from 1970 to 2003. At the same time, humanity's Ecological Footprint—the demand people place upon the natural world—has increased to the point where the Earth is unable to keep up in the struggle to regenerate. The International Union for Conservation of Nature (IUCN) notes in a video that: 1 bird out of 8, 1 mammal out of 4, 1 conifer out of 4, 1 amphibian out of 3, and 6 marine turtles out of 7, are all threatened with extinction

In addition, 75% of genetic diversity of agricultural crops has been lost, 75% of the world's fisheries are fully or over exploited, Up to 70% of the world's known species risk extinction if the global temperatures rise by more than 3.5°C, 1/3rd of reef-building corals around the world are threatened with extinction, Every second a parcel of rainforest the size of a football field disappears and Over 350 million people suffer from severe water scarcity.

The world environmental situation is likely to be further aggravated by the increasingly rapid, large scale global extinction of species. It occurred in the 20th century at a rate that was a thousand times higher than the average rate during the preceding 65 million years. This is likely to destabilize various ecosystems including agricultural systems. In a slow extinction, various balancing mechanisms can develop. Noone knows what will be the result of this extremely rapid extinction rate. What is known, for sure, is that the world ecological system has been kept in balance through a very complex and multifaceted interaction between a huge numbers of species. This rapid extinction is therefore likely to precipitate collapses of ecosystems at a global scale.

This is predicted to create large-scale agricultural problems, threatening food supplies to hundreds of millions of people. This ecological prediction does not take into consideration the effects of global warming which will further aggravate the situation. Industrialized fishing has contributed importantly to mass extinction due to repeatedly failed attempts at limiting the fishing. A new global study concludes that 90 percent of all large fishes have disappeared from the world's oceans in the past half century, the devastating result of industrial fishing. The study, which took 10 years to complete and was published in the international journal Nature, paints a grim picture of the Earth's current populations of such species as sharks, swordfish, tuna and marlin. The loss of predatory fishes is likely to cause multiple complex imbalances in marine ecology.

Another cause for extensive fish extinction is the destruction of coral reefs. This is caused by a combination of causes, including warming of oceans, damage from fishing tools and a harmful infection of coral organisms promoted by ocean pollution. It will take hundreds of thousands of years to restore what is now being destroyed in a few decades. According to the most comprehensive study done so far in this field, over a million species will be lost in the coming 50 years. The most important cause was found to be climate change.

Declining Ocean Biodiversity

It is not just fish in the oceans that may be struggling, but most biodiversity in the seas. This includes mammals such as whales, dolphins, polar bears, birds e.g. penguins, and other creatures e.g. krill.

In the past century, commercial whaling has decimated numerous whale populations, many of which have struggled to recover.

Commercial whaling in the past was for whale oil. With no reason to use whale oil today, commercial whaling is mainly for food, while there is also some hunting for scientific research purposes. Large scale commercialized whaling was so destructive that in 1986 a moratorium on whaling was set up by the International Whaling Commission (IWC). As early as the mid-1930s, there were international attempts to recognize the impact of whaling and try and make it more sustainable, resulting in the actual set up of the IWC in 1946. Many commercial whaling nations have been part of this moratorium but have various objections and other pressures to try and resume whaling.

Japan often claims its whale-hunting is for scientific research; the general population are often quite skeptical of such claims. Japan is the prime example of hunting whales for the stated aim of scientific research while a lot of skepticism says it is for food. Greenpeace and other organizations often release findings that argue Japan's whaling to be excessive or primarily for food, and for research as secondary. General public negativity of commercial whaling has also led to a difference between traditional whaling communities in the arctic region and conservationists. Traditional indigenous communities have typically hunted whale in far smaller numbers commercially, mostly for local food consumption, but the impacts of large-scale commercial whaling has meant even their hunting is under pressure.

The massive reduction in the local whale population meant the killer whales in that region that usually preyed on the younger whales moved to other animals such as seals.

– As seal numbers declined, the killer whales targeted otters
– As otter numbers were decimated, the urchins and other targets of otters flourished
– These decimated the kelp forests where many fish larvae grew in relative protection
– The exposed fish larvae were easy pickings for a variety of sea life

Fishermen's livelihoods were destroyed.

A 20-year study has shown that deforestation and introduction of non-native species has led to about 12.5% of the world's plant species to become critically rare. In fact, as an example, a study suggests that the Amazon damage is worse than previously thought, due to previously undetected types of selective logging and deforestation. A report from the World Commission on Forests and Sustainable Development suggests that the forests of the world have been exploited to the point of crisis and that major changes in global forest management strategies would be needed to avoid the devastation.

What also makes this a problem is that many of the endangered species are only found in small areas of land, often within the borders of a single country. New species of animals and plants are still being discovered. In Papua New Guinea, 44 new species of animals were discovered recently in the forests. Logging may affect these animals' habitats, though. The loss of rainforests around the world, where many species of life are found will mean that potential knowledge, whether medicinal, sustenance sources, or evolutionary and scientific information etc. could be lost. Brazil, which is estimated to have around 55,000 species of flora, amounting to some 22% of the world's total and India for example, which has about 46,000 and some 81,000 animal species amounting to some 8% of the world's biodiversity, are also under various pressures, from corporate globalization, deforrestation, etc. So too are many other biodiverse regions, such as Indonesia, parts of Africa, and other tropical regions.

Sustainable Forests or Sustainable Profits?

The overly corporate-led form of globalization that we see today also affects how natural resources are used and what priorities they are used for It is true that cutting down forests or converting natural forests into monocultures of pine and eucalyptus for industrial raw material generates revenues and growth. But this growth is based on robbing the forest of its biodiversity and its capacity to conserve soil and water. This growth is based on robbing forest communities of their sources of food, fodder, fuel, fiber, medicine, and security from floods and drought.

We hear more about sustainable forestry practices by the large logging multinationals. However, what does that really mean? Who is it sustainable for? Society and the environment, or for the logging companies? By replanting trees that will grow quickly and allow them to be felled for "sustained" logging sounds like a good strategy. However, the trees that are favored for this (eucalyptus) require a lot of water to grow so quickly. The eucalyptus trees achieve this rapid growth by tapping large quantities of groundwater, impoverishing surrounding vegetation and threatening to dry up local water courses.

Some government institutions even buy illegal timber from pristine forests. For example, it is claimed that UK buys all of its Mahogany from pristine forests in Brazil where 80% of all timber is traded illegally. Even though Brazil has now tried to introduce a moratorium on Mahogany logging for two years, this has been slammed by some as too little, too late.

People and Forests

Quite often we make blanket statements or generalized conclusions that people are the cause of deforestation. While that is true, unfortunately all people around the world are not equal, and it also also follows that some are more responsible for deforestation than others. Often, in forests of the Amazon, Africa, or Asia, forest protection schemes have been promoted that go against indigenous peoples and cultures, rather than work with them.

As Indian activist and scientist Vandana Shiva and others have shown in countless work, indigenous people often have their cultures and lifestyle structured in a way that works with nature and would not undermine their own resource base. For example, in her book *Stolen Harvests* she describes how their traditional knowledge has been beneficial to the environment and has been developed and geared towards this understanding and respect of the ecosystems around them.

Hopetoun falls, Australia; an example of trying to preserve nature while allowing tourism. Yet because of blanket

conclusions that humankind is responsible for deforestation, we risk assuming all types of societies are equally responsible for deforestation that is damaging to the environment.

Misuse of Land and Resources

How land is used to produce food can have enormous impacts on the environment and its sustainability. Intensive breeding of livestock and poultry for such restaurants leads to deforestation, land degradation, and contamination of water sources and other natural resources. For every pound of red meat, poultry, eggs, and milk produced, farm fields lose about five pounds of irreplaceable top soil. The water necessary for meat breeding comes to about 190 gallons per animal per day, or ten times what a normal Indian family is supposed to use in one day, if it gets water at all.

Overall, animal farms use nearly 40 percent of the world's total grain production. In the United States, nearly 70 percent of grain production is fed to livestock. In Indian Agriculture, women use up to 150 different species of plants as medicine, food, or fodder. For the poorest, this biodiversity is the most important resource for survival.

Because industrial agriculture promotes the use of monocultures, rather than a diversity of crops, the loss of biodiversity is leading to more resource usage, as described above. This as well as other political situations such as the motives for dumping surplus food on to developing countries to undersell the local farmers, leads to further hunger around the world.

Long Term Costs

If ecosystems deteriorates to an unsustainable level, then the problems resulting can be very expensive, economically, to reverse. Vanishing coral reefs, forests and other ecosystems can all take their toll and even make the effects of some natural events even worse. The cost of the effects together with the related problems that can arise like disease, and other illness, or rebuilding and so on is much more costly than the maintenance and sustainable development practices that could be used instead. As an example, and assuming a somewhat alarmist scenario, if enough trees and forests and related ecosystems vanish or deteriorate sufficiently:

- Then the oxygen-producing benefits from such ecosystems is threatened.
- The atmosphere would suffer from more pollution.
- The cost to tackle this and the related illnesses, problems and other cascading effects would be enormous as it can be assumed that industrial pollution could increase, with less natural ecosystems to "soak" it up
- Furthermore, other species in that ecosystem that would depend on this would be further at risk as well, which would lead to a downward spiral for that ecosystem.

Compare those costs to taking precautionary measures such as protecting forests and promoting more sustainable forms of development. Of course, people will argue that these situations will not occur for whatever reasons.

Amphibians

The class Amphibians is made up of frogs, toads, newts, salamanders and caecilians—little known legless organisms that resemble giant earthworms. Nearly one third of the approximately 6000 known amphibian species are threatened with extinction. These animals produce a wide range of novel substances, some of which are made only by amphibians living in the wild, not by those in captivity.

These include the:-

Pumiliotoxins, like those made by the Panamanian Poison Frog that may lead to medicines that strengthen the contractions of the heart and thus prove useful in treating heart disease. Alkaloids made by species like the Ecuadorian Poison Frog, which could be the source of a new and novel generation of pain-killers. Antibacterial compounds produced in the skin of frogs and toads such as the African.

Clawed Frog and South and Central American Leaf Frogs

Bradykinins and maximakinins, made in the skin glands of species like the Chinese Large-Webbed Bell Toad; Mexican Leaf Frog, and North American Pickerel Frog that dilate the smooth muscle of blood vessels in mammals and therefore offer promising avenues for treating high blood pressure. Frog glue, produced by species such as the Australian frog, could lead to natural adhesives for repairing cartilage and other tissue tears in humans. Many species of newts and salamanders, such as the Eastern Spotted Newt, can re-grow tissues such as heart muscle; nerve tissue in the spinal cord and even whole organs. As we are in evolutionary terms relatively closely related to these species, they are vital models for understanding how we might someday harness our own dormant regenerative potential. Some frogs, such as the Gray Tree Frog and the Chorus Frog can survive long periods of freezing without suffering cell damage—understanding how these frogs do this may yield key insights into how we might better preserve scarce organs needed for transplant.

Declining Amphibian Populations

Amphibians are particularly sensitive to changes in the environment. Amphibians have been described as a marker species or the equivalent of "canaries of the coal mines" meaning they provide an important signal to the health of biodiversity; when they are stressed and struggling, biodiversity may be under pressure. When they are doing well, biodiversity is probably healthy. Unfortunately, as has been feared for many years now, amphibian species are declining at an alarming rate.

The Golden Toad of Monteverde, Costa Rica was among the first casualties of amphibian declines.

Formerly abundant, it was last seen in 1989. Malcom MacCallum of the Biological Sciences Program, Texas A&M University calculated that the current extinction rate of amphibians could be 211 times the background amphibian extinction rate. He added that "If current estimates of amphibian species in imminent danger of extinction are included in these calculations, then the current amphibian extinction rate may range from 25,039–45,474 times the background extinction rate for amphibians. It is difficult to explain this unprecedented and accelerating rate of extinction as a natural phenomenon." Dwindling fish stocks *IPS* reports that fish catches are expected to decline dramatically in the world's tropical regions because of climate change. Furthermore, "in 2006, aquaculture consumed 57 percent of fish meal and 87 percent of fish oil" as industrial fisheries operating in tropical regions have been "scooping up enormous amounts of fish anchovies, herring, mackerel and other small pelagic forage fish to feed to farmed salmon or turn into animal feed or pet food." This has resulted in higher prices for fish, hitting the poorest the most. It is significant to note that, mass extinctions of marine life due to industrialized fishing has been a concern for many years. Yet, it rarely makes mainstream headlines. However, a report warning of marine species loss becoming a threat to the entire global fishing industry did gain media attention.

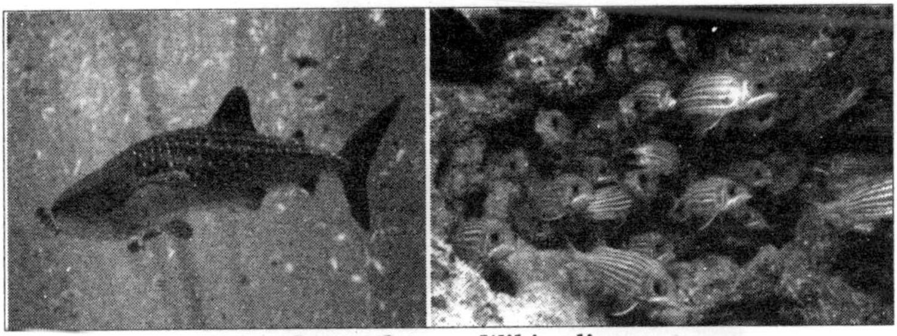

Image Source: Wikipedia

A research article in the journal, *Science,* warned commercial fish and seafood species may all crash by 2048. At the current rate of loss, it is feared the oceans may never recover. Extensive coastal pollution, climate change, over-fishing and the enormously wasteful practice of deep-sea trawling are all contributing to the problem. Each form of life works together with the surrounding environment to help recycle waste, maintain the ecosystem, and provide services that others—including humans—use and benefit from. For example, as Steve Palumbi of Stamford University noted, the ocean ecosystems can take sewage and recycle it into nutrients; Scrub toxins out of the water; Produce food for many species, including humans, Turns carbon dioxide into food and oxygen, With massive species loss, the report warns, at *current* rates, in less than 50 years and the ecosystems could reach the point of no return, where they would not be able to regenerate themselves.

Dr. Boris Worm, in ocean research, commented that: Whether we looked at tide pools or studies over the entire world's ocean, we saw the same picture emerging. In losing species we lose the productivity and stability of entire ecosystems. In an update to the above story, 3 years later, 2009, Dr. Worm was a bit more optimistic that some fish stocks can rebound, if managed properly. But it is a tough challenge "since 80 percent of global fisheries are already fully or over-exploited. "An example of over fishing that has a ripple-effect on the whole fish-food chain is shark hunting.

The Great White Shark is the largest predatory fish. Millions of sharks are killed each year from over fishing and trade. Many die accidentally in fishing nets set for tuna and swordfish, while others are caught for their meat or just for their fins. A demand for shark-fin soup in places like China and Taiwan is decimating shark populations. Shark fin soup

is considered a delicacy not even a necessity and can be extremely lucrative. So much money can be obtained just from the fin that fishermen hunting sharks will simply catch sharks and cut off their fins while they are alive, tossing the wriggling shark back into the ocean to die, as it cannot swim without its fin. This saves a lot of room on fishing boats. Sharks are known as the "apex predator" of the seas. That is because in general sharks are at the top of the food chain. Without sufficient shark numbers the balance they provide to the ecosystem is threatened because nature evolved this balance through many millennia.

Bears

Nine species of bear are threatened with extinction including the polar bear; the Giant Panda, and the Asiatic Black Bear. The threats to bears are similar to those amphibians face, but in addition many bears are at risk because they are killed for body parts, such as gall bladders, which can command high prices in black markets in places like China, Japan and Thailand. Several medical benefits have already arisen from the study of bears, including the development of ursodeoxycholic acid, found in the gall bladders of some bear species such as polar and black bears, into a medicine.

The substance is used to prevent the build up of bile during pregnancy; dissolve certain kinds of gallstones; and prolong the life of patients with a specific kind of liver disease, known as primary biliary cirrhosis, giving them more time to find a liver transplant. Some bear species, known as "denning" bears because they enter into a largely dormant state when food is scarce, are of tremendous value to medicine as they are able to recycle a wide variety of their body s substances.

Unlike people, who if „bed-ridden for a five-month period can lose up to a third of their bone mass, bears actually lay down new bone during the denning period. Bears appear to produce a substance that inhibits cells that break down bone and promote substances that encourage bone and cartilage-making cells. Currently, 740,000 deaths a year are the result of hip fractures worldwide, a large number of which are caused by osteoporosis. By 2050 there will be an estimated six million osteoporosis-linked hip fractures globally.

Denning bears can survive for a period of five months or more without excreting their urinary wastes, whereas humans would die from the build up of these toxic substances after only a few days. An estimated 1.5 million people worldwide are receiving treatment for end-stage renal disease, and more than 80,000 die each year in the U.S. alone from this disease. By studying denning bears, we may be able to learn how to treat them more effectively and help large numbers to survive. Denning bears may also hold clues to treating Type 1 and Type II diabetes as well as obesity. Worldwide there are an estimated 150 to 200 million cases of Type II diabetes. When produced in a non-invasive and ethically acceptable way, without pushing already threatened species further towards extinction, these substances are of great value to medicine.

Gymnosperms including Pines and Spruces

Close to 1,000 species of Gymnosperms have been identified. Evolutionary they are among the oldest of any plants alive but many groups, such as the cycads, are classified as endangered. Several pharmaceuticals, including decongestants and the anti-cancer drug taxol, have already been isolated from gymnosperms. The researchers believe many more are yet to be discovered and may be lost if species of Gymnosperms become extinct. Substances from one Gymnosperm, the Ginkgo tree may reduce the production of receptors in the human nervous system linked with memory loss. Thus they may play a role in countering Alzheimer s disease. They may also help in the treatment of epilepsy and depression.

Cone Snails

Around 700 species make up the cone snails, seven of which were identified only since 2004. While only four are now classified as vulnerable, no thorough assessment has been made in over ten years and thus current listings may underestimate the true number of endangered cone snail species. For example almost 70 per cent of some 380 cone snail species surveyed had more than half their geographic range within areas where coral reefs, their main habitats, are threatened.

Cone snail species may produce as many as 70,000 to 140,000 peptide compounds, large numbers of which may have value as human medicines, yet only a few hundred have been characterized. One compound, known as ziconotide, is thought to be 1000 times more potent than morphine and has been shown in clinical trials to provide significant pain relief for advanced cancer and AIDS patients. Another cone snail compound has been shown in animal models to protect brain cells from death during times of inadequate blood flow. It could prove a breakthrough therapy for people suffering head injuries and strokes and may even contribute to therapy for patients with Parkinson s and Alzheimer s. Other potential developments from cone snail peptides include treatments for urinary incontinence and cardiac arrhythmias.

Sharks

There are at least 400 species of sharks, which, as a group, evolved in ancient seas 400 to 450 million years ago. Many species are now threatened, with some species, such as the Scalloped Hammerhead, White Shark and Thresher Shark, falling in numbers by as much as 75 percent over the past 15 years. Over-fishing has been the main reason for the losses, and has been driven by: an increased demand for shark meat as a substitute for traditional commercial fish catches in foods like fish and chips; the rise in consumption of shark fin soup; increases in by-catch, for example, in tuna fisheries; and an increased market for shark cartilage products for a variety of unproved medical purposes.

Squalamine, a substance isolated from sharks such as dogfish, especially abundant in their livers, may lead to a new generation of antibiotics as well as treatments against fungal and protozoan infections.

Studies are also being undertaken with squalamine compounds as possible antitumor and appetite-suppressant substances. Trials are now also underway to see if squalamine can treat age-related macular degeneration which can lead to severe vision loss. The shark substance may halt the growth of new blood cells in the retina, which is linked to a loss of retinal function and blindness in these patients. The salt glands of some sharks are also being studied to gain insight into how the human kidney functions and how chloride ions are transported across membranes, which may shed light on two diseases—cystic fibrosis and polycystic kidney disease.

Sharks, having evolved as some of the first creatures with a fully functioning „adaptive immune system are irreplaceable models to help us understand human immunity. "What potential these creatures may still hold to further our knowledge of immunity is being rapidly depleted with the mass slaughter of sharks and the endangerment of sharks worldwide".

Horseshoe Crabs

There are four species of horseshoe crabs, with each organism possessing four eyes and six other light-detecting organs as well as blood that turns cobalt blue when exposed to the air. Because only around ten offspring survive out of the estimated 90,000 eggs produced by a female, they are highly sensitive to overfishing. Once harvested and processed to be used as fertilizer, they are now used as bait for eel and whelk fisheries. Horseshoe crabs are also important in the food chain, especially for Extinction-prone groups of species Human impacts on the environment–such as habitat loss, over-exploitation, species introductions, and pollution–do not threaten all groups of species equally. At greatest risk are species with small population sizes, species whose populations vary greatly and species with slow rates of population growth. More specifically, the following groups of organisms are particularly susceptible to extinction:

Species at Higher Trophic Levels

Species high in the food chain tend to be large, rare animals with slow rates of population growth. Particularly susceptible to over-exploitation or habitat loss, these species include the Caribbean monk seal, great auk (*Alca impennis*), Steller's sea cow (*Hydrodamalis stelleri*), thylacine (*Thylacinus cynocephalus*, also known as the Tasmanian wolf), and many endangered birds of prey, cats, wolves, foxes, and whales.

Local Endemics

Species with restricted ranges are often threatened by habitat loss. For example, the entire natural habitat of the Devil's Hole pupfish (*Cyprinodon diabolis*) is a spring measuring roughly 3 by 15 meters. Water development, pollution, or habitat alteration could easily drive the species to extinction in the wild. The extremely high rate of island species extinction further testifies to the threat that local endemics face.

Species with Chronically Small Populations

This category overlaps the first. Since many species at higher trophic levels have sparsely distributed populations, habitat restriction or fragmentation may reduce their populations to very small levels. However, the population sizes of species at lower trophic levels may also be extremely small in a given habitat or region. Many tropical forest tree species, for instance, have very low population densities and are thus threatened by habitat loss and fragmentation.

Largest Members of a Guild

Whether or not a species is on the top trophic level, large species have high metabolic demands, require large habitats, and tend to occur in low densities. Thus, the largest species within a group of species sharing similar food sources a guild tend to be a high risk of extinction. For example, all of the lemurs that have died out since Madagascar was colonized by human beings were as large or larger than the surviving species.

Species with Poor Dispersal and Colonization Ability

As with local endemics, species with narrow habitat requirements and species that can't disperse easily to new habitats are at high risk of extinction, even if their population is widespread. For example, in the face of a warming climate, some of the most threatened species will be those that can't disperse as fast as suitable habitat moves to higher elevations and latitudes.

Species with Colonial Nesting Habits

Colonial nesting species are particularly susceptible to over-exploitation or the loss of breeding habitat. Several species of marine turtles, for example, have been threatened by the development of nesting beaches for tourism and other purposes. Similarly, the vulnerability of the extinct passenger pigeon (*Ectopistes migratorius*) to hunters was increased by the birds' tendency to nest in enormous colonies.

Migratory Species

Migratory species depend upon suitable habitat in their summer and winter range *and* along the course of their migratory route. Thus, the potential for adverse effects of habitat changes on migrant populations is high. In North America, Kirtland's warbler (*Dendroica kirtlandi*), Backman's warbler (*Vermivora backmanii*), and the Whooping Crane (*Grus americana*) are endangered because habitat in both their breeding and wintering ranges has shrunk.

Species Dependent on Unreliable Resources

These species' populations fluctuate greatly, so they face increased threats of extinction when their population is low. Populations of frugivores and nectarivores, for example, may be reduced significantly during years when fruit or nectar crops are low.

Species with Little Evolutionary Experience with Disturbances

In regions where human beings have a longstanding presence, the species most sensitive to human disturbance have already been lost and many of the remaining species have adapted to the additional disturbance. In contrast, species are extremely vulnerable where human disturbance has no historical precedent. Thus, the loss of the great auk, passenger pigeon, elephant birds, and moas, and the endangerment of the musk oxen (*Ovibos moschatus*) were all related to the introduction of a predator where none had existed before or to the introduction of a predator with a new means of hunting – say, the rifle–that the prey had not previously experienced.

Biodiversity Loss : Cascade Effects

Cascade Effects

The loss of a species can have various effects on the remaining species in an ecosystem–what kind and how many depends upon the characteristics of the ecosystem and upon the species' role in its structure. Cascade effects occur when the local extinction of one species significantly changes the population sizes of other species, potentially leading to other extirpations. Such cascade effects are particularly likely when the lost species is a "keystone predator," a "keystone mutualist," or the prey of a "specialist predator."

Keystone Predators

How much the loss of a predator affects its prey population size depends on how much the predator limits the prey population. If the size of the prey population is determined by factors other than predation, it is said to be *donor-controlled.* For example, such bottom-dwelling invertebrates as mussels or barnacles consume only a small fraction of the ocean's plankton; consequently, removing these invertebrates would not affect the plankton's population size much. Similarly, although the passenger pigeon *(Ectopistes migratorius),* Carolina parakeet *(Conuropsis carolinensis),* and wild turkey *(Meleagris gallopavo)* were the dominant large seed-predators of eastern North America's forests, the extinction of the first two species and the near-extinction of the third apparently changed the forest little. The loss of the bird species increased seed survival, but this increase had no effect on the density of trees because what limited the tree populations was the availability of space, not seed predation.

In a *predator-controlled* system, the size of the prey population is determined by predation. In such systems, the impact of the loss of the predator can be substantial, especially if the predator is relatively high in the food chain. So-called keystone predators affect not only their prey's population size but also the community's species diversity. By limiting the population size of a species that would otherwise outcompete other species, keystone predators can help maintain high species diversity. When the keystone predator *Pisaster ochraceus* a starfish is removed from the intertidal zone in the Pacific northwest of the United States, the intertidal community changes from one with a high diversity of relatively large bottom-dwelling invertebrates to a virtual monoculture of the starfish's favorite prey–the mussel *(Mytelus edulis).*

Keystone predators represent a vexing problem for conservation. Many keystone predators, such as sea otters, face a high risk of extinction because they tend to be high in the food chain and relatively sparsely distributed. Yet, determining which species play keystone roles in a community may be extremely difficult. For example, the small moth *Cactoblastis catorum*, introduced into Queensland from South America, appears to be a minor component of the Queensland community since its population size and biomass aren't great. However, it is a keystone predator in Australia and probably also in South America. In the absence of the moth, *Opuntia* cactus also introduced from South America covered 62,000 square kilometers of Queensland but in its presence *Opuntia* has been reduced to small isolated patches. In general, determining whether a given species plays a keystone role requires considerable study and experimentation.

Although certain species have much more influence than others on an ecosystem's structure, not all ecosystems include a single species that exerts such a pervasive influence. In fact, most ecosystems are somewhat sensitive to the loss of any one of many species, though some losses have greater impact on the system than others. Nevertheless, determining which species function as keystone predators in an ecosystem can help managers find-tune conservation activities.

Mutualists and Keystone Mutualists

Species involved in mutualisms–mutually beneficial interactions such as pollination–are deeply affected if their partner mutualists are lost. For example, the loss of a fruit-eating species may severely affect the plants whose fruits it disperses. Similarly, rare tropical plants dependent on specific species for pollination would be threatened with extinction by the loss of the pollinator. Some orchids, for instance, rely on a single species of euglossine bee for pollination; without the bee, the plant would become extinct. By the same token, pollinators that specialize on a specific plant species could die out if the plant does. Most species of fig *(Ficus* spp.) are each pollinated by a different species of wasp, and each species of wasp pollinates only one fig, so the loss of a fig species could wipe out one of the specialist wasps.

Maintaining plant-pollinator interactions in tropical forests is of utmost importance since tropical forests are much more sensitive to extinctions of pollinators than are temperate forests. Whereas less than 30 percent of trees in the northern United States are pollinated by animals as opposed to wind, more than 95 percent of the trees in one dry forest in Costa Rica are believed to rely on animal pollination. Certain species–keystone mutualists–involved in mutualistic interactions may sometimes assume great importance in the community. For example, during the dry season in the tropical forest at the Cocha Cashu Biological Station in Manu National Park of southeastern Peru, only 12 of the approximately 2,000 plant species support the entire fruit-eating community of mammals and birds. During this period, fruit production drops to less than 5 percent of the peak production, so the food available to the fruit-eating species is severely restricted. The 12 plant species still producing fruit and nectar meet the food needs of as much as 80 percent of the entire mammal community and a major fraction of the avian species at the site. Clearly, the loss of one or more of these "keystone mutualists" could significantly harm forgiver populations.

While the identification of keystone predators may be extremely difficult, keystone mutualists can generally be identified through non-experimental observations of the resources that the species use. Once identified, keystone mutualists can be managed to ensure the survival of the species dependent upon them. By increasing the abundance of the keystone mutualists for instance, it might be possible to increase populations of dependent species.

Specialist Predators

In general, the loss of a species tends to have greater impacts on its own predators if the species is relatively high in the food chain. Populations of such species, including carnivores, tend to be limited by the amount of prey available, whereas species low in the food chain, such as herbivores, are more often limited by their predators or pathogens. Consequently, the loss of an herbivore tends to directly reduce the supply of the factor i.e. food limiting the carnivore population, while the removal of a plant species may have no effect on the herbivore population.

One exception to this pattern occurs when the plant species removed is the prey of a specialist predator–that is, a species with no alternative food source, For example, many insects eat just one species of plant because they have evolved mechanisms for detoxifying the plant's chemical defenses. If a such a plant were lost, the highly specialized insect would be too. Widespread and common prey species are more likely to have specialist predators that are rare or locally endemic species, so the loss of common species would tend to have a large impact on species higher in the food chain.

Habitat Loss Blamed for More Species Decline

Habitat loss is having a serious impact on Europe's butterflies, beetles and dragonflies. The release of the European Red List, commissioned by the European Commission, shows that nine percent of butterflies, 11 percent of saproxylic beetles i.e. beetles that depend on decaying wood) and 14 percent of dragonflies are threatened with extinction within Europe. Some species are so threatened that they are at risk of global extinction and are now included in the latest update of the IUCN Red List of Threatened Species.

"When talking about threatened species, people tend to think of larger, more charismatic creatures such as pandas or tigers, but we mustn't forget that the small species on our planet are just as

Nickerl's Fritillary (Melitaea Aurelia)

important, and are also in need of conservation action. Butterflies, for instance, play a hugely pivotal role as pollinators in the ecosystems in which they live".

According to new studies commissioned by the European Commission and carried out by IUCN, Butterfly Conservation Europe and the European Invertebrates Survey, nearly a third (31 percent) of Europe's 435 butterfly species have declining populations and nine percent are already threatened with extinction. For example, the Madeiran Large White Butterfly (*Pieris wollastoni*) is Critically Endangered (possibly extinct), having not been seen on Madeira for at least 20 years, and the Macedonian Grayling Butterfly (*Pseudochazara cingovskii*) in the Former Yugoslav Republic of Macedonia is also Critically Endangered because quarrying activities are reducing its habitat.

"Most butterflies at risk are confined to southern Europe; their main threat is habitat loss, most often caused by changes in agricultural practices, either through intensification or abandonment, or to climate change, forest fires and the expansion of tourism"'. For the first time, saproxylic beetles have been assessed for the IUCN Red List. These beetles are unique because they are highly dependent on decaying wood, particularly in forests, and play an essential role in recycling nutrients. A third of the 431 species assessed are unique to Europe. Almost 11 percent (46 species) are at risk of being lost from the region, and seven percent (29 species) are threatened with extinction at the global level. A further 13 percent (56 species) are listed as Near Threatened within Europe.

The main long-term threats to saproxylic beetles are habitat loss due to logging and the decline in the number of mature trees. The Violet Click Beetle (*Limoniscus violaceous*) is an Endangered species that typically lives in large tree cavities containing wood mould. It is under threat from changing woodland management practices. Dragonflies occur almost everywhere in Europe, with the highest numbers in southern France, the foothills of the Alps and parts of the Balkan Peninsula. Fourteen percent of the 130 dragonfly species assessed are at risk; five of these are threatened with global extinction. A further 11 percent are considered Near Threatened within Europe. Like butterflies, most of the threatened species are confined to southern parts of Europe. Increasingly hot and dry summers combined with intensified water extraction for drinking and irrigation are causing the dragonflies' wetland habitats to dry up.

Three of the most threatened dragonflies of Europe are endemic to the brooks and small rivers of Greece and nearby countries, including Albania, Bulgaria and Turkey. If no action is taken species like the Greek Red Damsel may become extinct during the first half of this century.

Mass Extinction Rates

Extinction rates have varied considerably over the history of life on earth. Paleontologists distinguish five episodes of "mass extinctions"–relatively short (1 million to 10 million year) periods during which a significant fraction of diversity in a wide range of taxa went extinct. The most significant mass extinction, as the end of the Permian (250 million years ago), may have eliminated 77 to 96 percent of species. Even apart from these mass extinctions, background rates of extinction are not constant. For example, for the past 250 million years, relatively high rates of extinction have occurred nine times–at intervals of approximately 26 to 28 million years. Two of these nine episodes were mass extinctions, one in the late Triassic, 220 million years ago, and one in the late Cretaceous, 65 million years ago.

Global biological diversity is now close to its all time high. Floral diversity, for example, reached its highest level ever several tens of thousands of years ago. Similarly, the diversity of marine fauna has risen to a peak in the last few million years.

Humanity's Contribution to Species Extinctions

Humanity's first significant contribution to the rate of global extinction may have occurred 15,000 to 25,000 years ago, when hunting of large mammals apparently caused or contributed to significant extinctions in North and South America and

Australia. These three continents lost 74 to 86 percent of the genera of "megafauna"-mammals greater than 44 kg-at a time. While the cause of these extinctions remains a matter of controversy, even if humanity is not wholly responsible, there is no doubt that for millennia, people have significantly altered the landscape with untold effects on native flora and fauna. For at least 50,000 years, intentional burning has occurred in the savannas of Africa. At least 5,000 years ago, in Europe, deforestation and the conversion of wildlands to pasture began and there is evidence in North America that for as long as 4,000 years indigenous peoples influenced the structure of forest communities, provided opportunities for weedy species and such herbivores as bison to expand their ranges, and caused at least local species extinctions. In Central America, forest had already been removed from large areas before the Spanish arrived. The prehistoric colonization of islands by human beings and their commensals substantially affected the diversity of island species. Fossil evidence suggests that 98 species of endemic birds were present in the Hawaiian Islands in A.D. 400 when the islands were first colonized by Polynesians. About 50 of these species became extinct before the first European contact in 1778.

Most experts believe that these extinctions resulted from a combination of the clearing of extensive tracts of lowland forest for agriculture, predation and disturbance by introduced species (the Polynesian rat, domestic pig, and domestic dog), and hunting. Similarly, following human colonization of New Zealand in A.D. 1000, the introduction of the domestic dog and the Polynesian rat, combined with the deforestation of large areas by fire and intensive hunting of larger birds, led to the extinction of 13 species of moas i.e. large flightless birds and 16 other endemic birds before the arrival of Europeans. Humanity is thought to have caused other extinctions following the colonization of Madagascar in A.D. 500 and the Chatham Islands in A.D. 1000. Early human colonization of oceanic islands may have led to the extinction of as many as one-quarter of the bird species that existed several millennia ago.

Endemic vs. Introduced Species

In the 15th and 16th centuries, the global spread of European cultures and their attendant livestock, crops, weeds, and diseases increased the loss of island flora and fauna and added to the threats to continental species. In later centuries, the growth of trans-oceanic human travel and commerce led to the spread of a tremendous variety of species to new regions of the world and to the human colonization of many uninhabited islands. Between 1840 and 1880, more than 60 species of vertebrates were released in Australia. Between 1800 and 1980, the number of introduced insect species in the United States grew from about 36 to more than 1200. Many of the European introductions and colonizations, like those of earlier colonizers, significantly influenced native flora and fauna. In Hawaii, the arrival of european explorers added cats, two new species of rats, the barn owl, the small Indian mongoose, and several avian diseases. In the next two centuries, habitat degradation, disease, and predation caused the loss of 17 endemic bird species, reducing the endemic avifauna to 31 percent of the diversity found in A.D. 400; several more species now verge on extinction.

Europeans first visited the uninhabited Mascarene islands such as Mauritius, Reunion, and Rodrigues in the early 1500s and released pigs and monkeys on the islands. In the mid-1600s, the Dutch settled the islands, and in the next 300 years twenty species of birds–including the Dodo–and 8 species of reptiles were lost. The extreme vulnerability of island endemics is exemplified by the fate of the flightless Stephen Island wren, driven to extinction by a single cat owned by a lighthouse keeper on an islet off New Zealand.

The relationship between biodiversity and ecosystem functioning has emerged as a central issue in ecological and environmental sciences during the last decade. Increasing domination of ecosystems by humans is steadily transforming them into depauperate systems. Because ecosystems collectively determine the biogeochemical processes that regulate the earth system, the potential ecological consequences of biodiversity loss have aroused considerable interest.

Recent experimental and theoretical work in this area has also led to animated debates and controversies. These controversies are currently in the process of being resolved. We now know that the productivity and nutrient retention ability of grassland ecosystems are affected by both local complementarily between species with different ecological niches and regional stochastic processes involved in community assembly. A moderate number of key complementarily species are probably critical to explain the level of primary production observed in grassland ecosystems at the small spatial and temporal scales considered in recent experiments.

These experiments, however, are very likely to underestimate the true extent of the impacts of biodiversity loss on ecosystem functioning. First, there is theoretical and experimental evidence that the diversity of functionally similar species can also buffer ecosystem processes against environmental fluctuations and enhance their long-term average magnitude. Second, heterogeneous environments are best used by an array of species with different specializations, and habitat destruction and fragmentation may prevent appropriate dominant species from being recruited in each community. The larger the temporal and spatial scales considered, the higher the diversity that is likely to be needed to maintain particular ecosystem processes.

Priorities in Conservation

Given that some species and genes are sure to be lost and some natural habitats converted to other land uses in the coming decades, how should scarce human and financial resources be spent to minimize biotic impoverishment and maximize biodiversity's contribution to human well-being? How can the most useful components of biodiversity be saved for current or future use? And how can such ethical considerations as our responsibility to other species and to future generations be incorporated into conservation priorities?

Clearly, these questions cannot be answered with scientific information on the nature, distribution, and status of biodiversity alone. Such factors as awareness of the problem and its solutions, perceived trade-offs with other basic human needs, and local political considerations will all influence conservation priorities.

Distinctiveness: Numbers Aren't Everything.

To conserve the widest variety of the world's life forms and the complexes in which they occur, the more distinctive elements of this diversity should receive high priority. For example, preserving an assemblage of plants and animals is comparatively more important if those species are found nowhere else or are not included in protected areas. Similarly, if the choice is between conserving a species with many close relatives or a species with few, the more distinctive species should be saved.

Maintaining the highest number of species without considering their taxonomic position makes little sense, as a comparison of marine and terrestrial environments shows. Terrestrial environments contain at least 80 percent of the world's total species, mainly because vascular plants and insects are so numerous on land-accounting for nearly 72 percent of all described species in the world-and so poorly represented in marine environments. However, the sea contains greater proportions of higher taxonomic units. Marine ecosystems contain representatives of some 43 phyla while terrestrial environments are home to only 28 phyla. The sea contains fully 90 percent of all classes and phyla of animals. Clearly efforts to conserve the widest array of biodiversity must give attention to all levels of the taxonomic hierarchy.

Threat: Saving the Most Beleaguered Species and Ecosystems First

In general, the threat to biological diversity is influenced by how widespread species are, how common they are over their range, and by such human pressures as harvesting, land conversion, and environmental pollution. The threats to species vary dramatically by region, and in each region some ecosystems are more threatened than others. For instance, Central America's tropical rainforests are less threatened than the remaining fragments of tropical dry forest in that region. In such cases, the dry forests should receive the most attention even though the rainforests contain more species.

These three factors–distinctiveness, utility, and threat–help make value judgments more explicit as conservation priorities are set. If a strictly global perspective is adopted, then preserving the most species-rich sites in the world seems most important. Indeed, this global perspective has led some organizations to focus on conservation in the "megadiversity" countries–Brazil, Colombia, Indonesia, Madagascar, Mexico, Zaire and others. In contrast, from a nationalist perspective, each country's biodiversity is worth roughly what every other country's is. Similarly, if today's needs for genes, species, or ecosystems are considered paramount, tomorrow's needs can't be too.

Biotic Impoverishment: The Impact of Agriculture

Because biodiversity incorporates differences at all levels of biological organisation, from individuals to ecosystems, and substantial change in a landscape will bring about a change in biodiversity. Such changes are of particular concern when patterns of landuse cause the loss of species or changes in their distribution. Mooney et al. (1995) noted that in agricultural regions throughout Australia trajectories of change are consistently towards declining natural diversity. This decline in diversity results in changes to the rates and pathways of ecosystem processes that are essential for conserving biodiversity and sustaining agriculture.

Changes in Biota

Briggs & Leigh (1996) pointed out that twenty four plant species are known to have been lost from the wheatbelt as a whole and the area now has one of the highest numbers of rare and/or endangered plant species in Australia. Of the 348 plant species listed as rare or endangered in the wheatbelt, only 79 occur in designated nature reserves. Of the remainder, 135 species occur in road verges and 53 are found on privately owned remnants.

Changes to Ecosystem Processes

Changes to ecosystem functions have resulted from changes in physical processes due to altered land-cover, as well as from changes in the diversity of species that participate in the movement of nutrients, water and energy throughout the system.

Physical Changes

Hobbs (1992) conducted a research in a range of agricultural regions throughout the world indicate that extensive land clearing can modify patterns of radiation and alter fluxes of wind, water and nutrients across landscapes. Radiation fluxes tend to be more variable in agricultural fields than they are in remnant vegetation both throughout the day and between seasons, resulting in greater local temperature extremes. Where natural bushland and agricultural fields meet, the higher radiation from the bare ground affects the microclimate at the remnant edge causing changes in plant composition and structure (Palik & Murphy 1990). Altered radiation levels in fields are also likely to modify soil micro-habitats for soil-dwelling invertebrates. Increased exposure to wind affects conditions at the edge of remnants, with increased frequency of tree blow-downs, increased evapotranspiration and increased transfer of weed seeds, nutrients and agricultural chemicals from farmland into vegetation remnants. Similarly, increased surface water flow brings nutrients, chemicals and weed seeds into remnants (Cale & Hobbs 1991).

Changes in vegetation cover have affected the hydrological balance in the catchment causing a rise in saline water tables which impacts initially on vegetation remnants in the lower valley floors but, as water tables continue to rise, vegetation located higher up the slopes is increasingly being threatened.

Biotic Changes

Hobbs (1992) pointed out that considerable attention has been paid to changes in ecosystem function associated with changes in biodiversity. Much of the debate revolves around the question of whether greater levels of diversity result in greater ecosystem stability or resilience in the face of environmental variability. MacArthur (1957) argued that a greater number of species provides more pathways for energy to reach a consumer. Consequently, the loss of any one pathway would be less likely to have a detrimental impact on that consumer. Elton (1958) also contended that the increasing complexity of an ecosystem that resulted from greater species diversity should generate greater stability. However, Cody (1986) found that high diversity in Mediterranean-climate regions tends to be associated with a greater probability of rarity and a consequent greater risk of extinction. However, the loss of a rare species from a diverse system is likely to have less impact on ecosystem processes than would the loss of a more abundant species from a less diverse system. This is because a more diverse system will have a greater probability of containing other species which play a similar functional role and hence can compensate for the loss of a particular species. Such compensation may not occur if a 'keystone' species is lost. This is a species which has an impact on ecosystem processes that is disproportionate to its representation in the community. In such circumstances, a small change in diversity can have a significant impact on ecosystem function. Mooney et al. (1995) noted that the impact of changes in biodiversity on function will therefore depend upon a number of factors including the number of species in the system, the relative abundance of those species, the functional role played by the species lost or gained and the relationships between these and other species in the system.

Protecting Biological Diversity: The Need for Clear Objectives

Before any attempt is made to address conservation issues in landscapes used for production it is essential that the objective of the exercise is clearly identified. Two broad approaches to nature conservation can be considered: General enhancement, which attempts to maximise the number of indigenous species retained or, alternatively, to minimise the number lost within constraints imposed by other land use objectives. Strategic enhancement, which aims to ensure the persistence of particular species, groups of species, or all species that currently occur in a landscape. This type of approach could be extended to include the reintroduction of species that have been lost from the landscape being managed. The objective of the first type of approach - to maximise the number of species retained, or to minimise the number lost - is an open-ended objective. It identifies a general trajectory along which to proceed, but does not specify targets which can be used to assess success or failure. This type of approach does not consider which species will be conserved or lost but simply aims to increase the probability that any given species will persist. It is a 'general enhancement' objective, in that it aims to 'make things better or 'minimise the impact' in an unspecified manner within the constraints of other land uses.

Approaches based on retaining or reintroducing specified components of the biota can be considered to be 'strategic' because they require specification of the landscape elements and management regimes that are required to meet a specific objective. They are more rigorous because they have quantifiable outcomes by which we can judge the effectiveness of our actions. If our objective is to retain all of the biota or a particular component of that biota, then the loss of any species, or of particular designated species would constitute failure. Similarly, if our goal is to reintroduce particular species that have been lost from an area, the failure of those species to become established is a clear indication that the objective has not been met.

Biodiversity conservation is a challenging organizational task which involves: 1) the diversity of ecological systems and their components which have multiple attributes, e.g., with regards to their nature, private or public goods, and 2) the diversity of social systems, i.e. multiple stakeholders and their behavior, interactions, and rules. Coping with complexity in biodiversity conservation not only requires an understanding of the nature of the good to be conserved but also of how people perceive it and organize to deal with the task of conservation. The argument we bring forth is, that in order to cope with this challenge we need to 1) apply valuation methods which are from the complexity stance, 2) take better account of intrinsic values and feelings, as well as 3) consider social rationality. We think that decision-support in biodiversity conservation is potentially of little value because it neglects the complexity of the task, which is defined by social rationality and social processes of value construction, including intrinsic values, and feeling. The case we refer to in order to demonstrate how to provide decision-support in complex problem settings is the loss of coffee genetic resources in the montane rainforests of Southwest-Ethiopia. The world's *Coffea arabica* stocks originate from wild populations growing in these rainforests in Ethiopia, which are their natural habitat. Deforestation heavily endangers their existence.

From Simple to Complex: Types of Biological and Institutional Diversity

Wood (2000) defined biodiversity as differences among biological entities and as such it is not only a resource but also a condition. Such differences can be, e.g., structural, functional, or behavioral. As a resource biodiversity can be governed in terms of efficiency. That is a strict anthropocentric view and we can attach anthropocentric and instrumental values to biodiversity as a resource. Seen as an environmental condition or state, biodiversity is as a precondition for the maintenance of biological resources and a necessary precondition for adaptive evolution and self-regulatory mechanisms of ecosystems. This conception of biodiversity "transcends the problems inherent in the allocation of scarce resources among competing interests". Wood (2000) distinguishes biodiversity from the diversity of biological resources. The former is a necessary condition for the continuing flow of biological resources and therefore "the conservation of biodiversity can be seen as a means for maintaining values that are universal and largely independent of the competition over scarce biological resources. While biodiversity as a resource refers to single biological entities with distinct instrumental values for humans, biodiversity as a condition or state refers to the diversity of initial conditions which are a precondition for the evolving, complex, adaptive, and unpredictable nature of ecosystems.

Ellis (2005) holds that complexity has no clear theoretical definition but that complex structures consist of hierarchically ordered "modules" which are more or less independent and whose structures can be studied in their own right, e.g., cells and bodies or individuals and groups of resource users "who are linked to each other and to multiple resources that occur across multiple scales through multi-level governance arrangements". Berkes et al. (2003) further state that "problems and solutions of biodiversity conservation at the generic level are considerably different from those at the species level or landscape level. Different groups of conservationists focus on different levels; they may use different research approaches and may recommend different policies. Biodiversity can be considered at different levels of scale. However, because there are strong feedbacks among the genetic species and landscape levels, there are coupling between different levels, and the system should be analyzed simultaneously across scale." Complex systems are not only defined by structure and scale but also by change. Causal relation within systems can change and the relative intensity of these relations can also change.

Common and Stag (2005) describe complex systems as systems in which causal relations between components change, and which are able to adapt and evolve. The change of species distribution and composition as response to changing from industrial to organic agriculture is an example. Complexity and change in ecosystems are largely a result of the initial differences among biological entities of the system; yet another reason to not merely view biodiversity as a resource. The efficiency stance neglects complexity to a large extent. Although Baxter (1974) recognizes that some components of the ecosystem may be important to humans, as soon as this relationship is not evident or known it becomes irrelevant. Given that human actors have imperfect knowledge about complex social- and ecological systems, uncertainty prevails.

Different possible outcomes of their actions may be known but probabilities can not be assigned to them. Lugo (1995), for example, suggests that trying to quantify sustainable harvest levels in tropical forests rarely leads to ecosystem sustainability. This is because the entire socio-ecological system is in a state of constant change. Change and disequilibrium, rather than equilibrium, are the dominant system conditions. Complex systems are also defined by diversity. In general, ecologists take the view that biological diversity promotes the resilience of systems. Different groups of species take over different functions in a system and therefore the persistence of ecosystem functions over time i.e. the resilience of ecosystem functions depends on the diversity of species within functional groups. Gunderson and Holling (2002) distinguish two forms of diversity within functional groups: *functional compensation* and *functional reinforcement*. Functional compensation occurs when one species goes extinct and is replaced by another species with similar functions but which is less sensitive to environmental change or disturbance. Such compensation occurs at the same or within a narrow range of scales. Functional reinforcement occurs when species perform similar functions across very different scales. Small birds, for example, that feed on larvae help to regulate insect infestation; and when the larvae becomes a worm, different large birds perform the same task, yet at a different scale. The two types of diversity effects can be described as follows: 1) Similar species at similar scales with similar functions but different sensitivities to disturbance, and 2) Different species at different scales with similar functions and different sensitivities to disturbance. This "within-scale and between-scale diversity produces an overlapping reinforcement of function that is remarkably robust" and which the authors call 'imbricated redundancy'. System resilience the ability of an ecosystem to re-establish itself after disturbance critically relies on this redundancy and it is a serious error to assume that minor or redundant species can afford to be lost, because their importance may only be detected when they are needed, i.e. following a disruption.

If we regard these two types of diversity and their effects on system resilience as organising principles, we find surprisingly similar effects in social systems. Similar to the idea of 'imbricated redundancy' of species within and across scales is the idea of institutional diversity and multi-level governance for biodiversity conservation, or 'a hierarchy of cascaded institutions'. Apart from the creation of a 'scattering of strong local jurisdictions' and the promotion of the economic value of biodiversity resources this institutional structure is required to match jurisdictional imperatives; and Martin (2003) argues that it is a condition for effective, stable, and equitable conservation at the national level. Ostrom (1998) argues along the same lines that in order to govern complexity in ecosystems appropriately, the respective nested and multi-layered institutions will be complex because of the 'Law of Requisite Variety'. This law which was developed by Ashby (1960), a psychiatrist and one of the founding fathers of cybernetics, states that any regulative system needs as much variety in the actions that it can take as exists in the system it is regulating.

Gunderson and Holling (2002) warn that seemingly redundant species can not be afforded to be lost. Despite this warning, one widely held argument is, especially among economists, that whatever way we choose to conserve biodiversity, it should be done efficiently; the criterion being defined with the below mentioned narrow understanding of individual rationality. Similarly, in the context of public service reform, individuals had argued on the ground of efficiency that many local jurisdictions should be merged into a single unit of government. Ostrom et al. (1961) challenge that presumption by stating that the consolidation argument need not hold "if agencies offer similar but differentiated services that impinge upon diverse communities of interest."

Other Important Findings

These statistics from the **IUCN Red List 2009** indicate that a minimum of 16,928 species are threatened with extinction: 21% of mammals, 12% of birds, 31% of reptiles, 30% of amphibians and 37% of fish globally are threatened. The

European Red List, published by IUCN and the European Commission, reveals that 23% of amphibians and 19% of reptiles (2009), 15% of mammals (2007) and 13% of birds (2004) in Europe are threatened with extinction. The latest assessment (2009) by the European Environment Agency led initiative **Streamlining of European Biodiversity Indicators** (SEBI 2010) warns that the number of invasive species in Europe continues to increase rapidly, with more and and more negative economic and ecological consequences. At European level, the **Mid-Term Assessment of Implementing the EU Biodiversity Action Plan** (BAP), published in December 2008, shows that 50% of species, over 40% of European bird species and up to 80% of habitat types have an unfavorable conservation status.

The first findings of the **Economics of Ecosystems and Biodiversity study** (TEEB) released in 2008 concluded that, in a "business as usual" scenario, the current decline in biodiversity and related loss of ecosystem services will continue and even accelerate. The annual welfare loss generated by the loss of ecosystem services by 2050 has been estimated at 6% of global GDP. As noted by the **Millennium Ecosystem Assessment** (MA) released in March 2005, human activities have taken the planet to the edge of a massive wave of species extinctions, further threatening our own well-being. For instance, current rates of loss are estimated at 100 to 1,000 times natural background levels. The global situation is alarming as pressures on biodiversity have intensified. Increasing demand for agricultural land, food and energy crops places even greater pressures on natural systems. At international level, the second **Global Biodiversity Outlook** (GBO2), launched in March 2006 in Brazil, assessed the current status and trends of biodiversity and the key drivers of biodiversity loss by putting a special emphasis on how ecosystem services contribute to human well-being, raising the fact that humanity's future depends on healthy ecosystems.

Current Trends in Biodiversity

Virtually all of Earth's ecosystems have been dramatically transformed through human actions and ecosystems continue to be converted for agricultural and other uses. The current loss of biodiversity and the related changes in the environment are now faster than ever before in human history and there is no sign of this process slowing down. Many animal and plant populations have declined in numbers, geographical spread, or both. Species extinction is a natural part of Earth's history. Human activity has increased the extinction rate by at least 100 times compared to the natural rate.

Comparing different types of measurements of biodiversity loss is not simple. The rate of change in one aspect of biodiversity, such as loss of species richness, does not necessarily reflect the change in another, such as habitat loss. Moreover, some aspects of biodiversity loss are not easily measured, for instance the fact that the same species are increasingly found at different locations on the planet and that overall biodiversity is decreasing. The Living Planet Index, compiled by the WWF, provides an indication of the declines in the overall abundance of wild species.

Factors Lead to Biodiversity Loss

Biodiversity is declining rapidly due to factors such as land use change, climate change, invasive species, overexploitation, and pollution. Such natural or human-induced factors–referred to as drivers–tend to interact and amplify each other. While changes in biodiversity are more clearly linked to direct drivers such as habitat loss, they are also linked to indirect drivers that are at the root of many changes in ecosystems. The main indirect drivers are changes in human population,

economic activity, and technology, as well as socio-political and cultural factors.

Different direct drivers have been critically important in different ecosystems over the past 50 years. For example, in terrestrial ecosystems, the main driver has been land cover change such as the conversion of forest to agriculture. In marine systems, however, fishing, and particularly over fishing, have been the main drivers of biodiversity loss. Overall, the

main factors directly driving biodiversity loss are: habitat change, such as fragmentation of forests; invasive alien species that establish and spread outside their normal distribution; overexploitation of natural resources; and pollution, particularly by excessive fertilizer use leading to excessive levels of nutrients in soil and water.

Recent changes in climate have already had significant impacts on biodiversity and ecosystems in certain regions. As climate change will become more severe, the harmful impacts on ecosystem services are expected to outweigh possible benefits, such as a longer growing season, in most regions of the world. Climate change is expected to exacerbate risks of extinctions, floods, droughts, population declines, and disease outbreaks.

Many drivers affecting biodiversity are stronger today than they were in the past and are also occurring together. Because exposure to one threat often makes a species more susceptible to another, multiple threats may have unexpectedly dramatic impacts on biodiversity. Drivers of extinction range from local to global in scope and from immediate to long-term in their effects. For example, the extinction of species due to habitat loss can be rapid for some species, while it may take hundreds of years for others.

Research Findings

Barbier, (1992) made a study of the economics of gum arabic in the Sudan, because of a decline in the real producer price of gum arabic relative to other crops such as sesame, groundnuts, sorghum and millet, the private profitability of gum arabic was lower than that of other crops except in the Tendelti system of the White Nile where field crop damage occurs frequently. But it also concluded that inclusion of the indirect benefits of gum arabic i.e. control of erosion/runoff, wind breaks, dune fixation, nitrogen fixation and the role of the gum belt in controlling desertification meant that its social profitability was much higher than its private profitability. In other words, the estimated private rate of return understated the value of *Acacia senegal* precisely because it does not take into account the indirect value of the resource in the provision of a range of ecological functions. More recently, the link between the diversity of species and the disruption of local ecological services has been clarified. Holling (1992) noted that many ecological functions are supported at any one moment in time by a relatively small number of species. The removal of these species can cause a fundamental transformation of the ecosystem-whether from forest to grassland, or grassland to a shrubby semi-desert. Where a forest or savanna is transformed in this way, most of the main ecosystem functions are disturbed. Examples of such species include the grass species that maintain productivity in savannas or the insectivorous birds that mediate budworm outbreaks in the eastern boreal forest.

Walker, (1993) pointed out that in both forests and savannas there is an element of redundancy in the system under given environmental conditions. Species redundancy, in this context, means that their removal has relatively few implications for the functioning and productivity of the system-at least under those environmental conditions. There is always a threshold of diversity below which their various functions cannot be maintained. Any self-organising living systems require a minimum diversity of species to capture solar energy and to develop the cyclic relation of fundamental compounds between producers, consumers and decomposers on which biological productivity depends. If the level of biodiversity drops below such a threshold the productivity of the system will fall. However, since not all species play a vital part at all times, it is often possible to delete some species without immediately affecting the functioning of the system.

The important point, though, is that even if the biodiversity needed to maintain ecological services under one set of environmental conditions is low relative to the actual level of biodiversity, a change in environmental conditions can alter the mix of organisms, populations and communities needed to maintain those services. That is to say, biodiversity has value both in supporting the flow of ecological services under given environmental conditions, and in assuring those services over a range of environmental conditions. It has an option or insurance value. The level of biodiversity in an agroecosystem determines its capacity to respond to external shocks, whether market or environmental. This is measured by its resilience. Holling et al (1995) pointed out that from an ecological perspective, biodiversity protects ecosystem resilience by underwriting the provision of ecosystem services over a range of environmental conditions. Certain species have greater ecological value under one state of nature than others, but species that are 'passengers' under one state of nature may have a key structuring role to play under other states of nature. The ecological impact of biodiversity loss depends on the link between the species and the functions of the system. Whether the deletion of some species affects a given function depends on the number of alternative species that can support the function if the ecosystem is perturbed.

Take the example of livestock farming. Increasing livestock density has led to two major changes in the biodiversity of savanna rangelands. The first involves the loss of perennial grasses and their replacement by annuals, which vary far more in response to fluctuations in rainfall. The second involves the reduction in phenological diversity in the grass sward that moderates inter-annual variation in production. Walker (1988) reported that rangelands subject to light grazing pressure tend to have roughly even amounts of early-, mid- and late-season growing grasses implying that there is roughly same amount of grass to respond to rainfall whenever it occurs in the season. But in rangelands subject to heavy grazing pressure, the loss of early season palatable species implies an increase in the relative amount of later growing species. There is a reduction in the functional diversity of the range, and this leads to an increase in inter-annual variation in fodder production. McNaughton (1985) showed that Serengeti grasslands in which there was greatest variation to biomass also varied least in annual production.

The insurance value of biodiversity has accordingly been argued to lie in its role in protecting ecosystem resilience. In the rangelands example, the market value of the liveweight gains secured by having a combination of grass species is a measure of the insurance value of that combination. From an economic perspective, biodiversity has insurance value because of its role in protecting the productivity of agroecosystems over a range of both environmental and market conditions.

The genetic diversity of cultivated species and environmental risks

Genetic resources are used in agriculture in two ways: selection of species for domestication and cultivation, and genetic 'improvement' of domesticated species. The vulnerability of agriculture due to the narrow genetic base of most major crop plants is a well-recognised source of risk. More than 90% of the food supply derives from a handful of grasses viz., wheat, rice, corn and oats), nightshades (tomato and potato, mammals viz., cattle, sheep and pigs, and birds viz., chickens and ducks. The narrow genetic base of these species is a 'cause' of disease and pest epidemics, leading to sharp variation in agricultural yields. Wild or traditional genetic stocks have value because they offer scope for breeding or engineering resistance to crop pests and pathogens. Most cultivated crop varieties and many livestock strains already contain genetic material incorporated from wild relatives, or from crops and livestock still used by traditional farmers. Indeed, it has been estimated that at least half of the increase in agricultural productivity this century is directly attributable to artificial selection, recombination and intraspecific gene transfer procedures. Examples include the following.

The use of Turkish wheat to develop genetic resistance to disease in western wheat crops is valued in 1995 at US $50million per year. Ethiopian barley has been used to protect Californian barley from dwarf yellow virus, saving damage estimated at $160 million per year. Mexican beans have been used to improve resistance to the Mexican bean weavil, which destroys as much as 25% of stored beans in Africa and 15% in South America. Wild rice strains have improved protection of cultivated varieties against grassy stunt virus. Heywood, (1995) stated that the susceptibility of sugarcane to the mosaic virus was solved by introducing resistance to it from wild relatives e.g. *Sacharum spontaneum*. The value of biodiversity in this case lies in the fact that it provides the raw material for desirable genetic traits in crops. Genetic resources have been used to improve productivity, induce better resistance to pests, or to improve adaptation to harsh environments. This value depends on the technology of genetic transfer. Changes in biotechnology have altered both the use value of genetic resources, and the potential indirect consequences or external costs of their use. The most important of these changes is that biotechnology allows gene transfer both within species and between unrelated species, increasing the flexibility of genetic resources. At the same time, however, the introduction of biological control agents or genetically modified organisms may have indirect effects that give rise to external costs/benefits. The best understood indirect ecological effects are those relating to species introductions, and include impacts on the phenotypic characteristics of Individuals; the genetics, abundance and dynamics of populations; the structure and functioning of communities and ecosystem processes such nutrient cycling and disturbance regimes.

These effects include the deletion of indigenous species through predation, browsing or competition; genetic alteration of indigenous species through hybridisation; and the alteration of ecosystem structure and function including biogeochemical, hydrological and nutrient cycles, soil erosion and other geomorphological processes. The associated external costs include the health and health control costs of the introduction of pathogens and their vectors; the productivity costs of changes in population abundance and species compositions; the cost of pesticides in agroecosystems; defensive expenditures and control costs of invasives such the water hyacinth.

Potentially more important still are the effects of biotechology on the incentive to conserve wild relatives and traditional crops. Productivity increases associated with the Green Revolution have been associated with the abandonment of traditional varieties that have been bred over thousands of years. These landraces have been a major source of genetic diversity in agriculture, but many have disappeared with the Green Revolution. Not only has the total number of rice varieties planted been substantially reduced, but also the proportion of the total crop accounted for by ten or less varieties has been substantially increased. Current trends in biotechnology threaten to displace even more traditional varieties, and to increase the vulnerability of farmers who choose to adopt genetically uniform crops. Perrings, (1998) reported that there are no realistic estimates of the insurance value of the displaced landraces. This would, in any event, explain variation in habitat loss.

To identify the relation between poverty, population growth and environmental change implicit in the Brundtland hypothesis requires a different approach. Dasgupta (1993) argues that poverty drives fertility through the microeconomic decisions of rural households. Where conditions are such that people are locked into particular technologies, population growth then leads directly to pressure on the resource base. This may be exacerbated by the failure of both markets and policies. Pearce and Warford, (1993) reported that in Sub-

Saharan Africa, for example, a combination of (a) cultural and institutional rigidities and (b) policy-induced declining real producer prices is argued to have led to a positive feedback loop between rural poverty, population growth, landlessness and pressure on soils, vegetation and water.

6

Climate Change and Biodiversity

Climate change is defined as a variation either in the mean state of the climate or in its variability, persisting for an extended period, typically decades or longer. It encompasses temperature increases; sea-level rises, changes in precipitation patterns, and increases in the frequency of extreme weather events. Biodiversity and climate change are closely linked, and each impacts upon the other: biodiversity is threatened by human-induced climate change, but biodiversity resources can reduce the impacts of climate change on population and ecosystems. Our planet's climate is changing rapidly

The Earth is undergoing profound changes to its climate. There is now little doubt that this results from human activities, mainly the burning of fossil fuels. Climatic changes have occurred throughout Earth's history. However, these recent changes are different because they are taking place faster and are unlikely to be reversed by natural processes.

Climate Change Affects all Biodiversity

The effects of climate change on biodiversity are far-reaching and operate at many different levels - from individuals to ecosystems. At the species level, climate change affects particular species in different ways. It may alter their distribution, abundance, behaviour, phenology, the timing of events such as migration or breeding, morphology i.e. size and shape and genetic composition. Many such effects have already been documented in a wide range of species. Birds provide some of the clearest examples, in both single species and multi-species studies.

Climate Change makes other Threats Worse

Climate change may affect species directly, for example through changes in temperature and rainfall. Often, however, the indirect effects are even more important. There may be increased pressure from competitors, predators, parasites, diseases and disturbances such as fires or storms. Climate change will often act in combination with major threats such as habitat loss and alien invasive species, making their impacts considerably worse.

The links between Climate Change and Biodiversity

In its Fourth Assessment Report, the Intergovernmental Panel on Climate Change (IPCC) states with 'very high confidence' that human activities since 1750 have caused global warming. For the next two decades, says the IPCC, about 0.2 °C of warming per decade is projected. Among the many knock-on effects are a sea level rise of more than 50 centimetres by 2100. It is also likely that some 20 to 30 per cent of plant and animal species assessed will be at greater risk of extinction if the rise in global average temperatures exceeds 2-3 °C.

Biodiversity, Land Use and Climate Change

Changes to biodiversity and natural systems are likely to have a major influence on the global climate. Land use changes leading to habitat and thus biodiversity losses can also boost greenhouse gas emissions. For instance forests, which are vital carbon sinks, release carbon dioxide (CO2) into the atmosphere when cut down or burnt. Land use changes, particularly deforestation in tropical regions—where forests tend to be very rich in biodiversity—are responsible for roughly 18 per cent of human-driven CO2 emissions.

Peatlands hold about a third of the carbon contained in soil worldwide, and greenhouse gases are released every time they are burnt, drained or converted to cropland. Many peatlands are important biodiversity reservoirs or stopover points for migratory species. So while the measurable effects of actual biodiversity loss on climate change may be highly variable, it is far more certain that conserving biodiversity will help in mitigating climate change.

Using Biodiversity to Cope with Climate Change Impacts

Scientists agree that even if greenhouse gas emissions were to stabilise—an unlikely scenario, given inertia on the part of governments and the public—global warming and sea level rise will persist for centuries. This is down to the timescales associated with climate processes and feedback. So the need to adapt to climate change impacts is inevitable. It is also already happening across the globe. Many people are using natural resources and biodiversity, including genetic diversity, as part of the adaptation process. For instance, wild relatives of food crops can be used to breed new varieties that can cope with changing conditions. In India, for instance, some farmers are accessing different crop varieties through traditional exchange systems, developing new varieties, and adapting farming practices to cope with hotter temperatures, pest infestations and disease.

In many regions of the developing world, the rural poor already rely heavily on wild food sources and medicinal Swap shop: agro-biodiversity and traditional knowledge in India It may be known in the North for tea, but Darjeeling district in West Bengal, India, is rich in traditional rice varieties. Here, seed purity is maintained through exchanges within and between villages. Women farmers adapt to changing conditions by procuring seeds of different varieties through traditional practices of seed saving, exchanges and networks. For example, a villager growing rice at 1200 metres obtained a different variety suited to higher temperatures from a villager at 750 metres. To cope with water shortages, farmers engaged in wet rice cultivation are trying to get native seeds suitable for dry paddy fields from others in neighbouring Sikkim.

Farmers in the region are also adapting agricultural practices and developing new varieties. One farmer is planting ginger a little later than usual to help tackle pest infestation. Another has used a wild relative of cardamom to breed a new, disease-resistant variety—which is such a success, it has attracted support from the state government for sale to neighbouring regions. Traditional farming systems and farmers' innovations are clearly at the cutting edge in adaptation to climate change. While scientists and policymakers work to find solutions, local farmers have already amassed considerable experience of how to cope, based on their observation and experimentation in the field.

Conserved ecosystems offer many other services vital for adapting to climate change. Wetlands are important reservoirs for floodwater. Vegetation such as hedges protects agricultural land from excessive water or wind erosion in times of heavy rainfall or drought. And by preventing erosion on hillsides, vegetation also reduces the risk of landslides when rain comes in heavy bursts. Watersheds with intact plant cover slow the movement of rainfall to rivers and so reduce flood risks downstream. Mangroves are well-known coastal buffers, reducing the strength of waves before they reach the shore and so protecting against cyclone damage to coasts and seaside communities.

How climate change affects biodiversity Climate change will have a number of impacts on biodiversity, from ecosystem to species level. The most obvious is the effect that flooding, sea level rise and changes in temperature will have on ecosystem boundaries. As a result of these shifts in boundary, some ecosystems will expand into new areas, while others will become smaller. Habitats will change as rainfall and temperatures change, and some species will not be able to keep up, leading to a sharp increase in extinction rates. The 2005 Millennium Ecosystem Assessment (MEA) estimates that by the end of this century, climate change will be the main driver of biodiversity loss. Along with predicting a higher risk of species extinctions, the IPCC says that temperature increases up to 3°C are also very likely to trigger substantial changes in the structure and functioning of all ecosystems.

The impacts of climate change on biodiversity will vary from region to region. The most rapid changes in climate are expected in the far north and south and in mountainous regions. These also happen to be the regions where species are more likely to be 'painted into a corner', with no alternative habitat to which they can migrate. Species with small populations, or populations restricted to small areas, are also especially vulnerable to any climatic shifts. Higher water temperatures have already caused devastating losses of biodiversity in coral reefs. Global warming is also causing shifts in the reproductive cycles and growing seasons of certain species, which can in their turn affect how ecosystems function. The equilibrium of ecosystems can also be upset when, for example, insect pests previously unknown in the UK survive the warmer winters. Migrating species may be affected dramatically by any changes to stopover sites key to their survival, or when seasonal availability of food sources is no longer synchronised with migration times.

How Responses to Climate Change Affect Biodiversity

The strategies used to counter climate change can also affect biodiversity adversely. Measures for reducing greenhouse gas emissions are one example. Some forms of renewable energy technology can lead to poor outcomes for biodiversity. Biofuel plantations may involve clearing areas of high biodiversity, such as tropical forests leading to greenhouse gas emissions on top of biodiversity loss, and introduce monocultures of alien species and damaging agrochemicals. By 'drowning' land and disrupting river flows, large hydropower schemes can cause the loss of terrestrial and aquatic biodiversity and inhibit fish migration.

Dammed water can also become a net emitter of greenhouse gases as submerged soils and vegetation decay and release CO2 and methane. Poorly located wind farms, on bird migration routes for example, can kill significant numbers of birds. By contrast, some renewable energy measures such as efficient stoves and biogas use can conserve carbon reservoirs and reduce pressure on forests. An increase in demand for freshwater, which is likely as the climate warms, could degrade wetlands, rivers and streams and so damage key ecosystem services.

Poverty, Climate Change and Biodiversity

Poor people are disproportionately vulnerable to the loss of biodiversity and ecosystem services. And although they are responsible for emitting the lowest levels of greenhouse gases, they suffer most from the impacts of climate change. Recognising this, the United Nations Framework Convention on Climate Change (UNFCCC) asserts that there are 'common but differentiated responsibilities' for tackling climate change. But along with the Convention on Biological Diversity (CBD) and the Millennium Development Goals (MDGs), these agreements do not specify the strategies and methods to be used by parties to each agreement to meet their stated aims. While links between climate change, biodiversity and poverty are clear, and there is an obvious.

Over the course of history, ecosystems have always had to adapt to changing climatic conditions. The current rate of climate change, however, is higher than ever recorded. It is now obvious that humans have impacted biodiversity loss–directly through the destruction of habitats, and indirectly through climate change. Current rates of climate change have already resulted in species composition changes. As the climate warms up or cools down, many local species have to shift from their current habitat to areas better suited to their needs. This means that with rising average temperatures many species will experience a range shift towards the poles, potentially pushing out existing species in those areas. In other cases warming will cause species to shift their ranges upwards along altitudinal gradients, with the same effect.

Changing temperatures will also influence their reproductive cycles, their growth patterns and, also as a result of range shifts, the interaction between species. All this may occur over time frames as short as decades. Affected species are likely to include those to which humankind has attached various values e.g. cultural emblems, indigenous crops, plants that typically mark the landscape. This will also include wild plants that are locally used, such as medicinal herbs, mushrooms and wild flowers. Examples of the urban impacts of such species changes are: Loss of trees in streets, gardens and parks, as changing temperatures add stress and reduce their resilience to pests; Loss of species and damage of forests, reducing economic gains as well as recreational values; Risk of collapsing waste water treatment systems as micro-organisms are vulnerable to temperature differences; Damages to flora and fauna through the immigrating of alien species; Risks to human health through the immigration of disease carrying insects such as mosquitoes.

Changes to local biological diversity and species composition poses tremendous challenges to conservation planning and implementation. Common conservation practices need to be revised in order to accommodate the current and future challenges resulting from climate change. If rapid and irreversible change in biodiversity is to be avoided, conservation strategies need to focus more on supporting the species' natural capacity to adapt to change. Beyond a morale obligation to act, helping species to adapt may also avoid the loss of important ecosystem services and the cultural and economic values attached to particular species.

At the local and regional level, conservation planning and action needs to include the creation of networks of protected areas. This will increase the 'connectivity' of the landscape and enable species to gradually shift from their current habitats to more suitable ones. Cities and towns in fact have an important function as habitat corridors for flora and fauna. If biodiversity can exist only in mere 'islands' within an otherwise inhos- pitable landscape, many species will not be able to survive climatic changes as they cannot migrate to other areas.

Local and regional authorities can restructure their urban development plans and zoning to allow for the gradual creation of biodiversity corridors, for example along linear infrastructure elements such as rivers, roads and railway lines. Some changes that will occur in ecosystems as a result of climate change: Accelerated habitat loss: Small islands of habitats may not be able to sustain themselves as the climate changes; Changes in range: The geographical range of many species will move towards the poles or higher in altitude; Increased extinction rates: Many species that are already threatened are likely to become extinct; Changes in physiology: For example reproduction timings, duration of growth season for plants.

Some changes that will occur in ecosystems as a result of climate change: Accelerated habitat loss: Small islands of habitats may not be able to sustain themselves as the climate changes; Changes in range: As the climate changes, extreme weather events such as storms, floods, droughts and heat waves are expected to become more intense and unpredictable. Healthy ecosystems play a crucial role in mitigating the impact of climate-induced disasters. For example, a biologically diverse and healthy forest ecosystem has a high capacity to absorb torrential rain. It provides tree cover and undergrowth that can reduce the erosive impact of intense precipitation, slow down the surface run-off and support the seepage of rain water into the soil and lower rock beds. These processes help reduce the risk of flash floods, mud slides and soil erosion. Conserving a patch of native vegetation may therefore be a key component of integrated disaster risk reduction in the face of climate change. Local and regional authorities will be called upon to take action to effectively use their regulatory capacity to help infrastructure systems and people adapt to a changing climate. The destructive Biodiversity management measures Active protection and management of existing habitats Increasing the size of existing habitats and creating new habitats Increasing the connectivity of habitats by creating ecological networks and connecting wildlife territories.

Maintaining critical ecosystem functions and aesthetic value; increasing human well-being Increasing the chances for survival of threatened species Maintaining the existing composition of ecosystems and enabling their survival. Healthy ecosystems can help reducing the impacts of extreme weather events. Increased overall ability of ecosystems to adapt; providing physical adaptation factors such as innercity shade Slowing down extinction rates of ecologically, culturally, economically significant species Biodiversity management measures and their contribution to climate change adaptation. Outcomes for cities and regions impact of climate-related disasters on human settlements depends considerably on the ability of local and regional authorities to effectively factor climate change risks in infrastructure planning and the design of municipal services.

Biodiversity management therefore needs to become integrated with disaster risk reduction planning and emergency response. For example, local and regional authorities can help reduce flood risks by protecting and restoring riverine and coastal vegetation. This will help regulate a more even flow of water in a catchment, reducing the impacts of extreme precipitation, storm surges or coastal wave action. In vegetated areas, the rate of water seepage into the ground is also higher, which reduces the impacts of droughts. All these measures will support local economic and infrastructure systems to adapt to a changing climate.

If carefully planned based on local indicators for climate change, even technologically geographical range of many species will move towards the poles or higher in altitude; Increased extinction rates: Many species that are already threatened are likely to become extinct; Changes in physiology: For example reproduction timings, duration of growth season for plants. Biodiversity also plays a crucial but often underestimated role in the mitigation of climate change. Mitigation refers to actions that reduce the extent of global warming. The biological diversity of ecosystems is a key factor in ensuring the natural processes undertaken by ecosystems. One of these processes is growth. As they grow, plants, fungi and soil bacteria work together to bind carbon dioxide gas from the atmosphere into the soil, wood and other organic matter. This process 'sequesters' carbon dioxide from the atmosphere, thereby helping to reduce the amount of this greenhouse gas, which is most responsible for global warming.

The release of CO_2 through deforestation and land use change accounts for as much as 25% of total human-induced greenhouse gas emissions. Maintaining local biodiversity and increasing urban green space, in particular forest areas, therefore are significant and effective contributions towards protecting the global climate. Apart from conserving green areas, local and regional authorities can provide the regulatory framework for encouraging vegetation growth on private properties.

Increasing the green areas in densely populated areas does not only contribute to global climate protection. Trees and other vegetation also help improve the local air quality, provide shade and lead to a cooler and more humid microclimate, which is particularly relevant in hot climates and seasons. Investing in global climate protection will quickly pay off locally and improve the quality of life in cities and towns. Simple and inexpensive measures can result in tremendous benefits. While national governments in many countries provide national adaptation plans, local and regional authorities have the challenging task to gather local data on climate change and making responsible decisions for local adaptation to climate change.

Such comprehensive 'anticipatory adaptation' will need to focus on technological, socio-economic and ecological support mechanisms alike. Conserving healthy ecosystems and promoting the adaptive capacity of species helps ensure that urban and rural environments can retain their 'fitness' and continue to provide the ecosystem services so vital for their inhabitants, despite the uncertainties of a changing local climate. Biodiversity management measures and their contribution to climate change adaptation Biodiversity management measures Active protection and management of existing habitats, Increasing the size of existing habitats and creating new habitats, Increasing the connectivity of habitats by creating ecological networks and connecting wildlife territories, Outcomes for cities and regions, Maintaining critical ecosystem functions and aesthetic value; increasing human well-being, Increasing the chances for survival of threatened species, Maintaining the existing composition of ecosystems and enabling their survival. Contribution to climate change Adaptation,

Healthy ecosystems can help reducing the impacts of extreme weather events Increased overall ability of ecosystems to adapt; providing physical adaptation factors such as innercity shade, Slowing down extinction rates of ecologically, culturally, economically significant species. Local and regional authorities are key players in coordinating and implementing biodiversity management measures that take climate change into account.

For helping species to adapt to climate change, they can: Adapt zoning and urban development plans to increase the connectivity of habitats to enable species to migrate more easily; Protect biologically diverse habitats and plan for an expansion of protected areas to increase the probability for species' local survival despite climate change; For protecting the climate and mitigate further climate change, local and regional authorities can: Participate in climate protection programs such as the Cities for Climate Protection Campaign that effectively reduce greenhouse gas emissions, for example through an extension of public transport, energy efficiency measures and the tapping of renewable energy sources; Invest in the planting of trees to increase carbon sequestration; Provide incentives for private and corporate stakeholders invest in renewable energy and energy efficiency.

Climate change can no longer be avoided, local and regional authorities need to regulate and plan for the adaptation of human activity to a changing climate. This can include: Increasing tree cover and green space in urban areas to cool local temperatures down during the hot season and create more liveable microclimates; Protect and restore riverine or coastal vegetation for reducing the risks of flooding as a result of extreme weather events; Rehabilitate and diversify municipal forests and wet lands to help regulate a more evenly distributed flow in watersheds. These are only a few examples of local and regional action for integrated biodiversity management that should help inspire creative planning and management processes suited for tackling the local and regional specific impacts of climate change.

Climate change is killing frogs finds new research The dramatic global decline of amphibians may be directly connected to global warming warns a new study published in the journal Nature.

Looking at a group of frogs found in biodiversity hotspots in Central and South America, scientists found links between higher temperatures and frog extinctions caused by a skin fungus. The infectious skin disease—a type of chytrid fungus-Batrachochytrium dendrobatidis—is now found in frog populations around the world and is the main suspect in the rapid disappearance of amphibians.

A survey of the planet's amphibian species, found that nearly a third (32%) of the world's 5743 known amphibian species are threatened and 129 species have gone extinct since 1980. Among the species to disappear is Costa Rica's Golden toad-Bufo periglenes and the Gastric Brooding Frog-Rheobatrachus silus of Queensland, Australia. Scientists believe there may be around 10,000 amphibian species on the planet, although this number is likely to be ailing fast.

Climate change is killing frogs finds new research Rhett A. Butler, mongabay.com January 11, 2006 The dramatic global decline of amphibians may be directly connected to global warming warns a new Panama golden frog study published in the journal Nature.

Using records of sea-surface and air temperatures, the researchers discovered strong correlations between changes in climate and the last known sightings of the frogs. According to the scientists, the Earth's rising temperatures enhance cloud cover on tropical mountains, leading to cooler days and warmer nights, both of which favor the chytrid fungus which grows and reproduces best at temperatures between 63 to 77 degrees Fahrenheit. The fungus kills frogs mostly in cool highlands or during winter, implying that low temperatures make it more deadly. Thus the disease flourishes in warm years and puts frogs living at mid-altitudes at greatest risk. In coming years, further changes in climate will likely produce conditions more favorable for the fungus and detrimental to frogs and other amphibians.

The new theory for the frogs' decline "leaps over a major roadblock in our understanding," said Sam Scheiner, program director for the National Science Foundation ecology of infectious diseases program, which funded the research. "This study demonstrates the complex nature of global climate change, including how climate affects the spread of disease, and why these must be integrated if we are to understand and reduce threats to species extinctions."

Ecologists fear that the global decline of amphibians may have broader implications for the world's environment. Because amphibians have highly permeable skin and spend a portion of their life in water and on land, they are sensitive to environmental change and can act as the proverbial "canary in a coal mine," indicating the relative health of an ecosystem. As they die, scientists are left wondering what plant or animal group is next.

Scheiner says global warming and the accompanying emergence of infectious diseases are a real and immediate threat to biodiversity and a growing challenge for humankind. The disappearance of amphibians is troubling on other grounds as well. Many frogs and toads secrete toxic alkaloids to protect themselves from predators. These alkaloids have been found to have medical applications as pain killers. Thus the loss of each amphibian species eliminates the possibility of finding compounds with potential curative properties.

The Kihansi Spray Toad, a small toad that is virtually extinct in the wild after its habitat was destroyed by a hydroelectric project in Tanzania, only exists today thanks to efforts by the Bronx Zoo of New York, the Detroit Zoo in Michigan, and the Toledo Zoo in Ohio. Nevertheless, the toad's captive population has fluctuated dramatically in the past few years due to its biology they breed in groups and small size, and difficulties typically associated with captive amphibians.

In a study published in Raffles Bulletin of Zoology, researchers confirmed the discovery of 35 new frog species in Sri Lanka over the past decade brining the number of frog species in the island country to 105. However, the survey found that 17 of these species have disappeared and at least another 11 face imminent extinction unless their habitat is protected.

Ward says that of the five major extinctions have occurred in the past 500 million years—the Ordovician, the Devonian, the Permian, the Triassic and the Cretaceous—, only one, the Cretaceous event which wiped out the dinosaurs, was likely caused by an asteroid impact. As evidence, Ward cites new research that suggests the Permian extinction of 250 million years ago was caused by huge volcanic eruptions in Siberia which triggered catastrophic climate change and wiped out 95 percent of lifeforms in the world's oceans and about 75 percent of land creatures. Further, says Ward, there is evidence that the Triassic extinction involved higher temperatures. Examining carbon isotopes in rocks dating from the Triassic extinction of 200 million years ago, Ward and his colleagues found that "the amount of carbon dioxide in the atmosphere was up to 100 times what it is today, and that the levels fluctuated wildly over tens of thousands of years." This suggests a gradual increase in global temperatures doomed many of Earth's species to extinction. "The Triassic event isn't something that happened overnight,". Ward's conclusion has significant implication in today's world where climbing carbon dioxide levels are producing higher temperatures.

While Ward is well-respected in his field, not everyone agrees with his interpretation of ancient data. Many paleontologists still finger planetary impact as the primary culprit of past mass extinctions.

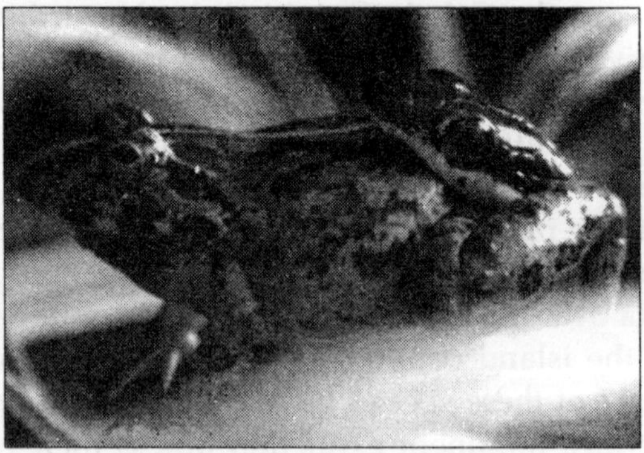

Regardless, it is evident that we are currently in the midst of a sixth great extinction event. One that is almost entirely of our making. Habitat destruction, introduction of alien species, and overexploitation has resulted in the documented extinction of more than 800 species since 1500, although the actual toll will never be known. Presently. extinctions are running at 100 to 1,000 times the background rate, according to the U.N.'s Millennium Ecosystem Assessment.

Past extinctions have shown it takes at least 5 million years to restore biodiversity to the level equal to that before the extinction event. Our actions today will determine whether Earth will be biologically impoverished for the 500 trillion or more humans that may inhabit the Earth during that future period. It urges parties to adopt measures on a global, regional and national basis to ensure that potential risks of ocean fertilization and other geo-engineering schemes have been carefully considered in advance and if allowed to proceed are subject to permits based on prior environmental impact assessments, advance notification and consultation, and requirements for monitoring of effects and reporting to verify results; and it urges Parties to fully assess the likely risks and benefits of sub-seabed emplacement which is dumping in the terms of the London Protocol of carbon dioxide on marine organisms and related ecosystems in advance of permitting any such activity. IUCN urges Parties to address the issue of purity of any carbon dioxide streams in advance to ensure that they are not mixed with other materials that may also be harmful in a marine environment.

Urges Parties to adopt an ecosystem approach that integrates efforts to reduce emissions from deforestation and degradation, in coherence with both the CBD and the UNFCCC; In terms of biodiversity and climate change adaptation Encourages Parties and other States to ensure that conservation measures to reduce the vulnerability of ecosystems and livelihoods are given high priority in global discussions, not only in the context of climate change, but also in those on biodiversity conservation within the framework of the CBD and other biodiversity-related multilateral environmental agreements;

CBD calls on Parties to integrate invasive species management in the context of measures to address climate change and biodiversity loss; it urges Parties to adopt adaptation measures to safeguard agricultural productivity and water supplies, and conservation measures that reduce the vulnerability of ecosystems and the livelihoods of people who depend on them;

Many of the flagship species such as the polar bears, North Atlantic Right whales, frogs, tigers, African elephants and coral reefs are already facing the burnt of climate change in reducing populations, breeding pressures and migrations. In addition, islands such as the Yasawa island in Fiji, Fualopa in Tuvalu, several atolls in Maldives are all under imminent threat of their very existence due to changes in climate causing sea level rise and increased incidence of storms and cyclones.

Threats to species conservation, so far, have primarily been the effects of habitat loss or fragmentation, introductions of alien species, hunting and pollution. However, global warming and climate change are emerging as a significant threat to species and ecosystems. Due to climate change related stresses coral bleaching and resulting die-off is reported from the Indian Ocean at 50% levels, hibernation times of some species like gothic marmots are changed by about a month, the average weight of an adult female polar bear is reported to have been reduced by 20% in the past 25 years. The effects of climate change are pervasive and transnational, whereas most other threats are relatively localized. The potential consequences of climate change are sufficiently great and immediate.

Scenarios for the potential human impact on the biodiversity of terrestrialand freshwater biomes have been developed by several researchers. They model the consequences of predicted changes in land use, climate, nitrogen deposition, biotic exchange species introductions and invasions, and CO2 concentration on the biodiversity of different biomes, making assumptions about the sensitivity of each biome to each of these main drivers of change. Across all biomes, changes in climatic conditions and land use are predicted to have the greatest effect on biodiversity. However, there is considerable variation among biomes, with, for example, Arctic biomes being most sensitive to climate change, tropical forests most sensitive to land use change.

Climate Change Challenge to Biodiversity

Natural levels of atmospheric gases such as carbon dioxide and methane - green house gases are increasing at an alarming rate due to human induced pressures. Such pressures include burning of fossil fuels such as petrol, diesel and coal, farming activities such as intensive rice farming in irrigated conditions, land use changes such as deforestation and clearing of land to developmental activities. Some of the changes that can be observed clearly include: Increase of mean temperatures by 0.6 C since 1800's Global mean sea level risen by 10-20cm Reduction of overall volume of glaciers in Switzerland by

two-thirds. A 40% reduction in the thickness of ice in Artic during late summers and early autumn In addition, loss of about 40-60% of total available water in catchment basins of Niger, Lake Chad and Senegal, retreat of upto 70% of sandy shorelines are causing heavy burdens to people of Africa who already are affected by famine, food shortages and disease outbreaks.

Climate change also affects species at the level of cells and genes Changes in the genetic makeup of species are expected as organisms adapt to new climatic conditions, and increases in temperature can also lead to increases in the rate at which cells use energy causing growth related problems.

Other observed impacts of climate change include changes in the timing of reproduction in certain species; in the length of the growing season in many regions; in the abundance of different species; and in the frequency of pest and disease outbreaks. For example, higher temperatures have led to an increase in the number of eggs laid by the spruce budworm, already one of the most devastating pests in North America's boreal forest. This could in turn contribute to more severe outbreaks of this pest.

How can Biodiversity Conservation Help?

Conservation of biodiversity can reduce the negative impacts of climate change and variability to a great extent. Conservation of habitats can reduce the amounts of carbon-di-oxide released into the atmosphere. Current deforestation rates are estimated to be responsible for about 20% of human induced carbon emissions. Conserving certain species such as mangroves and drought resistant crops can reduce adverse impacts of climate change such as flooding and rainfall shortages. In addition, biodiversity offers a range of options for adaptation to climate change as well. Terrestrial and ocean ecosystems play a significant role in the global carbon cycle. Managing these ecosystems properly can reduce green house gases significantly. Each year upto 60 gigatons (Gt) of carbon is taken up and released by terrestrial ecosystems while ocean ecosystems take up and release upto 90 Gt of carbon. On average terrestrial ecosystems store about 3 Gt of carbon and oceans store about 1.7 Gt. The average emission rate of carbon from burning of fossil fuels and industrial processes is about 6.3 Gt and through deforestation there is an emission of 1.6 Gt of carbon. The result is a net build up of 3.2 Gt of atmospheric carbon per year.

Biodiversity conservation and appropriate management of ecosystems provides a significant option to deal with

sequestering carbon and to reduce global warming. In addition to mitigating the effects of climate change, biodiversity also provides opportunities for adaptation to climate change and variability. The ecosystem approach of the Convention on Biological Diversity provides a flexible management framework to address climate change mitigation and adaptation activities in a broad perspective. Just as climate change alters the state of biodiversity, changes in biodiversity can also affect the world's climate. Perhaps most significant one is the change in land use that leads to loss of biodiversity contributing to increased greenhouse gas emissions.

Forests are a major store of carbon. Carbon dioxide is released into the atmosphere whenever there are forest fires, or when forests are cut down. It is estimated that one-third of the carbon dioxide released into the atmosphere between 1850 and 1998 came from land-based sources, predominantly the destruction of forests. Forests also help to regulate moisture in the atmosphere and can reduce temperatures. Peatlands or mires are another major store of greenhouse gases. As they hold roughly one-third of the carbon contained in soil worldwide, greenhouse gases are released into the atmosphere every time wetlands or peatlands are burned, drained, converted to agriculture or degraded. A dramatic example of this was the peatland forest fires that occurred in Indonesia during 1997. These fires released amounts of carbon

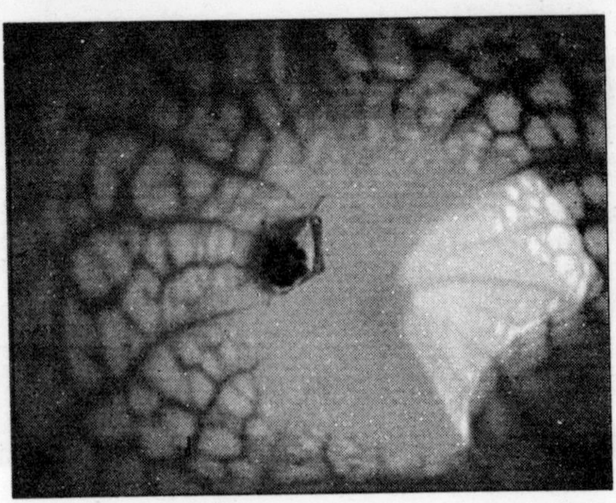

dioxide equivalent to 40 per cent of the world's average yearly carbon emissions from fossil fuels.

There are also feedback mechanisms at work between biodiversity and climate change. For example, some species of ocean algae release a chemical into the atmosphere known as dimethyl sulfate (DMS). When ocean temperatures rise in response to global warming more DMS is released. But DMS is also associated with the formation of clouds. This means that a boom in populations of some algal species due to increased water temperatures and a resultant boost in DMS levels, may actually help reduce temperatures because extra cloud cover reduces the amount of heat that reaches the Earth's surface. Land use, land-use change and forestry (LULUCF) activities can play an important role in reducing green house gas emissions into the atmosphere. Such Land use, land-use change and forestry activities could include conservation of existing carbon pools, sequestration by increasing the size of carbon pools and substitution of fossil fuel energy by use of modern biomass. Plantation of native tree species will support not only enhanced biodiversity conservation but also contribute to better accumulation of carbon. In addition, agroforestry systems, if properly designed and managed, could contribute to substantial sequestration of carbon and reduce soil erosion and water usage. Adoption of agricultural management practices such as conservation tillage, erosion control practices, irrigation contribute to better management of local biodiversity at the same time reduce green house gas emissions.

Similarly, improved management of grasslands can enhance carbon storage in soils and conserves biodiversity. Conservation of biodiversity and proper maintenance of ecosystem structure and function constitute an important climate change adaptation strategy as genetically diverse populations and species rich ecosystems have greater potential to adapt to climate change. Such management practices also contribute to significant provision of goods and services from such ecosystems to people living in and around such areas, including direct economic benefits.

Climate Change, Conservation and Development

The Johannesburg Declaration from the World Summit on Sustainable Development (WSSD) notes that, "...the adverse effects of climate change are already evident, natural disasters are more frequent and more devastating and developing countries are more vulnerable....".

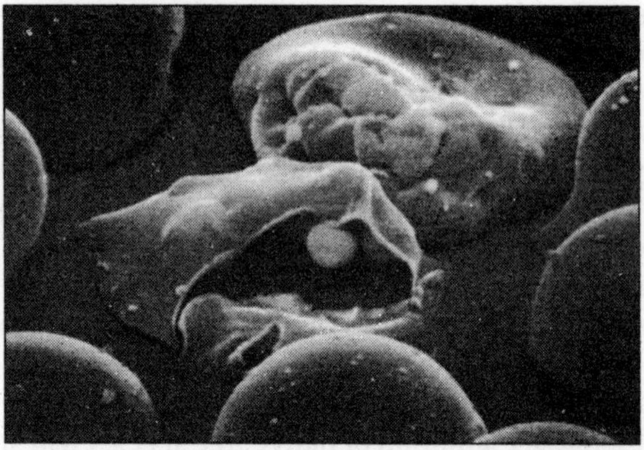

While climate change will have global impacts, poor countries and poor people will be most vulnerable because of their high degree of dependence on natural resources that are directly impacted by climate change. Limited human, institutional and financial capacities to cope climate variability pose a major hurdle to these people. Climate change threatens to undo decades of development and poverty reduction efforts. Climate change is also expected to contribute to severe water and heat stresses that have profound impacts on the livelihoods and health of the poor, including their assets and quality of life. This enhanced vulnerability of the poor limits the effectiveness of development interventions and raises the questions of how to most effectively integrate climate change concerns into development planning.

In addition to threats to biodiversity, climate change will also causes severe damage to human health and could contribute to negative growth at national levels. Modeling based on IPCC scenarios suggest that temperature rise by

2010 could lead to significant increase in potential breeding grounds for malaria in parts of Brazil, Southern Africa and the horn of Africa.

Biodiversity and Climate Change

What Changes might we Expect in the Future?

Computer models predict an average global temperature increase of 1.4 to 5.8 degrees C by the year 2100. Predicted impacts associated with such a temperature increase include: a further rise in global mean sea level of 9 to 88 cm, more precipitation in temperate regions and Southeast Asia, associated with a higher probability of floods, less precipitation in Central Asia, the Mediterranean region, Africa, parts of Australia and New Zealand, associated with a greater probability of droughts, more frequent and powerful extreme climatic events, such as heat waves, storms, and hurricanes, an expanded range of some dangerous "vector-borne diseases", such as malaria, and further warming of the Arctic and Antarctic, leading to more sea-ice disappearance.

Why Act Now?

Climate change is already a reality. Even if all anthropogenic emissions were to stop now, changes would continue into the future. We must, therefore, increase climate change mitigation and adaptation efforts.

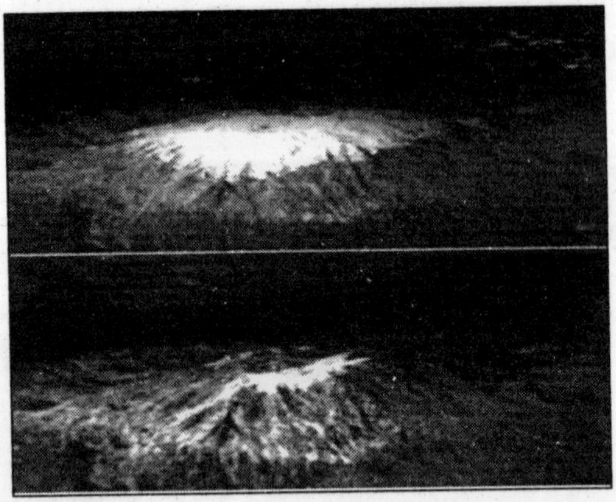

Although ecosystems have adapted to changing conditions in the past, current changes are occurring at rates not seen historically. In general, the faster the climate changes, the greater the impact on people and ecosystems. Reductions in greenhouse gas emissions can lessen these pressures, giving these systems more time to adapt.

In addition to mitigation, however, there is an urgent need to develop and implement climate change adaptation plans. People and the natural environment have become particularly vulnerable to the impacts of climate change. Indeed, activities that lead to the degradation of the environment, such as deforestation and overgrazing, can exacerbate the consequences of climate change. In many countries, more people, particularly those at lower income levels, are now forced to live in exposed and marginal areas i.e. floodplains, exposed hillsides, arid or semiarid lands, putting them at risk to the negative impacts of climate change. For these people, even minor changes in climate can have a disastrous impact on lives and livelihoods. The same can be said of many species, which are adapted to very specific climatic conditions. A small change in these conditions could mean that we lose these species forever. While there is still more to understand about climate change, enough is known about the range of impacts, the magnitude of risks, and the potential for adaptation to act now.

Biodiversity: To Reduce the Impacts of Climate Change

The resilience of ecosystems can be enhanced and the risk of damage to human and natural ecosystems reduced through the adoption of biodiversity-based adaptive and mitigative strategies. Mitigation is described as a human intervention to reduce greenhouse gas sources or enhance carbon sequestration, while adaptation to climate change refers to adjustments in natural or human systems in response to climatic stimuli or their effects, which moderates harm or exploits beneficial opportunities. Examples of activities that promote mitigation of or adaptation to climate change include: maintaining and restoring native ecosystems, protecting and enhancing ecosystem services, managing habitats for endangered species, creating refuges and buffer zones, and establishing networks of terrestrial, freshwater and marine protected areas that take into account projected changes in climate.

Polar ecosystems are home to an array of plants and animals that survive in some of the most extreme conditions in the world. The seas surrounding the Antarctic are rich in plankton, which support a rich marine food chain, while the Arctic supports many mammals and has an important role in the annual cycle of migratory birds. The biodiversity of the Arctic is fundamental to the livelihoods of Arctic peoples.

Vulnerability to Climate Change

Polar regions are now experiencing some of the most rapid and severe climate change on Earth, which will contribute to environmental and socio-economic changes, many of which have already begun. Polar regions are highly vulnerable to climate change because: During the 20th century, Arctic air temperatures increased by approximately degrees C. This increase is ten times faster than the observed global-mean surface temperature. An additional warming of about 4-7 degrees C in the Arctic is predicted for the next 100 years.

Polar species and societies have developed very specialized adaptations to the harsh conditions found at the poles, thus they are extremely vulnerable to dramatic changes in these conditions. Low resilience to changing environmental parameters, including fluctuations of air temperatures and precipitation dynamics.

Walruses, polar bears, seals and other marine mammals that rely on sea ice for resting, feeding and breeding are particularly threatened by climate change. For example, studies reveal that in 1980, the average weight of female polar bears in western Hudson Bay, Canada, was 650 pounds. In 2004, their average weight was only 507 pounds. It is believed that the progressively earlier breakup of the Arctic sea ice is responsible for the fall in the polar bears' average weight. Reduced sea-ice extent is also believed to have caused a 50% decline in emperor penguin populations in Terre Adélie.

Populations of krill and other small organisms may also decline as ice recedes. Due to the high importance of krill in food chains, the entire marine food web could be adversely affected. The livelihood of indigenous people in the Arctic is already being affected by climate change. Losses in biodiversity affect the traditional practices of indigenous people, particularly fishing and hunting. For example, the Saami people have observed changes in reindeer grazing pastures, and the Inuit people of Canada have observed reductions in the ringed seal population, their single most important source of food.

Contribution to Climate Change

The warming of polar regions has repercussions in the rest of the world. Indeed, melting of highly reflective snow and ice uncovers darker land and ocean surfaces, increasing absorption of the sun's heat and further warming the planet. Moreover, snow and ice melt raises global sea level. The melting of ice sheets in Antarctica and Greenland is estimated to account for one third of the sea level rise. This addition of fresh water in the ocean is also slowing ocean circulation, affecting global and regional climate.

Adaptation Options

The reduction of other stressors, such as permafrost degradation, chemical pollution, overfishing, land-use changes, and habitat fragmentation could improve polar ecosystems' resilience to climate change. Adaptation activities can make use of local and indigenous knowledge and participation. Indeed, indigenous people can contribute to the understanding of changes in the Arctic through their observations and perspectives on changes in biodiversity and ecosystem functioning. For example, the Inuvialuit Hunters and Trappers in Canada's High Arctic, along with the International Institute for Sustainable Development (IISD), initiated a year-long project to document Arctic climate change and communicate it to Canadian and international audiences. During the initiative, a video and several scientific journal articles were produced to communicate the negative impacts of climate change observed in the Arctic and to understand the adaptive strategies that local people are using in response.

Remedial Action

Action to halt or reduce climate change, most obviously by reducing CO2 emissions from fossil fuel burning has been proposed through global mechanisms such as the Kyoto Protoco. Furthermore, even if CO2 emissions were to stop immediately an average warming of 1-2°C would still occur. Mitigation strategies such as the possible widespread planting of forests to sequester carbon would be of more immediate interest to conservation planning. Much will depend on where forests are planted whether in existing degraded forest, or via conversion of grassland or agricultural land, what types of trees are planted mixed stands of native trees may be expected to have greater conservation value than monoculture plantations of exotics and the intensity of management. Action to halt the decline of old-growth forests would produce the greatest immediate benefit to conservation.

Research Priorities

For conservation biologists to incorporate climate change into planning, the following could be some starting points: w Climate change scenarios are better made at the regional/ ecosystem/biome scale. Therefore it is important for biodiversity experts to work closely with climate change specialists in designing options for conservation and climate change mitigation as well as adaptation. While climate change models are useful for conservation purposes, they cannot be considered independent of other ecosystem and environmental variables.

Understanding the variation in climate experienced by species in an assemblage and an ecosystem in a biome in the past will help predict their response to future change. It is therefore important to understand evaluate species and ecosystem dynamics. Improved predictions of the frequency and magnitude of extreme events will help conservation specialists to design appropriate interventions aimed to strengthen conservation practice.

Policy Linkages

The many ways in which climate and biodiversity interact are gradually leading researchers and policymakers to the conclusion that working on the two issues together will be more effective than dealing with them separately. At a practical level, this will mean that the initiatives emanating from the UN Framework Convention on Climate Change (UNFCCC) and Convention on Biological Diversity (CBD) will need to be put into practice in a more integrated way. The climate convention aims to stabilise atmospheric concentrations of greenhouse gases, primarily through negotiation of global agreements such as the Kyoto Protocol, but also by providing guidance to governments on practical ways to adapt to a changing climate. The goals of the biodiversity convention are the conservation of biological diversity, the sustainable use of its components and the fair and equitable sharing of the benefits arising out of the utilisation of genetic resources.

Despite their separate objectives, members of the two conventions have been considering ways of working together. The biodiversity convention, for example, has an Ad Hoc Technical Expert Group, which has identified a number of possible areas where links between the two conventions could be strengthened. Similarly, the secretariats to the climate and biodiversity conventions, as well as the UN Convention to Combat Desertification, have established a joint liaison group to help link up initiatives related to climate change and to biodiversity.

One mechanism that will catalyse stronger links between policymakers working on climate change and biodiversity, and that could also have positive implications for biodiversity, is the Clean Development Mechanism of the Kyoto Protocol. This allows developed countries to invest in emission is reducing projects in developing countries to obtain credits, which they put towards meeting their own emissions targets. Projects to establish plantations as carbon sinks could be added to the activities under this mechanism. If established in degraded areas and composed of indigenous species, such plantations could become valuable habitats that help to conserve biodiversity.

Overall, all environmental issues concern natural systems are linked in myriad ways. Before long, research and policymaking both at the national and international level will need to reflect this interconnectedness. Unless we focus our attention on actions that link conservation, adaptation and mitigation achieving the CBD targets of reducing biodiversity loss by 2010, the Kyoto Protocol targets of reducing GHG emissions by 2012 and achieving sustainable development by 2015 as per the MDGs will all remain a distant dream.

In practice, however, getting those responsible for implementing the two conventions to work together will not be easy. The conventions have separate constituencies, administration arrangements, negotiators and guiding scientific bodies.

In recent years, several alternative models for international environmental governance have been suggested. One possible

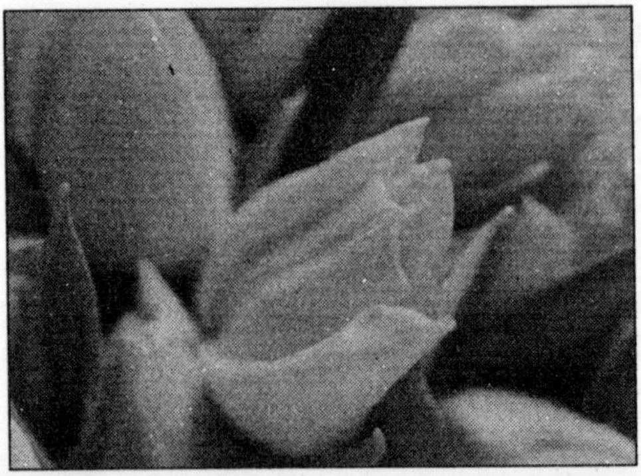

reform at the national level would be to encourage countries to set up a single body to deal with their obligations under all international environmental agreements. At present, many countries have a separate national organisation handling each concern that have little contact with each other, despite the fact that they may report to the same ministry.

Another area for national-level reform could be in the area of disaster management. For example, countries that are drawing up plans to deal with climate-induced disasters could identify not just vulnerable human settlements, but also the local ecosystems on which they depend. An important element of this highly integrated approach is that it would protect ecosystems not just as economic mainstays of local peoples, but also as havens of biodiversity.

7

Biodiversity Convention and Conservation

The Convention on Biological Diversity (CBD), known informally as the Biodiversity Convention, is an international legally binding treaty that was adopted in Rio de Janeiro in June 1992. The Convention has three main goals: Conservation of biological diversity or biodiversity, sustainable use of its components and fair and equitable sharing of benefits arising from genetic resources. In other words, its objective is to develop national strategies for the conservation and sustainable use of biological diversity. It is often seen as the key document regarding sustainable development. The Convention was opened for signature at the Earth Summit in Rio de Janeiro on 5 June 1992 and entered into force on 29 December 1993. 2010 is the International Year of Biodiversity. The Secretariat of the Convention on Biological Diversity is the focal point for the International Year of Biodiversity.

Convention Meaning

The convention recognized for the first time in international law that the conservation of biological diversity is "a common concern of humankind" and is an integral part of the development process. The agreement covers all ecosystems, species, and genetic resources. It links traditional conservation efforts to the economic goal of using biological resources sustainably. It sets principles for the fair and equitable sharing of the benefits arising from the use of genetic resources, notably those destined for commercial use. It also covers the rapidly expanding field of biotechnology through its Cartagena Protocol on Biosafety, addressing technology development and transfer, benefit-sharing and biosafety issues. Importantly, the Convention is legally binding; countries that join it are obliged to implement its provisions.

The convention reminds decision-makers that natural resources are not infinite and sets out a philosophy of sustainable use. While past conservation efforts were aimed at protecting particular species and habitats, the Convention recognizes that ecosystems, species and genes must be used for the benefit of humans. However, this should be done in a way and at a rate that does not lead to the long-term decline of biological diversity. The convention also offers decision-makers guidance based on the precautionary principle that where there is a threat of significant reduction or loss of biological diversity, lack of full scientific certainty should not be used as a reason for postponing measures to avoid or minimize such a threat. The Convention acknowledges that substantial investments are required to conserve biological diversity. It argues, however, that conservation will bring us significant environmental, economic and social benefits in return. Issues under the convention

Some of the many issues dealt with under the convention include:

- Measures and incentives for the conservation and sustainable use of biological diversity.
- Regulated access to genetic resources and traditional knowledge, including Prior Informed Consent of the party providing resources.
- Sharing, in a fair and equitable way, the results of research and development and the benefits arising from the commercial and other utilization of genetic resources with the Contracting Party providing such resources governments and local communities that provided the traditional knowledge or biodiversity resources utilized.

- Access to and transfer of technology, including biotechnology, to the governments and local communities that provided traditional knowledge and biodiversity resources.
- Technical and scientific cooperation.
- Impact assessment.
- Education and public awareness.
- Provision of financial resources.
- National reporting on efforts to implement treaty commitments.

The Cartagena Protocol on Biosafety of the Convention, also known as the Biosafety Protocol, was adopted in January 2000. The Biosafety Protocol seeks to protect biological diversity from the potential risks posed by living modified organisms resulting from modern biotechnology.

The Biosafety Protocol makes clear that products from new technologies must be based on the precautionary principle and allow developing nations to balance public health against economic benefits. It will for example let countries ban imports of a genetically modified organism if they feel there is not enough scientific evidence the product is safe and requires exporters to label shipments containing genetically altered commodities such as corn or cotton.

The required number of 50 instruments of ratification/ accession/approval/acceptance by countries was reached in May 2003. In accordance with the provisions of its Article 37, the Protocol entered into force on 11 September 2003. Six international conventions focus on biodiversity issues: the Convention on Biological Diversity year of entry into force: 199), the Convention on Conservation of Migratory Species, the Convention on International Trade in Endangered Species of Wild Fauna and Flora (1975), the International Treaty on Plant Genetic Resources for Food and Agriculture (2004), the Ramsar Convention on Wetlands (1971) and the World Heritage Convention (1972).

Each of the biodiversity-related conventions works to implement actions at the national, regional and international level in order to reach shared goals of conservation and sustainable use. In meeting their objectives, the conventions have developed a number of complementary approaches such as site, species, genetic resources and ecosystem-based and operational tools viz programmes of work, trade permits and certificates, multilateral system for access and benefit-sharing, regional agreements, site listings, funds.

The Six Biodiversity-related Conventions

Convention on Biological Diversity the objectives of the CBD are the conservation of biological diversity, the sustainable use of its components, and the fair and equitable sharing of the benefits arising from commercial and other utilization of genetic resources. The agreement covers all ecosystems, species, and genetic resources. Convention on International Trade in Endangered Species of Wild Fauna and Flora (CITES) the CITES aims to ensure that international trade in specimens of wild animals and plants does not threaten their survival. Through its three appendices, the Convention accords varying degrees of protection to more than 30,000 plant and animal species.

Convention on the Conservation of Migratory Species of Wild Animals the CMS, or the Bonn Convention aims to conserve terrestrial, marine and avian migratory species throughout their range. Parties to the CMS work together to conserve migratory species and their habitats by providing strict protection for the most endangered migratory species, by concluding regional multilateral agreements for the conservation and management of specific species or categories of species, and by undertaking co-operative research and conservation activities.

The International Treaty on Plant Genetic Resources for Food and Agriculture the objectives of the Treaty are the conservation and sustainable use of plant genetic resources for food and agriculture and the fair and equitable sharing of the benefits arising out of their use, in harmony with the Convention on Biological Diversity, for sustainable agriculture and food security. The Treaty covers all plant genetic resources for food and agriculture, while it's Multilateral System of Access and Benefit-sharing covers a specific list of 64 crops and forages. The Treaty also includes provisions on Farmers' Rights. Convention on Wetlands popularly known as the Ramsar Convention the Ramsar Convention provides the framework for national action and international cooperation for the conservation and wise use of wetlands and their resources. The convention covers all aspects of wetland conservation and wise use, recognizing wetlands as ecosystems that are extremely important for biodiversity conservation in general and for the well-being of human communities.

World Heritage Convention (WHC) the primary mission of the WHC is to identify and conserve the world's cultural and natural heritage, by drawing up a list of sites whose outstanding values should be preserved for all humanity and to ensure their protection through a closer co-operation among nations. Each convention stands on its own—with its own specific objectives and commitments—inter-linkages between the issues each addresses, and potential complementarities in their monitoring and implementation processes, provide a basis for cooperation. With the target of achieving by 2010 a significant reduction of the current rate of biodiversity loss, set by the Strategic Plan of the Convention on Biological Diversity, and later endorsed by the World Summit on Sustainable Development and incorporated into the Millennium Development Goals, the need to promote cooperation among the biodiversity-related conventions while reducing duplication of effort has become increasingly relevant.

The governing bodies of each Convention have set out specific mandates for cooperation among the biodiversity-related conventions. In line with these mandates, reference to cooperation is made in a number of decisions from these governing bodies, and has led to the development of memoranda of cooperation and joint work programmes between the conventions. These provisions have resulted in a wide range of cooperative activities being undertaken by the Conventions in support of shared goals. To further enhance cooperation, a Biodiversity Liaison Group comprising the executive heads of the six biodiversity-related conventions was established in 2002. Options for advancing cooperation have been put forward at the meetings of the Biodiversity Liaison Group and other meetings. India biodiversity conventional The country is divided into 10 biogeographic regions. The diverse physical features and climatic situations have formed ecological habitats like forests, grasslands, wetlands, coastal and marine ecosystems and desert ecosystems, which harbour and sustain immense biodiversity. Biogeographically, India is situated at the tri-junction of three realms-Afro-tropical, Indo-Malayan and Paleo-Arctic realms, and therefore, has characteristic elements from each of them. This assemblage of three distinct realms makes the country rich and unique in biological diversity.

The country is also one of the 12 primary centres of origin of cultivated plants and domesticated animals. It is considered to be the homeland of 167 important plant species of cereals, millets, fruits, condiments, vegetables, pulses, fibre crops and oilseeds, and 114 breeds of domesticated animals.

About 4,900 species of flowering plants are endemic to the country. These are distributed among 141 genera belonging to 47 families. These are concentrated in the floristically rich areas of North-East India, the Western Ghats, North-West Himalayas and the Andaman and Nicobar Islands. These areas constitute two of the 18 hot spots identified in the world. It is estimated that 62 per cent of the known amphibian species are endemic to India of which a majority is found in Western Ghats.

Approximately 65 per cent of the total geographical area has been surveyed so far. Based on this, over 46,000 species of plants and 81,000 species of animals have been described by the Botanical Survey of India (BSI) established in 1890 and Zoological Survey of India (ZSI) established in 1916, respectively. This list is being constantly upgraded, specially in lower plants and invertebrate animals. The Forest Survey of India established in 1981 assess the forest cover with a view to develop an accurate database for planning and monitoring purposes. Conservation and sustainable use of biological resources based on local knowledge systems and practices is ingrained in Indian ethos. The country has a number of alternative medicines, like *Ayurveda, Unani, Siddha* and Homeopathic systems which are predominantly based on plant based raw materials in most of their preparations and formulations. Herbal preparations for various purposes including pharmaceutical and cosmetic purposes form part of the traditional biodiversity uses in India.

The strategies for conservation and sustainable utilisation of biodiversity have comprised providing special status and protection to biodiversity-rich areas by declaring them as national parks, wildlife sanctuaries, biosphere reserves, ecologically fragile and sensitive areas. Other strategies include offloading pressure from reserve forests by alternative measures of fuelwood and fodder need satisfaction by afforestation of degraded areas and wastelands and creation of *ex-situ* conservartion facilities such as gene banks. For example, the Tura Range in Garo Hills of Meghalaya is a gene sanctuary for preserving the rich native diversity of wild citrus and musa species. Approximately, 4.2 per cent of the total geographical area of the country has been earmarked for extensive *in-situ* conservation of habitats and ecosystems. A protected area network of 85 national parks and 448 wildlife sanctuaries has been created. The results of this network have been significant in restoring viable population of large mammals such as tiger, lion, rhinoceros, crocodiles and elephants.

The Indian Council of Forestry Research and Education (ICFRE) has identified 309 forest preservation plots of representative forest types for conservation of viable and representative areas of biodiversity. Out of these plots, 187 area in natural forests and 112 are in plantations, covering a total area of 8,500 hectares. A programme "Eco-development" for *in-situ* conservation of biological diversity involving local communities has been initiated in recent years. The concept of ecodevelopment integrates the ecological and economic parameters for sustained conservation of ecosystems by involving the local communities with the maintenance of earmarked regions surrounding protected areas. The economic needs of the local communities are taken care under this programme through provision of alternative sources of income and a steady availability of forest and related produce.

Programmes have also been launched for scientific management and wise use of wetlands, mangroves and coral reef ecosystems. Twenty one wetlands, and mangrove areas and 4 coral reef areas have been identified for intensive conservation and management purposes. Mangroves conservation is one of the thrust areas of the Ministry of Environment and Forests. Six significant wetlands of India have been declared as "Ramsar Sites" under the Ramsar Convention. Under the World Heritage Convention, five natural sites have been declared as "World Heritage Sites". To conserve the representative ecosystems, a Biosphere Reserve Programme is being implemented. Twelve biodiversity rich areas of the country have been designated as Biosphere Reserves applying the UNESCO/MAB criteria. These reserves aim at conserving the biological diversity and genetic integrity of plants, animals and microorganisms in their totality as part of the natural ecosystems, so as to ensure that self-perpetuation and unhindered evolution of the living resources.

The Ministry of Environment and Forests constituted the National Afforestation and Eco-development Board (NAEB) in August 1992. NAEB has evolved specific schemes for promoting afforestation and management strategies which help the states in developing specific afforestation and management strategies and eco-development packages for augmenting biomass production through a participatory planning process of joint forest management and microplanning. To complement *in-situ* conservation, attention has been paid to *ex-situ* conservation measures. According to currently available survey, central government and state governments together run and manage 33 botanical gardens. Universities have their own botanical gardens. There are 275 zoos, deer parks, safari parks and aquaria. A Central Zoo Authority was set up to secure better management of zoos. A scheme 'Assistance to

Botanical Gardens' provides one-time assistance to botanical gardens to strengthen and institute measures for *ex-situ* conservation of threatened and endangered species in their respective regions.

Formal policies and programmes for conservation and sustainable utilisation of biodiversity resources dates back to several decades. The concept of environmental protection is enshrined in the Indian constitution in articles 48a and 51a(g). Major central acts relevant to biodiversity include Forest Act, 1927, Wildlife (Protection) Act, 1972, Forest (Conservation) Act, 1980, and Environment (Protection) Act, 1986. The various central Acts are supported by a number of state laws and statutes concerning forests and other natural resources. The policies and strategies directly relevant to biodiversity include National Forest Policy amended in 1988, National Conservation Strategy and Policy Statement for Environment and Sustainable Development, National Agricultural Policy, National Land Use Policy, National Fisheries Policy, National Policy and Action Strategy on Biodiversity, National Wildlife Action Plan and Environmental Action Plan.

India is a party to the Convention on Biological Diversity (CBD). The main objectives of this convention are; conservation of biological diversity, sustainable use of the components of biodiversity and fair and equitable sharing of benefits arising out of the utilisation of genetic resources. The main implementation measures for the CBD are through national strategies, legislation, and administrative instruments to be developed in accordance with the country's conditions and capabilities.

Adopting a consultative process with the stakeholders, a national policy and action strategy on biological diversity has been drawn up as a macro-level statement of strategies. This plan aims at consolidating the on-going efforts of conservation and sustainable use of biological diversity, identifying gaps in various sectors and providing a policy and programme regime to ensure attainment of the three objectives of the CBD. This famework document provides the basis for developing detailed action at micro and regional levels.

The Ministry has accessed funding from GEF for preparing detailed National Biodiversity Strategy and Action Plan (NBSAP) using participatory planning approach and is currently being implemented.

Identifying the critically important areas and gaps in taxonomic work, an All India Coordinated Project on Taxonomy (AICOPTAX) has been launched for prioritised purpose-oriented capacity building in taxonomy. To encourage taxonomic work, Janaki Ammal National Award in taxonomy has been instituted. India's richness in biological resources and related indigenous knowledge is well recognised. One of the major challenges before the country lies in adopting an instrument which help realise the objectives of equitable benefit sharing enshrined in the convention. India has worked out a bio-diversity legislation. It aims at regulating access to biological resources. The legislation has been prepared after extensive consultations with concerned people and agencies.

Biodiversity Conservation Programmes

There are two methods of biodiversity conservation, in situ and ex situ. The former envisages conservation within the natural ecosystem such as protected areas such as wildlife sanctuaries, national parks, biosphere reserves, heritage sites, etc.), and the latter is a method of conservation outside natural habitats viz botanical and zoological gardens, gene banks, seed banks etc. In case of domesticated or cultivated species, conservation means conservation in the surroundings where they have developed their distinctive properties. In addition to this there are special projects envisaged by Government of India to protect threatened species such as Project Tiger, Project Elephant, Project Rhino, Project Hangul, etc. Global Initiate In addition to the aforesaid conservation programmes, laws related to intellectual property rightly from part of the global initiates in biodiversity conservation. A product derived by the activity of the productive human mind or intellect is referred to as intellectual property. Cultural and artistical knowledge, industrial designs, logos that come in the ambit of trade marks pictorial or verbal, music related products, sound recording, biotechnology products, industrial secrets, etc comes with in the framework of intellectual properties.

Intellectual Property Acts

Patent	Patent Act 1970
Copyright	Copyright Act 2000
Trade mark	Trade mark Act 1999
Design	Design Act 2000
Circuit layout	Semiconductor Intended layout Design Ac
Geographical indicators	GI of Goods (Registration Protection) Act
BiodiversityNew Plant varieties	Biodiversity Act 2002New Plant varieties and Farmers RightsProtection Act 2001

Activities in India

Several activities related to biodiversity conservation are going on in India. Many such activities, including formation of biosphere reserves, are centered in biodiversity-rich forests ecosystems. In addition to this, many ecosystems such as wetlands, mangroves and sacred groves have also been brought under conservation schemes. As mentioned, Project Elephant, Project Tiger, etc are some of the national efforts in India towards conservation of selected individual species. National Botanic Gardens have been established at Lucknow and Kolkata for the conservation of plants. Further, there are Orchidaria at Yercard and Meghalaya for the conservation of orchids. The Government of India has also enacted several laws aimed at conserving our rich biodiversity.

Conservation of Biodiversity

A. In situ conservation. The best method of in situ conservation of areas rich in bioresources is the designation of areas as. Biosphere Reserves. (BR) as well as natural sanctuaries, such as of orchids. The regions of BRs are protected from human interference. Those areas can be considered as nature.s playground where dynamics of evolution can be studied. Biosphere Reserves Several areas so far have been identified in India as potential BRs. These include: (i) Mysore Plateau

Wynad in the Nilgiris with the Silent Valley (ii) Kanchendzonga National Park on Sikkim (iii) The Nanda Devi Sanctuary (iv) Valley of Floowers (v) Kedarnath in Uttar Pradesh (vi) Simlipal and Jeypore Hill Forest stretch in Orissa (vii) Kanha National Park, Madhya Pradesh Biodiversity need December 11, 2003 I Proceedings of Round Table Discussion (viii) North Island of Andaman and Jarawa Tribal Reserves (ix) Sundarbans in West Bengal (x) Namdapha, Lalichopra, Pakku, Tawang and Welong in Arunachal Pradesh (xi) Gulf of Mannar in Tamil Nadu, and (xii) Tura Ridge in Meghalaya (a) Monitoring of BRs Biosphere Reserves, because of the absence of human interference, remain undisturbed and the environment with its ecosystem reflects the dynamics of evolution.

The origin and death of genotypes is a common feature of such natural reserves. Through struggle for existence and natural environmental changes, genotypes adapted to changed niches survive with the gradual elimination of the unfit. As such, documentation of the flora and fauna of the biosphere should be a continuing process to record the dynamicity of species change in evolution. Along with establishment of Biosphere Reserves, which has already been achieved, the continuous documentation and monitoring with exploration of resources is essential.

In situ conservation other than in BRs In addition to Biosphere Reserves, germplasm centers as well as orchid sanctuaries, which normally harbour a wide variety of genotypes, certain species with high medicinal and economic value with strict ecological preferences, including stress environments, such as alpine, semi-alpine or even desert regions need to be preserved in situ. In view of their wide importance, coupled with the fact that their preferences are very specific in a restricted set-up of nature, their conservation in situ is essential for propagation and sustainable utilization. In order to increase the number of individuals of such species, tissue culture may be resorted to along with hardening of plants. Reserves can be created with such plants in one or more than one areas representing special ecological niches. The taxa requiring growth in such special areas in West Bengal (in situ) should include Daphne cannabina, Allium wallichii, Allium strachyii, Aconitum sp., E himalaya and Vaccinium sp. B. Ex situ conservation . Attention is to be focused on a large number of medicinals, which due to their therapeutic value are gradually becoming rare and facing extinction as a result of over-exploitation from natural habitats and absence of systematic cultivation.

Wild and other species of these genera need to be preserved for widening of their gene pool. Fortunately for legumes, the focus is widespread and conservative programmes are attended. (a) The Cucurbits serve both as vegetables and medicinals, and have wide distribution in West Bengal. Concerted effort is needed to collect wild varieties of Cucurbits in this zone - the group, which undoubtedly is going to be one of the most important families yielding drugs and vegetables for the future. Of the tropical resources, Cucurbits now occupy a prominent position in the western (b) Other medicinals include Withania somnifera, Hemidesmus indicus, Centalla asiatica, Evolvulus alsinoides, Phyllanthus sp., Zingiber sp. (Ginger family), Amaranthus sp. and Piper sp.

Under the targets for conservation, over and above, the categories listed, there are several other categories of plants in this part of the world, the genotypes of which need to be conserved, such as, (c) Plantation crops. Like tea, coffee, rubber, palm, cardamom, jute, Cinchona and others. All these crops are being cultivated on a wide-scale; and research and development of these categories of plants are supposed to be under the jurisdiction of different government departments and agencies. However, excepting the National Bureau of Plant Genetic Resources (NBPGR), where germplasms are being maintained, majority of the other research stations are principally dealing with cultivated crops without necessarily having the wild species and genotypes being maintained and conserved, as in allied species of Corchorus, Cinchona, Camellia and others. In fact, conservation of the entire gene pool of these crops needs to be mandated from all our bioresources.

Enrichment of Biodiversity

With the powerful tools at hand, scientists are now engaged in moulding genetic architecture of plants thus leading to enrichment of genetic vis-à-vis biodiversity, in addition to nature.s own mechanism of enrichment through mutations and recombinations. Genetic manipulation. both through mutations and genetic engineering i.e., vertical and horizontal gene transfer have paid very high dividends for agricultural crops. Despite such a strong base of genetic and gene manipulation in India, no serious attempt has so far been made for the enrichment of genetic vis-à-vis biodiversity of medicinal crops in this subcontinent. Only in the recent years, attention has been focused to improve herbal plants conferring resistance to biotic and abiotic stresses through horizontal or direct gene transfer and simultaneously augmenting the active principle. The medicinal plant resources of India offer enormous scope for the enrichment of genetic diversity in India. The potential of genetic manipulation in other economic and plantation crops, too, are enormous. The powerful tools of manipulation both for vertical and horizontal transfer of genes can conveniently be geared towards enrichment of diversity, which is the natural heritage of humanity.

A large number of countries are amending their laws to incorporate the potential of new technology and also to comply with the stipulations of the Convention on Biological Diversity (CBD) and the Agreement on Trade-Related Aspects of Intellectual Property Rights (TRIPS) of the World Trade Organisation (WTO). Developing nations feel threatened by the inclusion of an option for patent protection for all life forms, such as plants and animals in Article 27(3)b of TRIPS, which states: Article 27. Patentable Subject Matter 3. Members may also exclude from patentability:

(b) Plants and animals other than microorganisms, and essentially biological processes for the production of plants or animals other than non-biological and microbiological processes. However, Members shall provide for the protection of plant varieties either by patents or by an effective suigeneris system or by any combination thereof.

This option intensifies the debate concerning IPRs for biological and genetic resources between the technology-rich North and the gene-rich South. Firms in the technology rich north have the technical capability to fully exploit the commercial potential of biological resources. Policy makers in developed economies want to implement international patent protection for genetically engineered products, because their firms face high research and development (R&D) costs. On the other hand, majority of bioresources and related knowledge, both crucial inputs to genetic R&D, are found in a small number of developing countries. Governments in the developing world are trying to achieve two goals:

- To acquire a fair share of the profits generated by foreign firms from commercial exploitation of new products developed using national biological resources and knowledge; and
- To protect national interests while conforming to international agreements, which now stipulate that national governments must protect the rights of those who develop new products using biological resources (e.g. Article 27(3)b of TRIPS).

This exercise is being carried out at a time when firms in the developed world are accused of pirating and patenting biological and associated TK of indigenous communities from the biodiversity-rich developing world for profit, without equitably sharing benefits or appropriate transfer of new technologies. The Indian Biodiversity Act December 11, 2003 station that Protection for biological resources through property rights legislation, such as patents, have traditionally been excluded on grounds of public morality, the need to encourage everyone from the smallest farmer to multinational companies to foster innovations and the need to prevent commercialization of critical sectors like food and healthcare. Patents have long been associated with technological innovations and their application to life forms has triggered debates all over the world. With regard to biodiversity, IPRs mete out differential treatment to human and natures creation. With the latter traditionally being excluded from patentability, genetically engineered life forms are being progressively patented with plants and more recently even animals coming under the patent net.

While IPRs could arguably be extended to cover agro-biodiversity, voices have been raised against this trend from different quarters. Protecting biological resources is only part of the problem. The commercial utility of biological resources increases if its uses are known and have been tested. This knowledge is often held as TK by indigenous communities. Traditional knowledge is the knowledge held by and acquired by members of a distinct culture through the means of inquiry peculiar to that culture, and concerning the culture itself or the local environment in which it exists TK has a number of different subsets, some of them designated by expressions such as .indigenous knowledge,. .folklore,. .traditional medicinal knowledge and others. TK is dynamic in nature and is changing its nature and evolving all the time through interactions with the social and physical environment. Its commercial use raises IPR questions, whose importance is heightened by international trade, communications and cultural exchange. Specific definition and protection of the TK is necessitated by the push from commercial sector to control, manage and market the knowledge.

Patent applicants, through documentary evidence, must show that their innovation is the result of a single act of discovery. Indigenous communities cannot protect information relating to TK or protection of biodiversity, if it is not the result of specific historic act of .discovery.. Indigenous knowledge, being part of the public domain and being trans-generational and communally shared, is therefore not protectable. Patents also motivate commercialisation and distribution. Indigenous communities may however, be largely concerned with prohibiting commercialisation and restricting use and distribution. Patents recognise only market economic values and ignore spiritual, aesthetic, or cultural-or even local economic-values. Indigenous people may value such information as they are linked to their cultural identity and symbolic unity.

Copyrights

Copyright necessitates identification of an author thus posing problems in many traditional societies. Copyrights are meant to benefit society through the granting of exclusive rights to.natural. And .juridical. Persons or creative individuals. not collective entities such as, indigenous people. In other words, creative expressions and collective innovations of traditional communities being ineligible for protection may legally be treated as free inputs for industrial R&D and the copyright industries. Copyright normally requires works to be documented. However, this is not so with folkloric expressions as they are usually passed on orally from one generation to the next. Moreover, the entity asserting its copyright - or indeed to claim any other IPR - must have legal standing which collective groups like rural communities usually lack according to national legal system. Trade Secrets While many traditional societies commonly share knowledge, it cannot be discounted that healers and other specialist knowledge-holders may possess knowledge that they would want to conceal from the world. Legal action may be used to force a company to pay compensation if it obtains such information illicitly. Conceivably, a considerable amount of TK could be protected under trade secret law.

However, the problem here is that if the inventor chose to commercialise his innovation as a trade secret, the growth of ideas may be hindered. Enforcing legal rights would also require the holder of the knowledge to prove that steps had been taken to ensure that the knowledge was protected as a secret. Further, simultaneous development of innovations similar to ones own may emerge and be protected affecting the interest of the company or the inventor having the trade secrets. Opponents of patenting often overlook the fact that useful information would become trade secrets if patenting is disallowed. Patenting forces an invention to become public knowledge, thus allowing other researchers to investigate knowledge from the new invention. An effective regime for protecting biodiversity and traditional knowledge should provide for the following, if it seeks to effectively uphold the interests of developing countries. Establishment of property rights for all actors involved in biodiversity management. Instead of protecting only the interests of corporate biotechnology firms, it should also uphold the interests of indigenous communities who have maintained the biological resources and also hold knowledge associated with it. This would involve assigning proper rights both for the resource held as well as associated knowledge.

Non-monopoly rights that recognises that biodiversity related innovations are the handiwork of various actors and each needs property rights over his inventions. So while firms can commercialise products produced using TK associated with biological resources, the indigenous community that provided the knowledge would get a share of the benefits. This allows different property rights systems to exist simultaneously. Recognition of non-commercial interests-This respects the desire of communities to refrain from commercializing their TK. It recognises that not all communities see monetary gains as the sole driving force behind the use of biological resources and TK is effective. In implementation this takes into account the broader context of a nations interests as well as its international obligations and requires that policy is effective not just on paper, but in actual practice.

The Indian Biodiversity Act

Responding to the new international regime of national sovereignty over biological resources ushered in by the CBD; the Indian Government formulated the Biological Diversity Act, 2002. The Act acknowledges the importance institutions of local governance, such as Panchayats, in managing biodiversity and affecting benefit sharing. However, in spite of the above features, the Act has several flaws. One such significant deficiency is that it lacks focus and makes excessive claims. It aims at regulating use of all biological resources through the entire expanse of the country. This would be a stupendous task anywhere in the world; it is quite impossible in a biomass-based civilization such as ours.

The Act further stipulates .no person who is a citizen of India or a body corporate, association or organization, which is registered in India shall obtain any biological resource for commercial utilization or bio-survey and bio-utilization except after giving prior intimation to the SBB concerned. It further clarifies that the provisions of this section shall not apply to the local people and communities of the area, including vaids and hakims, who have been practicing indigenous medicine. However, it does not define local people and local communities, leaving it to the bureaucracy and the rules to be formulated. It also leaves it to the discretion of the Government authorities to exempt other normally traded commodities. These arrangements are clearly inadequate and could lead to exploitation by parties with commercial interests.

The most prominent aspect of the Act is that it brings the countries biodiversity and associated TK under the regulation of the proposed NBA. Experts consider it .totally unfair. As it amounts to government control, read bureaucratic control, over knowledge. Since a large part of the knowledge is in the public domain and is acquired through the process of research, experimentation and observation, hence it .cannot be brought under the purview of any authority, leave alone NBA. It is further pointed out that the Bill .is based on a framework in which the government acts as a monopoly power, the sole regulator and benefactor.

The Act also overlooks the applicability of other legislation to access to biological resources and their sustainable use and development and does not seek to harmonise its provisions such that they do not contradict other relevant laws like the Drug and Cosmetics Act, Plant Varieties Protection (PVP) Act and Farmers. Rights Bill, Geographic Indicators Bill, etc. Moreover, it does not provide any mechanism for recognising assertions of benefit claimants and granting them rights. It does not specify whether they can adopt measures against grant of IPRs on genetic resources or knowledge taken from them. An analysis of the Indian Biodiversity Act, in terms of the four criteria mentioned above, reveals some shortcomings in the Act that would need to be rectified, either by amending it, or while rules and regulations are framed for its implementation.

In terms of proper establishment of property rights, the Act does not lay down detailed guidelines for assessing and assigning property rights related to biological resources and traditional knowledge. These tasks are left to the NBA, which will be required to establish detailed rules and regulations in this regard. To be really effective, the Act must assign property rights in a manner that protects the rights of all those who have contributed to the development of a commercially significant product. The Act must be discriminate enough, first to recognise the contributions that various parties make to a new invention or product, and then assign commensurate rights.

There is also no proper recognition of non-commerical interests in the Act. The Act calls for prior intimation to the SBB before obtaining resources for commercial use or bio-survey (Article 7). Equitable sharing of benefits arising out of utilisation of genetic resources is to be ensured by the NBA (Article 21.3). The Act does not recognise noncommercial interests of traditional communities. This is unlike legislation in other parts of the world, which recognise such rights. Costa Rica and the Organisation of African Unity (OAU) recognise the non-commercial interests of their people and acknowledge that receiving monetary returns is not the only force that drives innovations. The Biodiversity Law of Costa Rica grants the local communities and indigenous people the .Right to Cultural Objection and allows them to resist access to their resources and related knowledge for cultural, spiritual, social, economic or other motives. (Article 66).

Effective implementation of the Act is another area of concern. The Act requires the setting up of a NBA, which would advise the Central and State Governments on matters relating to the conservation of biodiversity, sustainable use of its components and equitable sharing of benefits arising out of the utilisation of biological resources. The law also provides for prohibition or restriction of such activities that may prove.detrimental or contrary to the objectives of conservation and sustainable use of biodiversity. (Article 24.2). To be really effective, the NBA needs to ensure that the rules and regulations that it frames are implemented fully. For example, documentation of TK could be counterproductive if it allows foreign firms to gain insights into the uses of biological resources without due acknowledgement or benefit sharing. Documentation of knowledge should be accompanied by strict monitoring of its use in order to ensure that benefits are shared. In the absence of such monitoring, it might be better not to document the knowledge at all.

It is necessary that the Indian Biodiversity Act be made more comprehensive so that it can take up all possible aspects of protection of TK and conservation and sustainable use of biodiversity. It is essential that the Biodiversity Bill recognise the rights of traditional communities and acknowledge the enormous contributions that they have made towards the conservation, development and sustainable use of plant and animal genetic resources. It should also provide for protection of biological resources as per customary practices and laws of the concerned communities. Communities should be given the right to protect their TK related to plant and genetic resources and receive equitable share of benefits arising from their use. Their participation in local and national level decision making on matters related to conservation and sustainable use of such resources should be encouraged.

It is important that they be granted the right to refuse access to such wealth, if they feel this permission would have detrimental effects on their cultural, spiritual and social heritage. There should also be scope for withdrawing consent or restricting activities that exert such harmful influences. They need to incorporate provisions that would allow traditional communities to refrain from accessing and prevent access to their genetic resources and knowledge even if it meant monetary losses.

In order to encourage conservation and sustainable use of biodiversity, both the pieces of Indian legislation need to put in place a comprehensive framework of incentives. Provisions must be added to award fiscal, scientific-technological and other kinds of incentives to activities or programmes carried out by people who contribute to achieving the above objective. In addition to establishing training programmes and research projects that encourage conservation and sustainable use of biodiversity, the laws should promote investment in this direction. Attempts to reform current Indian legislation so that it effectively upholds national interest and encourages sustainable use of resources could benefit from the experience of other developing countries to protect their biodiversity and traditional knowledge. It is evident that the effective implementation of the Indian Biodiversity Act will prove to be a major challenge and there is much work that lies ahead. Relationship between the TRIPS and the CBD Article 27(3)(b) of the TRIPS Agreement provides that plant varieties shall be protected under a patent or an effective sui generis legislation or a combination of both. India opted for adoption a sui generic legislation by the passing of the Protection of Plant Varieties and Farmers. Rights Act, 2001. However, there is no explanation in the TRIPS Agreement as to what constitutes an effective sui generis system.

The reasons for the inherent conflict, as pointed out by Kenya is as follows: 1 TRIPS reference to Article 27(3)(b) provides for the appropriation by the private parties of the genetic resources in a way that is inconsistent with the sovereign rights of the countries over those resources under the CBD; and TRIPS provides IPR protection of genetic material without ensuring the requirements of Prior Informed Consent (PIC) and benefit sharing under the CBD. The reasons for no conflict, as projected by European Community, Japan and the United States of America are as under: TRIPS and the CBD refer to different objects, purposes and subject matter.

International Treaty on Plant Genetic Resources for Food and Agriculture, 2001 The Commission on Genetic Resources for Food and Agriculture is a permanent forum within the Food and Agricultural Organisation (FAO) where governments discuss and negotiate matters relevant to genetic resources for food and agriculture. The FAO Conference established the Commission in 1983, mandate of which was broadened in 1995 to cover all components of biodiversity of relevance to food and agriculture. The FAO adopted an International Undertaking on Plant Genetic Resources in 1983. The Undertaking was revised with the aim of working in harmony with the CBD. It is known as the International Treaty on Plant Genetic Resources for Food and Agriculture, 2001. The Treaty incorporates the following objectives:

Conservation and sustainable use of plant genetic resources for food and agriculture fair and equitable sharing of benefits arising out of their use apart from the objectives stated above, the Treaty acknowledges the contribution of farmers and their rights in the area. It further emphasises on the need to develop synergy among agriculture, environment and commerce to achieve the objectives. Part three, Article 9 enumerates the farmer's rights. The treaty recognises the enormous contribution of local, indigenous and farming community for the conservation and development of plant genetic resources throughout the world. The responsibility for realising farmers rights vests with the national governments. Hence, each contracting party has to adopt suitable measures in this direction, which include protection of traditional knowledge, equitable participation in benefit sharing, and the right to participate in the decision-making at the national level on matters related to conservation and sustainable use of the resources. As per Part 3, the Article shall not be interpreted to limit the rights of farmers to save, use, exchange and sell farm-saved seed/propagating material. This provision is subject to national law as appropriate.

A multilateral system is provided in the Treaty for access to the genetic resources and benefit sharing. The Authority to determine access to the resources rests with the national governments and is subject to national legislation. Access to plant genetic resources for food and agriculture protected by intellectual and other property rights shall be consistent with relevant international agreements and with relevant national laws. Access shall be provided pursuant to a standard material transfer agreement (MTA), adopted by the Governing Body. Benefit sharing under the multilateral system is provided in Article13. The following mechanisms are provided in the Treaty:

Exchange of information; Access to and transfer of technology; Capacity building; and, Sharing of monetary and other benefits of commercialization, Convention for Protection of new Varieties of Plants. Article 17 states about the restrictions on the breeders. Accordingly, no contracting party may restrict the free exercise of a breeder's right for reasons other than of public interest. In any such case of restriction, contracting party shall take all necessary measures to ensure that the breeder received equitable remuneration. Article 15 of the Convention deals with exceptions to the breeder's right. Breeders Right shall not extend to the following cases: Acts done privately and for non-commercial purposes; Acts which are experimental; and Breeding other varieties. An optional restriction is placed on the breeders rights to permit farmers to use the variety for propagating on their own holdings, the product of the harvest, which they have obtained by planting on their own holdings. However, contracting party may exercise this exception within reasonable limits and subject to the safeguarding of the legitimate interests of the breeder. Any state, which is a member or is seeking membership, shall have necessary laws within its territory to give effect to the provisions of the Convention. States can seek the advice of UPOV Council to make its laws in conformity with the Convention.

8

People's Biodiversity Register

Biodiversity loss is a major environmental concern and there are several reasons for this. The magnitude of the loss is large and is occurring mainly for anthropogenic reasons connected with economic growth. Balmford et al., (2002) points out that there is uncertainty about the exact rate of species loss, natural scientists agree that the net loss is well in excess of normal background rates of extinction and the rate of extinctions globally appears to be increasing. While Dietz and Adger (2003) came to the conclusion that higher income countries expend more effort than lower income countries on biodiversity conservation, they find that this extra effort only partially decelerates biodiversity decline. Overall, the prospects of halting biodiversity decline in the foreseeable future seem to be bleak. Tisdell, (2003) hold the view that the decline not only includes wild species but also varieties of many domesticated animals used for productive purposes, such as livestock, and varieties of plants used for production.

This loss results in a diminution in natural capital and threatens the long-term sustainability of economic production. Material production based on the use of biological resources is threatened. In addition, to the extent that biodiversity adds to human satisfaction via the enjoyment of variety, novelty and intangible values, biodiversity loss impoverishes humankind. Furthermore, apart from any anthropocentric advantages from biodiversity conservation, there are ethical concerns, such as the view.

Basic Behavioral Issues

Early approaches to evaluating the economic desirability of environmental change continued to rely heavily on the individualistic rationality assumptions of neoclassical economics, and vestiges of this path dependence continue. Such evolutionary path dependence is not surprising. However, neoclassical economics was primarily concerned with providing an abstract theory of how markets for private goods work. It adopted an atomistic reductionist approach using the notion of ideal or representative types of consumers and producers and ignores biophysical conditions. Variations in human behavior and knowledge were ignored and economic preferences and decisions were assumed to be free of any complicating social influences. For private goods in a near-stationary world, this probably was a useful abstraction and undoubtedly neoclassical theory provided valuable insights into the way markets operate. However, in a world of significant change and one in which public goods and mixed goods are important, neoclassical economics is of reduced relevance. This is especially so because neoclassical economics pays no significant attention to the processes involved in preference formation and in decision-making. Because public goods, mixed goods, and uncertainty are so frequently encountered in the valuation of biodiversity changes, this limits the scope for applying the absolute rationality assumptions of neoclassical economics.

In evaluating attitudes to biodiversity loss and in using social benefit cost analysis, the features listed below occur. These considerations are not allowed for in neoclassical economics but it is desirable to allow for them in evaluating preferences about biodiversity. Individuals often lack knowledge of or have limited knowledge of species, the payoffs from their conservation, and the effectiveness of strategies to conserve them. This knowledge varies widely between individuals and for species.

Increased knowledge and experience often systematically alter valuations of or preferences for commodities. The dynamics of preference variation resulting from such changes is behaviorally important. Social influences on values can be important. This includes the type of moral dimensions mentioned by Etzioni (1988) and bandwagon effects.

The actual observed preferences of individuals and their behavior may be inconsistent with those predicted by neoclassical economics theory. This has, for example, been highlighted by the endowment effect. It could be noted that observed behavior is not necessarily informed behavior and, therefore, need not reflect an individual's 'true' preference for a commodity. This limits the usefulness of revealed preference techniques, such as the travel cost method, for economic valuation of biodiversity change.

In applying SCBA to decisions about nature conservation, researchers should be guided by the goal of optimally imperfect decision-making along the lines suggested by Baumol and Quandt (1964). SCBA should only be refined up to the point where the extra anticipated benefit from this equals its extra anticipated cost. The above issues will now be considered in relation to stated preferences, revealed preferences, and implications for conservation policy will be examined. They will be illustrated by recent research results.

Stated Preferences, especially Contingent Valuation, and Biodiversity conservation. In assessing stated preference approaches to eliciting preferences.

In view of the limited capacity of individuals to consider a large set of possibilities simultaneously, the process of asking individuals to state their preference for a particular object may focus their concentration on the object being evaluated and reduce their concentration on alternative objects. Compared to the situation prior to elicitation, this is likely to alter the individuals' stated contingent valuation of the object to be valued, probably elevating it if the object is considered to be desirable and depressing it if the object is believed to be undesirable. If this effect is present, it means that the individual's prior value is unobservable.

Economics and Biodiversity

From the perspective of an economist, biological diversity or 'biodiversity' is of interest for two fundamental reasons. First, biodiversity is valuable to society. That is, the greater the biodiversity we have, the better off we are and if we lose some biodiversity, we consider ourselves to be worse off. Second, choices made by society have made and are continuing to have effects on biodiversity. That is, some of the resource use decisions made by society–albeit inadvertently–have left us with less biodiversity. Clearing land for agriculture, harvesting timber from forests, draining wetlands for housing estates for example, have caused depletions in biodiversity.

Putting these two reasons together lead the economist to conclude that biodiversity is a scarce and valuable resource. And for an economist, that means their discipline has something to contribute to the biodiversity debate, simply because the focus of economics is on the analysis of the ways societies make choices about their scarce and valuable resources.

There is a need to provide an understanding of what economics has to say about biodiversity. Because the overriding goal of economics is to deliver choice solutions that make society better off, a strong emphasis will be given to the value of biodiversity because it is the creation of value that makes society better off. Specifically, three questions will be addressed:

1. Why is biodiversity valuable?
2. How can the value of biodiversity be estimated? and,
3. How can the value of biodiversity be delivered to society?

The goal is not to advocate or promote any particular type of analysis.

We begin with an attempt to define biodiversity. 'Attempt' is used advisedly here because it is immediately apparent that biodiversity is a complex concept that generates definitional conundrums. The importance to society and the policy significance of biodiversity are addressed to answer the question: 'Why is biodiversity valuable?' What economists have done to estimate the value of biodiversity. While the lists of valuation techniques developed and applications performed internationally are extensive, it is also apparent that the discipline does not find itself standing on 'safe' ground in this area. Some of the specific issues that have caused concerns both within and outside the economics profession are overviewed, especially in the light of observations made regarding the relevance of applications to policy making. Briefly, the third question of how the value of biodiversity can be delivered to society is tackled.

Revealed Preference and Biodiversity Conservation

There seems to be a widespread view that, since revealed preferences are based on observed behavior, they are likely to be more accurate indicators of individuals' actual preferences than stated preferences. However, the scope for using revealed preference methods is restricted because markets are often absent or incomplete, as in the case of pure public goods and mixed goods, and suitable surrogate markets do not always exist. In such cases, stated preference methods remain an essential means for gathering information about demand for these types of goods, such as the demand for nature conservation and biodiversity, although revealed preference methods can play a supplementary role in their economic evaluation. However, revealed preference methods are based on assumptions about human behavior. These are essential to interpret the behavior observed, and often they are the same behavioral assumptions as those that underpin neoclassical economic analysis.

Take the travel cost method, for example. It assumes that individuals are well informed about the economic choices available to them and are utility maximizers. This also implies that they are well informed about sites that they may visit. While this may be reasonable for sites previously visited by recreationists, many sites, such as national parks in remote areas, are poorly known to many visitors and can be regarded as experiential goods. In some cases, the majority of visitors will only visit these places once in their lifetime.

People's Biodiversity Register

Objectives of the People Biodiversity Register

The People's Biodiversity Register is a programme of organising information available in the domain of slow science on the status, uses and management of the living resources. This is information with practical ecologists, people who deal with living resouces as a part of their daily subsistence activities-as graziers, as fisherfolk, as basket weavers, etc. It is information with simple folk, who are nevertheless specialists of slow science, such as tribal medicinemen, or expert honey gatherers. It is also information with more sophisticated practitioners of slow science such as physicians trained in Ayurvedic colleges. The information may be purely orally transmitted or contained in palm leaf manuscripts or in printed Ayurvedic texts. The information may be guarded as secret, as certain herbal remedies are or may be very widely known-as the use of turmeric as an antiseptic is all over India. This information may be documented in full, or as claims that only partially reveal the information. It needs to be collated, verified, synthesized and made available to the society at large, as well as fed back to the village communities.

The objectives of the programme would thus embrace the creation of decentralised countrywide databases on :

- Status of biodiversity resources such as populations of medicinal plants, cultivars of fruit trees or freshwater fishes.
- Various factors such as harvests from natural populations, changes in agricultural practices or discharge of industrial effluents, affecting the biodiversity resources.
- Ongoing involvement of local communities/ individuals in sustainable use and conservation of biodiversity resouces, such as systems of regulated grazing on pastures, maintanence of varieties of fruit plants like jackfruit on farm bunds or protection of fish in sacred ponds.

- Local knowledge, widely shared and publically disclosed of properties and uses of biodivesity resources e.g. drought-resistance of a certain cultivar, methods of preservation of food or use of certain plants in treating human or livestock diseases.
- Local knowledge only partially disclosed; for instance, a claim that a paticular medicine woman knows of a cure for asthma.

The programme would also attempt to:

- Validate the information thus recorded; for instance, through crosschecking with other published information sources.
- Create a network of decentralized databases, ultimately linking them to a consolidated national data base which would give full credit to the origin of information at the level of an individual, a community or a village panchayat.
- Feed pertinent information, such as volume of trade and prices of medicinal plants back to the local communities.

We hope that the creation of such a countrywide decentralized yet networked system of information of slow science stream will serve several important purposes. It will for the first time create a machinery for monitoring the fate of a variety of biodiversity resources, be it medicinal plants, land-races of crops, breeds of regional livestock, or wild relatives of cultivated plants throughout the country. Such information could then form the basis of a stratergy for the conservation of these resources.

Such an information system giving full and proper credit to informants would serve to give recognition and encouragement to the practical ecologists everywhere, many of whom are formally uneducated, yet have a deep fund of knowledge about the living world and its human uses.

This would also be a system of properly recording and establishing links amongst the many different strands of more formal systems of slow science, such as the Ayurvedic traditions of Kerala and Western Himalayas or the Siddha medicine of Tamilnadu and the Tibetan medicine of Sikkim.

Such an information system is also essential to begin a process of drawing on a part of the commercial profits made from further development of this broad base of slow science. There could be several ways of tapping such profits. Biodiversity-based enterprises may be charged some fees for accessing these information bases. They may pay a cess on their sales. Such income may be deposited into a National Biodiversity Fund which may be used to reward individuals, communities or village panchayats for their contributions towards:

(a) Ongoing conservation of biodiversity resources on their private lands/waters or on public lands/ waters within their jurisdiction.

(b) Recording of practical ecological knowledge.

(c) Recording of more systematised slow science, such as locally available manuscripts on Siddha system of medicine.

Additionally, the information system could help individual entrepreneurs establish contacts with claimants who do not fully diclose their knowledge. The two may then negotiate and decide on equitable sharing of benefits of possible commercial exploitation of such knowledge by entering into an appropriate Information or Material Transfer Agreement.

Benefit Sharing

These suggestions involve making important choices in the context of sharing of benefits flowing from commercial exploitation of the biodiversity resource base and knowledge of its uses in the "slow" science stream. The first choice relates to rewarding for creation and maintenance of biodiversity, particularly in terms of cultivars of crops and land races of livestock, but also through systems such as protection to sacred groves, ponds or individual species. We may either focus on contributions made in the past or on the ongoing efforts. To assess ongoing contributions is easier; it may be based on recording of current stocks of biodiversity resources through the People's *Biodiversity Registers.*

In near total absence of information, the local collectors receive very low value for such resources, the public exchequer realises nothing in terms of collection fees or other taxes, and the resource is rapidly overused and depleted. Often the only information available with the public pertains to the identity of species in demand and the quantities harvested is at the ground level, with the local communites.

The intention is not only to tap from local communities information on ongoing harvests, but to involve them in sustainable management and reward them for their services and their knowledge. The PBR exercise would therefore explore in depth the motivations and the capabilities of different components of the peoplescape in sustainable management.

Nurturing Diversity

India's ecosystem people have rich traditions of sustainable use of biological resources and conservation of biodiversity, although many of these have been severely eroded over the last two centuries. In spite of this erosion, the Indian countryside remains dotted with millions of sacred trees of the keystone resource genus *Ficus,* and harbours tens of thousands of langurs and macaques and peafowl thanks to our cultural traditions. In the village of Masur-Lukkeri is a large roost of the gaint fruit bats on a sacred *Ficus* tree. While people do hunt these bats away from the roost at night, they are fully protected at the roost in the day. In the same village is a small patch of natural evergreen vegetation, protected as a sacred grove surrounding the Ishwara temple.

Socio-Economic Profile

An outlined socio-economic profile of the study area will help to understand people's dependency on the biological resources, either for internal use or for external supply. The following information will be required to get the socio-economic profile: Population - to get the total population of the study area individual household survey is to be carried out with reference to the following parameters Total number of members

Male	-	below 18 years, 18 - 65 years, above 65 years
Female	-	below 18 years, 18 - 65 years, above 65 years
Literacy rate		
Male	-	4-18 years, 18-65 years, above 65 years
Female	-	4-18 years, 18-65 years, above 65 years
Drinking water	-	Source, Number (in case of dug well and tube well)
Sanitation	-	Sanitary latrine present or absent Land holding - Agricultural land - Upto 5 cottah, 5 cottah - 1 bigha, 1 - 5 bigha, Above 5 bigha (1 cottah - 720 sq.ft; 1 bigha - 14400 sq.ft)

Occupation, Agriculture, Fisheries, Animal husbandry, Trading, Manufacturing, Other unorganized worker, Organized worker, Family wise monthly income Healthcare and dependency on traditional medicine Fuel use and source, Daily food habit, Staple food, Pulses, Vegetables, Oilseeds, Fish, egg, meat, Domesticated animals–number, types and breed (e.g., cow, goat, chicken, duck), Type of housing–Kutchha, Pucca

- Respondent's awareness to biological diversity, Special focus should be given to the more knowledgeable individual, men and women. During the survey, the following format may be used for recoding of information of knowledgeable individuals:

Natural Resources: Survey and Documentation Natural Resources include non-living elements like air, water, soil, minerals and also living elements like, plants, animals and microorganisms. Together the nature provides the life sustaining system.

Non-Living Natural Resources

Land Resource: Land and soil is a vital resource supporting biological diversity. Soil is not an inert substance. It provides shelter to tiny microbial organisms like bacteria and fungi as also micro arthropods and earthworms, not easily visible to naked eyes.

In order to understand land and soil, it is therefore essential to prepare a basic target area including urban, non-forest rural, forest, mountain, coastal, island etc. based land use map. To prepare such a map one has to study the following: *For Urban Area:* (I) Land for human habitation a. How much land is used b. What type of land is preferred (II) Roads a. Length of the Pucca road b. Length of the Kutchha road (III) Institutional and Commercial Area a. Small scale industries b. Markets c. Institutions like office, educational institution (IV) Parks, Gardens and Open Space a. Total park and garden area b. Total open space area (V) Wetlands a. Name, location, ownership, area and current uses of wetlands (Details are given under Water Resource) (VI) Rivers / Canals / Creeks- to understand the flowing water profile (if any) Length of the river/canal b. Current uses of river/canal water

Agricultural Land

- How much water is accumulated in low, medium and high land?
- Whether such land is used for single crop, two crops or more?
- What type of irrigation is currently practiced?
- What are the major crops used for cultivation?

Wetlands

- Name, location, ownership, area and current uses of wetlands in the village (Details are given under Water Resource)

Land for Human Habitation

- How much land is used?
- What type of land is preferred?

Fallow Land

- Area of fallow and their location
- Vegetation in the fallow land
- Fauna in the fallow
- Whether it is used for any religious purposes?
- Whether permanent fallow or current fallow?

Rivers / Canals / Creeks – to understand the flowing water profile

- Origin of the river/canal
- Length of the river/canal
- Current uses of river/canal water

Roads

- Length of the Pucca road
- Length of the Kutchha road

Jungle / bushes / forest patch

- Location
- Area

Sacred Grove / Pond

- Location
- Area
- History
- Salient features

General Features

- Forest types (Deciduous, evergreen, dry, wet etc.)
- Area under closed forest
- Area under open forest
- Category of forest (protected, reserved, unclassed)

Wetlands

- Name, location, area and current uses of wetlands (Details are given under Water Resource)

Rivers / Canals / Creeks – to understand the flowing water profile

- Origin of the river/canal
- Length of the river/canal
- Current uses of river/canal water

Forest Village, If Any

- Location of the village
- Size of the village
- Road length and type

Sacred Grove / Pond

- Location
- Area
- History
- Salient features

For Mountain Area

General Features

- Altitude
- Slope

Land under Forest

- Forest types (Deciduous, evergreen, dry, wet etc.)
- Area under closed forest
- Area under open forest
- Category of forest (protected, reserved, unclassed)

Barren Land

Landslide Zone

Agricultural Land

- How much water is accumulated in low, medium and high land?
- Whether such land is used for single crop, two crops or more?

– What type of irrigation is currently practiced?
– What are the major crops used for cultivation?
– Area under abandoned Jhum land
– Area under terrace cultivation

Wetlands

– Name, location, ownership, area and current uses of wetlands in the village (Details are given under Water Resource)

Land for Human Habitation

– How much land is used?
– What type of land is preferred?

Rivers / Canals / Creeks–to understand the flowing water profile

– Origin of the river/canal
– Length of the river/canal
– Current uses of river/canal water

Roads

– Length of the Pucca road
– Length of the Kutchha road

Jungle / bushes / forest patch

– Location
– Area

Sacred Grove / Pond

– Location
– Area
– History
– Salient features

For Coastal Area

General Features

- Location
- Length of coastline
- Current uses:
  - Human habitation
  - Area without habitation
  - Coastal industries
  - Coastal fisheries
  - Coastal forest
  - Coastal agriculture
  - Coastal road
  - Coastal wetland

Note: Details under each category can be followed as given above.

After studying the above, a Land Use Map is to be prepared on the basis of Mouza map to

- Demarcate target areas
- Identify parcel of land under each category (as shown above under I , II, III......)
- Color code the land uses on the map

Water Resources

Water is a vital resource sustaining life forms. Collect the information on ground water (from dug well data and hand pump data) and surface water and document

Surface Water Resource

- River, Canal, Wetland - location / name
- Area
- Uses - drinking / cooking / irrigation
- Produces of wetlands - fishes / vegetables / flower / aquatic plant
- Birds in the wetland
- Does the wetland acts as receptacle during flood / serve as water recharging area?

– Recreational use

Groundwater Resources (except in Mountain ecosystem)

– Number of hand-pumps used for lifting drinking water and approx. depth of each

– Number of shallow and deep tube wells used for irrigation-horsepower of motors

– used for each, hours of operation per day

Living Natural Resources

How to observe animals in the field?

In every target area, diverse groups of animals can exist in tropical climates. These may include animals without backbones invertebrates like earthworm, insects, spiders, scorpions, mollusks etc., or animals with backbones like fishes, toads and frogs, snakes, lizards, turtle, tortoises, birds and mammals such as squirrels, rat, mole, civet cat, fox etc. Since the animals occur at different period of time and season, they have to be observed both seasonally and also diurnally day and night. Some animals are best observed during early morning e.g., Birds or at the dusk time e.g., Bat while others may be observed before the day temperature increase e.g. Butterflie. It has to be remembered that animals inhabit different habitats viz., upper branches of the tree, grassland, banks of river, deep foliage, holes in the tree trunk, under the bark, on the ground or underground. Selection of sites for observation It will be useful to select the best possible site for observation where diverse biological species may be available. These may include the agricultural land, cultural fallow, village orchards, village ponds, forest patch, canal side and roadside. A complete information base can only emerge if year round, seasonal observations and recording is carried out.A brief outline of methods of observation for selected group of animals is given below:

Invertebrate Animals

Invertebrate animals can occur both in land and water in specific habitat condition under the stone, in rotten logs, in the flowers, on fruit trees, in bushes, in agricultural fields etc. The smallest invertebrate animal called protozoa body with single cell cannot be seen by naked eyes but they may exist in soil, water even within the body of the human being or other animals. Normally the visible animal species are recorded in PBR. These include:

Earthworms

Earthworms normally live in mineral rich soil. To observe and collect earthworm specimen: ÀÛƒÜ Select agricultural and other land; make 1 ft x 1 ft x 1 ft deep cavity at least in 5/10 places and collect earthworm sample in small plastic glass vial containing 70-80 percent alcohol spirit.

Note the date, time and place of collection in the field notebook. Local names can be further validated by scientific names once identified by subject specialists.

Insects and Spiders Insects comprise the largest number of living organisms in the world. So far nearly 8,50,000 species of insects have been documented in the world. Scientists believe this is only 4-5 percent of the total insects species that may exist in today's world; that means 95 percent of the insects are yet to be described and named. Obviously, the largest numbers are expected from southern tropical countries including India. So far only 6.5 percent of the insect species of the world have been recorded from India. Out of 29 orders of insects, representatives of 27 orders have been located in the country.

Usually insects are considered as enemies to the human society because they cause considerable damage to the agricultural crop and many species can act as vectors and transmit pathogens causing serious diseases in plants, animals and human being Cholera, Malaria, Kala Azar, Dengue are some of the insect borne diseases. On the other hand, a number of insects offer useful services to the plant kingdom and to the human society as pollinators, as natural enemies for pests, as providers of silk, honey, lac etc. Insects are the largest group of pollinators in the natural world.

Spiders may occur in several forms. Spiders act as predators for many of the insects and provide food to the higher animals like lizards and birds. In general, the goods and services from insects and the spiders can be listed as follows: Provide honey, lac and silk. Act as pollinators in agriculture, forestry and horticultural crops. Act as food in many parts of the world. Act as predator for pest insects larva of Lady bird beetle Help to increase fertility of the soil-soil inhabiting insects Provide aesthetic pleasure and lessons in social behavior e.g., butterflies and honey bees Many insects and spiders are used for medicinal purposes by the tribal communities Play an important role in the food chain in nature.

General Characteristics of Insects

Insects and spiders have jointed legs; the same character can also be seen in crustacean like Prawn, Shrimp and Crabs, centipedes and millipedes. Insects have normally a pair of antenna and an exoskeleton. Adult insects have three distinct parts–head, thorax and abdomen. Insects have three pairs of legs and usually one or two pairs of wings. Male and female insects may vary in colour, shape and size.

Collection and Observation

To collect insects and spiders, several methods can be used viz., Hand collection with a soft camel hair-brush (soft body insects) or with a pair of forceps. Such insects are normally preserved in 70 percent alcohol in glass vial. Net collection-Butterflies and other flying insects can be collected with the help of a specially prepared aerial net with a long handle or with a sweeping net attached to a short handle.

Traps–Insects can also be collected by setting up different types of traps, viz., Light trap: Light attracts insects at night especially at dark nights. An illuminated electric bulb or a petromax lamp can be used as a light source in the open field; a piece white cloth (3 ft X 3 ft) may be placed on the grass near the light source; when the insects are attracted to the light and settle on the white cloth they can easily be collected.

Pitfall Trap: This is a simple devise of using a plastic cup (9 cm X 8 cm), which is placed in a dug out pit at the floor level. Such cups are covered by a wire mesh and then by leaves and twigs. The cups should contain some attractant food or liquid. Such containers should be kept for 3-days and then taken out for observation.

To observe and collect insects: Select a site and demarcate an area measuring 50m X 10m. Observe and collect the insects and spiders by hand, net and traps. This can be repeated depending on the total area of the selected site. To preserve the collected specimens, first kill the insects and spiders using chloroform or benzene. Preserve butterflies and moths in envelops in dry condition and other insects in plastic or glass viol containing 70% alcohol spirit. Put serial number in the collected specimens and write it in the field notebook. Also note the date, time, description of the site, type of insect collected and number of each type. Identification of collected insect specimen:

Firstly, it should be identified whether the collected specimen is insect or not. For this following steps to be adopted: Presence of two pairs of antennae and not less than five pairs of legs–crustacean Presence of four pairs of legs and no antenna–arachnid Presence of three pairs of legs and adult specimen possessing wings except in some cases like Springtails, Lice, Earwigs etc. – insect Characteristics of insects of different orders: Insects are divided into 29 orders. The characteristics of most common insects are given below:

<table>
<tr>
<td>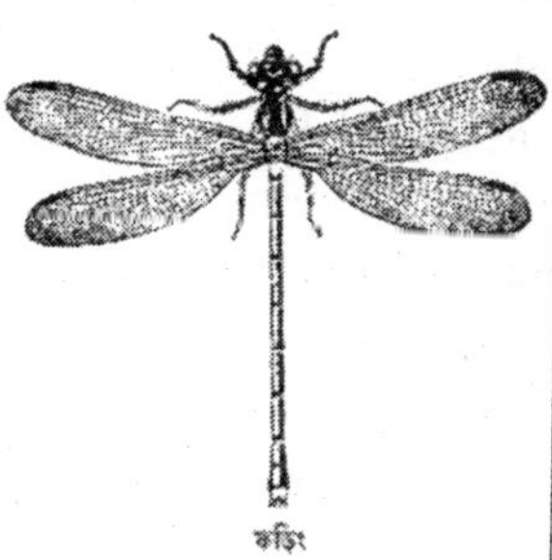</td>
<td>

- Two pairs of wings with several or many cross veins
- Large compound eyes present on the head
- Three pairs of legs with exoskeleton

e.g. Dragonflies, Damselflies
</td>
</tr>
</table>

B. Orthoptera 	• Two pairs of wings, fore wings are thin, hind wings are large with cross veins • Hind legs are thick and long, used for jumping, e.g. Grass hoppers, Crickets
C. Phasmida 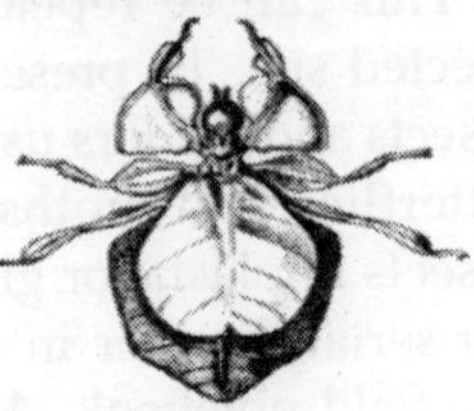	• Body flat leaf like or long and slender, stick like • Wingless e.g. The Walking Sticks,
D. Dictyopetra 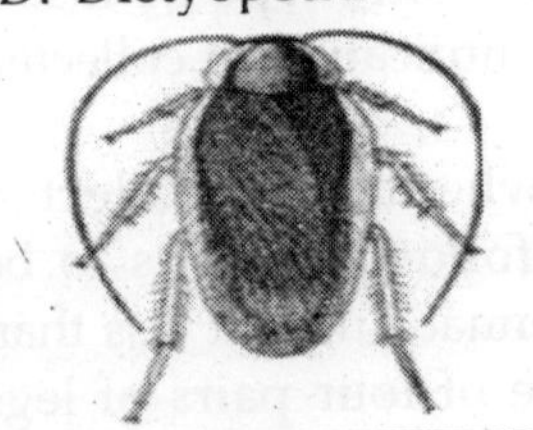	• Body flattened, swift on foot • Upper wings are bright and thin • Lower wings with cross veins e.g. Cockroaches
E. Mantodea 	• Two pairs of wings present • Front legs adapted for catching insect preys e.g. Praying Mantis
F. Isoptera 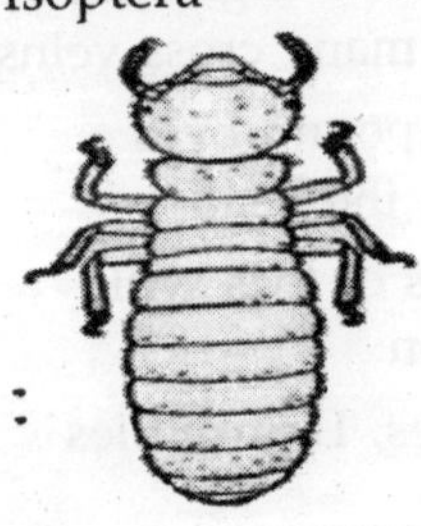	• Soft body • Wings are equal in size and with indistinct veins • In some forms wings are absent. • Winged insects normally comes out during rainy season. e.g. Termites, White Ants

G. Thysanoptera 	• Small in size • Pigmented body • Thin wings with veins and hairs e.g. Thrips
H. Hetroptera 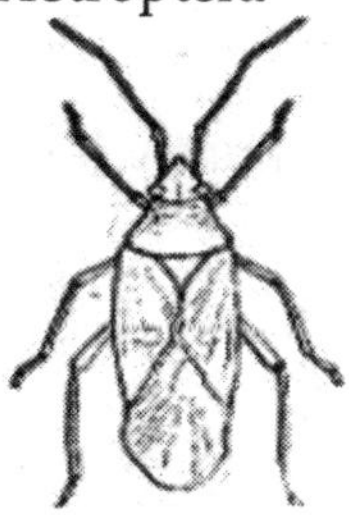	• Large, hard body • Fore wings thick, sometimes needle like • Hind wings with cross veins • Suck blood from other animal's body e.g. Bug
I. Homoptera 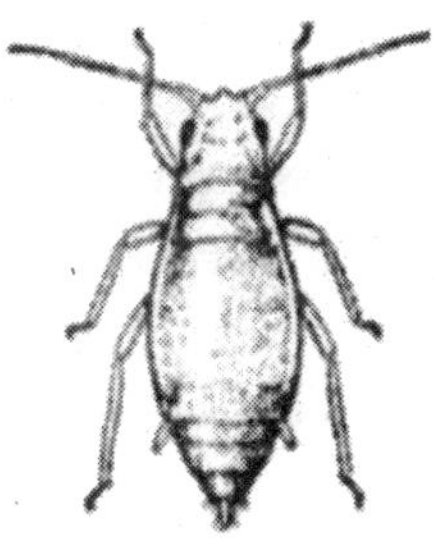	• Generally small insect • Soft legs, two pairs of wings with cross veins • Anterior part of head with sucking mouthparts • Wing less insects are also found in some stage of life cycle e.g. Aphids, Cicada, White flies, Leaf hoppers
J. Coleoptera 	• Two pairs of wings, fore wings hard called elytra, hind wingswith cross veins • Bright in colour e.g. Lady bird beetle, Dung beetle

<table>
<tr><td>K. Hymenoptera
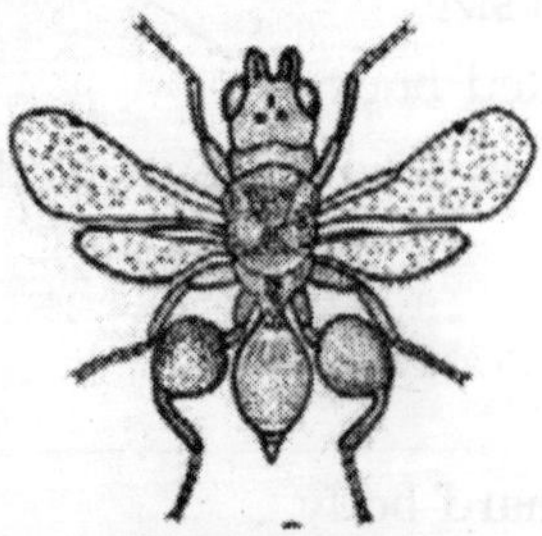</td><td>• Two pairs of wings with cross veins
• Hind wings are always smaller than fore wings
e.g. Honey bee, Ant, Wasps</td></tr>
<tr><td>L. Lepidoptera
</td><td>• Two pairs of wings with fine scales
• Body and wings are thin and flat
• Wings are bright in colour
e.g. Butterflies, Moths</td></tr>
<tr><td>M. Thysanura
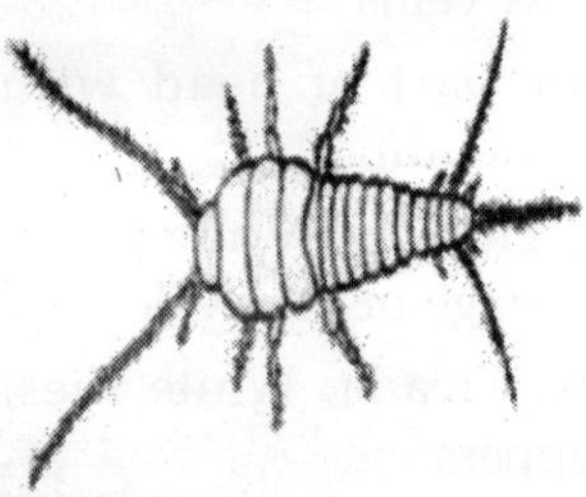</td><td>• Small, wing less, soft body
• Whole body is covered with silvery hairs
e.g. Silverfish, Bristletails</td></tr>
<tr><td>N. Collembola
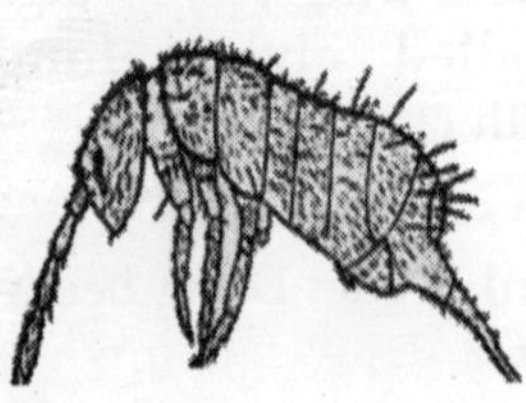</td><td>• Small, wing less, soft body
• Spring like structure present at the end of the tail, help in jumping
e.g. Springtail</td></tr>
</table>

Mollusks are represented by snails, mussels etc. They can be found in marshy areas as well as in the high land. Observation place for mollusks may be marshy land, agricultural field during monsoon, bund area of the field etc.

Vertebrate Animals

Fishes can be found from small ponds, beels and rivers in the village. Fishes can be distinguished based on the structure, scales, colour, size etc. A total of 2586 species of fishes have been recorded from India. To observe fish diversity of the particular place :

Collect the names of fishes cultured in the village wetlands

Collect the names of non-cultured fishes

Also collect the information regarding the fishes which are not found nowadays from elderly people Amphibia The group amphibian is represented by frogs, toads and newts. They are the cold blooded animals. Night time is the best time to observe Amphibia. In India about 209 species of amphibians have been recorded. Frogs and toads eat harmful insects of paddy field and therefore beneficial for the human being. To observe and collect Amphibia specimen Pit fall trap can be used near the water bodies or paddy field. The trapped specimen can be collected and observed later on. Amphibia can also be collected by cloth nets and by hand.

Reptiles include snakes, lizards, crocodiles, turtles etc. Snakes may be poisonous or nonpoisonous. A total of 485 species of reptiles are found in our country. These animals can be found during day time as well as night time. Reptiles may be observed at the selected sites by direct sighting (on trees, on ground, under stone, in crevices, near water) by the molt of the snakes from the elderly people about the common reptiles found in the area

Birds

Birds are colorful feathered animals. The male and female birds can be easily distinguished as the males are more colorful than females. Early morning and dusk time are the best for observing birds. The call, colour, structure of wing, beak, legs etc. are important for observing a bird species. Birds can be observed at the places where insects, amphibia, reptiles etc are observed. Select 50 m., long site. Imagine a circle of 10 m., radius at the two end points of the 50 m., long site. Stand at the centre of the circle and observe the birds for 10-15 minutes. Note the name of the known birds. Record major features of the unknown birds.

Mammals

Mammals are placed at the highest level of animal kingdom. They are characterized as warmblooded animals with hairy body and by the presence of mammary glands. In India, a total of 372 species of mammals are found. The smell, call, foot-print, excreta etc., are important for observing mammalian species. Generally most of the mammalian species come out during night. Therefore for direct observation, night time is preferred. Indirect information may be collected from their foot-print, call, habitat etc.

Domesticated Animals

Domesticated animals include cow, buffalo, sheep, goat, poultry chicken, duck etc.

Figure: Different Types of Legs of Birds

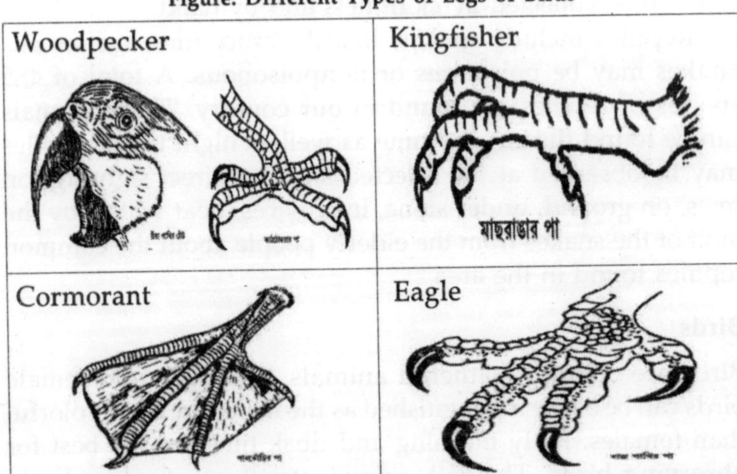

Figure: Different Types of Beaks of Birds

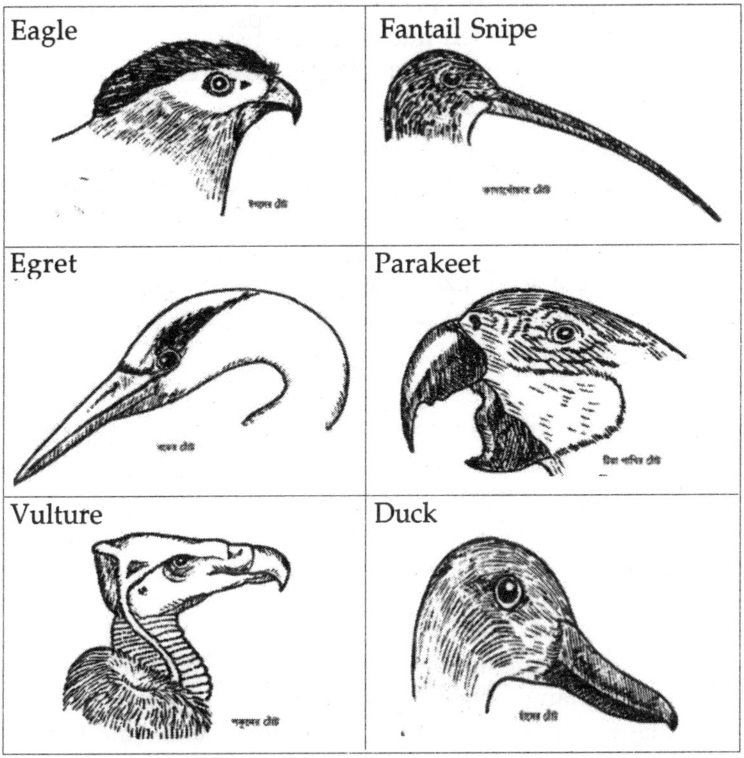

Plant Diversity (Flora) Plant kingdom can be divided into Gymnosperms and Angiosperms.

Gymnosperms

Gymnosperms are non-flowering plant group and reproduces by spores. This group can be divide into three sub-groups:

Thalophyta: Generally once-celled or multi cellular body. In case of multi-cellular species, body parts cannot be differentiated. e.g. Algae, Fungi Bryophyte: Small, shoot, leaves present, generally grows on marshy land. e.g. Moss Pteridophyte: The plant is having distinct root, shoot and leaves. Vascular bundle present. e.g. Fern

Angiosperms

These are the flowering plants and reproduce by seed. This group is further divided into two groups:

- When the seeds are open without any external coating. e.g. Pine, Cycas
- When the seeds have outer seed coating. This may again be divided into two groups called Monocot and Dicot. The first one having only one undivided seed, like rice and wheat and the second one has got divided seed, like gram, mango etc. The dicot group is again divided into following three sub-groups:

Herbs

Plants with soft shoot herbs can be on the basis of duration of lifecycle. Life cycle completes in one year. e.g. Paddy, Wheat Life cycle completes in two years. e.g. Radish Life cycle completes in more than two years. e.g. Ginger Herbs can also be divided in to climbers and creepers Shrubs: The plants larger than herbs but smaller than trees. Woody stem, medium height, branched. e.g. China rose, Trees: Plants with long, woody stem. Branching out after a certain height from the ground. e.g. Mango, Banyan Observation Procedure for Plants Wild Plants: Demarcate 50 m., long area Demarcate five 10 m., X 10 m., quadrant, one at the middle of 50 m., length and two each at the two sides of end two points Write description of the site Write down the uses of the plants especially medicinal use, abundance, and plants, which were present and used in the earlier times but no longer available. This data can be collected from the help of local villagers specially the elderly people.

Agricultural Crop Plants

Agricultural crops include cereals, pulses, vegetable, spices, oil seeds, fiber yielding plants, sugar yielding plants, cultivated fruit and flowers, green manure etc. [S=Summer; M=Monsoon; PM=Post Monsoon; W=Winter] Paddy: Paddy can be of many varieties, both indigenous as well as high yielding. Farmers are the best source for getting the information regarding cultivation of rice varieties. The morphological characters to be noted are:

Total plant: seedling height, plant height Culm: culm number Leaf: leaf length and width, leaf angle, flag leaf angle and colour Panicle: panicle type, length, weight, secondary branching pattern, number of grains, proportion of sterile grains, panicle axis Grain: grain length and width, 100-grain weight, brown rice length and width, brown rice colour, aroma etc Disease and pest resistance of plants may be observed directly by the rate incidence of pest insects and diseases.

Horticultural Plants

This includes flowering and fruit-bearing plants like Mango, Guava, China Rose, Jasmine etc. Timber yielding Plants This type includes woody plants grown for timber. e.g. Teak, Mahogany etc. Preparation of herbarium sheets for identification of plant species Herbarium sheets are important tools for identification of plants species. Specimen can be kept for many years as herbarium sheet for future research. Herbarium sheets must be prepared in case the plants cannot be identified at the site. Following materials are needed for herbarium sheet preparation: Bag (jute, paper, polythene) for keeping the specimen collected Knife for cutting the plant specimen Note book and pencil for note down the information required Weight for pressing thc plant specimen between newspapers Procedure:

Collect the plant specimen with leaves, flower and fruit. The size of the specimen must be 30 cm. x 20cm. Put necessary label with number, location, date of collection and name of the collector. Now write down these information in the notebook for future reference Put the specimen in between two newspapers and place the leaves, flowers and fruit properly Put the specimen within the paper on a plain surface and place some weight on it Change the newspaper as and when required for complete drying the specimen. Place the dried specimen outdoor for sun-drying Take a hard art paper of size 41.7 cm., X 26.5 cm. Place the dried specimen at the center of the paper and glued or stitched it with the paper. Put herbarium label at the lower right hand side of the herbarium sheet with information like scientific name, family, location/place, date, collects name, altitude from msl etc. This complete herbarium sheet can be preserved for many years with proper care.

Medicinal Plants Survey

In case the area in rich in medicinal plants, special focus is to be given to document the resources and its current use pattern. Based on the results of the documentation the BMC will be in a position to determine the resource potential, prospects of cultivation, sustainable use and trading.

Ritualistic and Social Use of Biodiversity

The population in the study area may use different biological resources for religious and other ritualistic purposes. In the rural areas, such materials are normally collected from nature but in the urban areas such materials are traded through commercial stores.

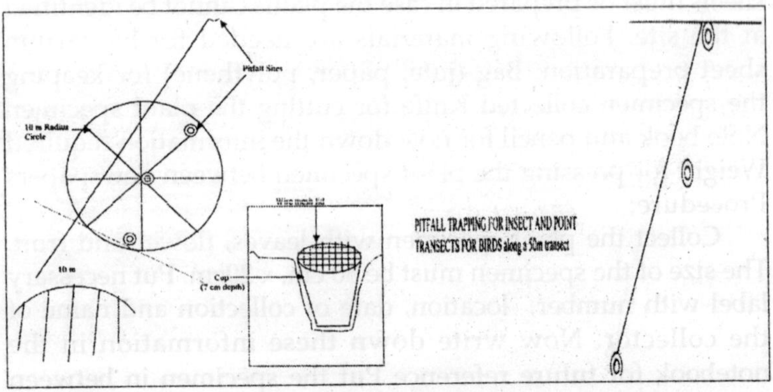

9

Biodiversity Economic Valuation

The economic values discussed in this chapter are anthropocentric. The word anthropocentric means that values are driven by human perceptions, or that are derived from values that humans hold. Although philosophers may argue that biodiversity resources have intrinsic or inherent value, economic values obtained through the nonmarket valuation techniques discussed in this lesson are based on values that humans place on them. It is important to understand this distinction. For many biodiversity resources where economists values, they are estimating values based on evidence drawn from human experience. These values therefore reflect human decisions and perceptions. Markets for most goods and services determine prices, and these prices reflect value. For example, automobiles are more expensive than toothpicks, and therefore and judge to be more valuable. The additional value of automobiles derives from the many materials and hours of labour that go into producing automobiles relative to toothpicks.

Most environmental resources, like water or air, are not traded in markets, and markets fail to account for the fact that many of us value clean water and clean air. These market failures are caused by externalities, or un-priced effects. While we discussed environmental policy for dealing with these environmental damages in the last several lessons, we have not talked about how to estimate marginal damage or marginal benefits curves. This section attempts to do just that.

In this section, we discuss valuation techniques for environmental resources. Understanding these values provides important information for setting environmental policy, as marginal damage and marginal benefit curves can provide information on the relative importance of alternative environmental goals (i.e. should we make our air cleaner by 20%, or should we focus on cleaning our water by 20%).

Economic Valuation: Concepts

The economic approach to valuing an environmental change (an improvement or a degradation) is based on individual' (or households') preference for that change. Preferences are assumed to be reflected in people's willingness to pay to secure the improvement, or avoid the degradation. Willingness to pay (Willingness to pay), in turn, is defined as the amount of goods, services or money individuals are willing to give up within a given time period to secure or avoid the change.

Preferences are reflected in people's willingness to accept compensation to forgo the improvements or endure the degradation. Willingness to accept compensation (Willingness to accept), in turn, is defined as the amount of goods, services or money individuals are willing to accept to forgo or endure the change.

In actual markets, people's preferences are reflected in their consumption and production behaviour so that market price reflect the Willingness to pay of buyers and Willingness to accept of sellers. Generally an individual will only consume a good or a service when its price is less than or equal to the individual's Willingness to pay. When the prices is less than Willingness to pay, the difference between the price and Willingness to pay is the surplus value that accrues to the individual beyond the amount he has to pay in the market. This surplus is known as consumer surplus in the economic literature. This implies that is most cases, the market price is only a lower-bound estimate of the individual's Willingness to pay.

For environmental goods and services which are not traded in actual market typically because of their open access and / or non-excludable property right characteristics, the task of finding out about people's preferences and estimating the economic value in question is even more complex since when there is no price information, there is not even a lower-bound estimate of Willingness tc pay. Note that the concept of consumer surplus is valid even for those goods and services that are not traded in markets.

The lack of markets and prices for many environmental goods and services means that the challenge for economists it two fold. The first task is to identify the reason why individuals may have preferences for or against an environmental change, or in other words, the ways in which an environmental change affects individual's well-being. The second task is to estimate the value of the environmental change or consumer surplus through a variety of economic valuation techniques. This section outlines the first task. The economic valuation techniques discussed can be used to estimate the consumer surplus both for marketed and non-marketed environmental goods and services.

There are a number of motivations behind people's preferences for biodiversity resources and the services these resources provided. They can be grouped as those that are related to the actual or future use of resources and their services (know as use values) and those that are not related to any use (passive use or non-use value).

Use Values Include

- Direct use values, where individuals make actual use of resource either in consumptive way, (e.g. harvesting forest timber and abstraction of water for drinking or commercial use) or a non-consumptive way (e.g. bird watching and trekking).
- Indirect use values, where society benefits from ecosystem functions (e.g. watershed protection or carbon sequestration by forests); and
- Options values, where individuals are wiling to pay for the option of using a resources in the future (e.g. future

visits to a wilderness area, or possible future pharmaceutical uses of biological resources).

A convenient classification of non-use values is in terms of

- Existence values, which reflect Willingness to pay to keep a resource in existence in a context where the individual expressing the value has no actual or planned use for his/herself or for anyone else.
- Altruistic values, which might arise when the individual is concerned that the resource in question should be available to others in the current generation: and
- Bequest values, which reflect concerns that the next and future generations should have the option to make use of the source.

These categories of motivations or preferences or value are collectively known as total economic value (Total Economic Value). Total Economic Value of a change, then is the sum of both use and non-use values:

= use values + non-use values

Total Economic Value

= direct use + indirect use + option + existence + bequest values

To arrive at the measure of society preferences for an environmental change (or change in societal well-being due to the environmental change), Total Economic Value expressed by individuals need to be aggregated across the population affected by the change in question. Note that some may be affected positively, while other may be affected negatively and both parties should be accounted for in estimating the effect of the change to the society.

Determining the 'affected' population is a crucial step of the valuation of environmental change and choosing the restoration option. Evidence from valuation exercises to date show that individual use values tend to be larger than individual non-use values while the aggregate non-use values tend to be larger than aggregate use values. This is mainly due to the fact that non-use values are likely to be held by a larger population than those who may be making actual use (or planning future use) of the natural resource in question.

A frequent question concerns how Total Economic Value is related to the notion of intrinsic value. Intrinsic value is often regarded as being a value that resides 'in' the asset in question, and especially environmental assets, but which is independent of human preferences. By definition, Total Economic Value relates to the preferences of individual human beings, so that if intrinsic value is defined to be independent of those preferences, Total Economic Value cannot encompass intrinsic values. However, notions of intrinsic value may well influence Willingness to pay/Willingness to accept, and some valuation techniques are particularly well-suited to revealing the motives for stated Willingness to pay/Willingness to accept. These motives vary and may well include notions such as a 'right to exist' for the asset in question. This is fairly common motive when the asset is, for example, a living creature. Hence, although Total Economic Value cannot embrace a measure of intrinsic value, some valuation techniques may help to make the motivations for Willingness to pay explicit, and those motives may well involve a concern on behalf of the object being valued.

Total Economic Value expressed by an individual for an environmental change is determined by a number of factors including but not limited to:

- Price paid for marketed resources if available, (e.g. fees and other costs incurred to facilitate recreational use of resources)
- Characteristics of the individual or household (e.g. income, age, gender, employment status, education, recreational choices, location, membership of an environmental group etc);
- Characteristics of the resource (e.g. whether use is made of the resource, availability and quality of the resource prior to the change in question, accessibility to the resource, the availability and characteristics of substitutes for the resources etc).

- Characteristics of the environmental change (e.g. reversible / irreversible, improvement / degradation, a voluntary change chosen by or an involuntary change imposed on those affected etc); and
- Other characteristics (e.g. weather, culture characteristics etc.).

Economic valuation techniques discussed are designed to identify and collect information on these factors in order to estimate the Total Economic Value of an environmental change. In doing this, the techniques not only generate an estimate for Total Economic Value but other information which can be valuable in designing restoration options.

Valuation Techniques

This section presents three approaches to valuing damage to natural resources: (1) economic valuation techniques, (2) benefits transfer and (3) scoring and weighting techniques. The first two approaches can generate both monetary and non-monetary expressions of people's preferences for or against damage to natural resources. The third approach is usually based on expert opinion and can generate non-monetary estimates of damage to natural resources.

Economic Valuation Techniques

There are three main types of economic valuation techniques

Conventional market techniques: these techniques rely on readily observable market prices as measure of value. They are appropriate for valuing damage where there are observable impacts to commercial operations that depend on the natural resource e.g. tourism, forest products, fisheries, etc. These are currently used in determining compensation e.g. fishery damage due to the Sea Empress oil spill. This technique is limited as it may as it may only be applied to marketed goods, i.e. a subset of use values. Because damage to commercial operations is usually site and resource-specific and straightforward to value (basically, the price-net of taxes and subsidies times the quantity of resource lost or damaged), they are dealt with only briefly in this study.

Revealed Preference Techniques

These techniques attempt to find 'surrogate' or hidden markets for natural resource, where through the price or costs of other goods and services, individuals implicitly express preferences for environmental resources. Methods of most relevance for the current context are travel cost models mainly to estimate recreational value of a site which may be of interest where damage affects only certain aspects of a site. Other techniques include hedonic property pricing (estimate the effect of environmental characteristics on property values) and averting behaviour using the expenditure on aversion as a measure of damage cost. Revealed preference techniques are preferred in some context because of their explicit link with actual, observed market prices. However, these techniques are useful only in the context of estimating use values.

Stated Preference Techniques

These are survey methods in which hypothetical markets are created by way of carefully structured questionnaires for respondents to express their preferences. While these techniques may be used to estimate use and/or non-use values for a resource, they are the only techniques available for estimating non-use values. Non-use values have been shown to be a significant portion of total economic value in the context of many natural resources, especially where the resources concerned is unique or the impacts is irreversible. Two variants of stated preference techniques exist: contingent valuation (which focuses on the natural resource as a whole) and choice modeling (which focuses on the individual attributes or characteristics of a given natural resource).

The second feature of concern is benefits transfer. This is the process of taking information about benefits from one context (the study side) and applying it to another context (the policy site) and is the subject of a rapidly growing literature. The reason is obvious: if benefits transfer is a valid procedure, then the need for original (or primary) studies could be vastly reduced. In an ideal world, values would be taken off the shelf and applied to new contexts. Less than ideally, this is how many actual cost-benefit studies have proceeded. It is less than ideally since various requirements for a valid benefits transfer exercise are rarely met in practice.

Finally, note that restoration costs are omitted as a valuation technique in the figure. In some circumstances it may be legitimate to estimate benefits by the cost of replacing an asset (e.g. a particular habitat). Strictly, however, restoration cost is not a proper valuation technique: for (i) it relates to costs, not preferences; (ii) full restoration may not be possible, (iii) restoration only occurs after a time lag and (iv) restoration cost ignore psychic cost as a result of an incident. But where it is clear that the asset in question is unique and that benefits greatly exceed costs even on a limited inspection of the information available, then restoration cost becomes a minimum estimate of benefits. As a general rule, however, restoration costs as the measure of the 'value' of a resource should only be used in exceptional circumstances.

Revealed Preferences Techniques

Reveled preferences techniques (RP) are appropriate whenever the relevant Willingness to pay information can be inferred from individuals' decisions in actual markets. They assume that purchases in markets are reliable indicators preference ; if a person actually pays x to buy a good, it is inferred that her Willingness to pay for that good is at least x. Inferring current actions is a relatively simple process. However, predicting actions through RP techniques under a new set of conditions not yet experienced is not so straight forward.

The type of data and the absence of direct enquiry into individual preferences for environmental goods restrict RP techniques to the estimation of use values only. This suggests RP techniques are appropriate for valuing biological resources when they provide current and potential future use values, such as habitats for recreation, fisheries for commercial or recreational used and so on. However, considering that the use values represent only a part of Total Economic Value, RP are likely to underestimate the Total Economic Value of a particular resource. Estimation of non-use values is of particular importance in the current context so RP techniques are covered only briefly in this report.

The variants of RP are the travel cost method, random utility models, hedonic property pricing, and averting behaviour. The most relevant of these in the current context are travel cost method and random utility models. Restoration cost techniques are also covered here, although it should be noted that these are not true valuation techniques.

Travel Cost Method

Economists have long debated what costs to include in the travel cost model. Do we include only the direct costs of transportation to the site, or do we also include the costs of giving up time to do something else. The opportunity cost of time away from work or from some other recreational activity may be substantial, and should be considered in travel cost models. Yet, short of knowing each persons wage rate, we may have little information to determine the value of time. Furthermore, people may value their recreational time more or less than their work time, so it may not even be proper to use the wage rate to determine the value of time spent at the recreational site.

Economists typically include the following in the travel costs (or price of accessing a site):

Machinery Costs

Costs of driving. This is usually a fixed value (i.e.$0.37 per mile) multiplied times the miles. Alternative modes of transportation will have different costs.

Time Costs

The opportunity costs of not working while traveling to visit the site. Taking time out to recreate is important, but also costly. These costs are an important component of the overall costs. Opportunity costs of recreation are often calculated as 30% of the wage rate, based on estimates from the leisure literature.

Valuing Changes in Environmental Quality at a Site

One important aspect of travel cost models is that they can potentially be used to value changes in the nonmarket value of sites when environmental amenities change. For example, suppose the water quality at a beach declines suddenly so that people cannot swim in the water. What effects would this have no value? There would likely be two effects: (1) All individuals would likely take fewer trips; (2) Some individuals will quit taking trips altogether. Suppose on the other hand, that water quality suddenly improves dramatically. Two similar effects would occur: (1) Many individuals would take more trips; (2) some new individuals would begin taking trips, perhaps from a longer distance.

Both gains in environmental quality and losses in quality will have an effect on the number of trips taken to the beach. This should have a measurable impact on the value of the beach. If we are fortunate to have data to estimate two travel cost functions, one before the change, and one after the change, we can compare consumer surplus in each case to obtain the value of the environmental change as the difference between the two. This can be shown.

Valuing Changes in Environmental Quality Across Sites

Valuing sites is useful for many policy decisions where resource managers must decide which sites to invest resources in developing or preserving. It also is a good tool when we have natural experiments such as discrete losses or gains in environmental quality at sites to estimate the value of changes at that site. However, it is of limited value for more general questions such as "What is the value of improved water quality in Lake Erie?" This type of change will affect a host of recreational sites, and travel cost models of individual sites will provide little knowledge of the broader effects. These results from the factors we noted above, such as unique features, which may also affect visitation. It is difficult to translate the effects of discrete changes at one or two sites to the remaining sites, and furthermore, we may not even have natural experiments with discrete changes to use.

Nevertheless, to accommodate increasing demands for information on the value of improving or reducing environmental quality, economists have developed methods for using the travel cost model for this type of valuation. In some instances, these models can be used to value amenities, such as water quality, across sites, rather than value the sites themselves. In the literature, these approaches are known as the Hedonic Travel Cost approach or the Random Utility Approach. Rather than providing explicit detail on what these approaches attempt to do, we provide general background on how the basic travel cost model can be extended for these types of valuation exercises.

To understand what these techniques entail, let's think about the problem of environmental quality from the perspective of the individual rather than from the perspective of the site, as we discussed above. Except for the few sites that have particular unique features, individuals have many sites to choose from when they decide to recreate. For instance, if I wish to hike in the forest, there are many trails to choose from, including those on National Forests, State Forests, National Parks, State Parks, Country Parks, and even City Parks. I am likely to choose to go to the trail that provides the mix of environmental characteristics which provides me with the most value for the time I have available. Travel costs still provide the information on the price of visiting the site I chose, but instead of just considering the chosen site, we now consider information on all other sites that I could have chosen.

By comparing data on individual decisions to travel to a range of sites, we can determine the value of the amenities at the sites rather than the sites themselves. Thus, rather than looking at one side and a bunch of visitors to that site only, our survey must now include many sites, and visitors to all of these sites, as well as the levels of the environmental amenities at each of the sites. With this information, we can compare the costs of travel to the different sites to estimate the value of the characteristics. Instead of focusing on the relationship between the number of trips and travel costs, we are interested in the relationship between travel costs and environmental quality. We can estimate a similar function to that in Figures 1-3 above, but use environmental quality on the x-axis. The importance of these methods is that we use information on many different sites, and many different visitors to estimate the value of environmental quality rather than each individual site. Consumer surplus can still be measured and used to estimate the value of environmental quality.

Travel Cost

Background

This method is usually applied to three valuation problems:

- Recreational services flows of an existing site;
- Ex-ante value of a new recreational site; and
- Change of the environmental quality of a recreational site

By observing how visitation rates to a site change as the environmental quality of the site changes, the methods also provides values for environmental quality itself. In order for this to be true, a weak complementary relation is assumed between the environmental good (such as scenery) and the private good (such as visits to forests). In this case the method measures how the demand curve for the market good shifts if the associated environmental attribute changes.

The fundamental insight that drives this model is that if a consumer wants to use the recreational services of a site he has to visit it. The travel cost to reach the site is considered as the implicit or the surrogate price of the visit, and changes in the travel cost will cause a variation in the quantity of visits. Observation of these variations across individuals will permit the estimation of demand functions and the derivation of the welfare measure.

Two main approaches are used, a zonal TC and individual TC. With the former, visitors to sites are invited to provide information on the trip (cost, length, purpose, other sites visited, etc.) as well as on other socio-economic features (income, age, sex, etc.). The entire area from which visitors originate is divided into a set of visitor zones. The analyst then defines the dependent variable as the visitor rate (the number of visits made from a particular zone in a period, divided by the population of that zone). With the individual TC, the analyst defines the dependent variable as the number of site visits made by each visitor over a specified period.

In mathematical terms the trip demand curve will be defined as:

$V_{ij} / N_i = f(C_i, X_i)$ with : Vij as the total number of trips by individuals from zone i to zone j per unit of time; N_i as population of zone I; C_j as visit cost from zone i to zone j[11]; X_i as socio-economic explanatory variables in zone i[12]. The visitor rate V_{ij} / N_i is generally calculated as visits per 1,000 population in zone h. In each zone the household consumer surplus for all visits to the site is calculated by integrating the equation of the type :

$$V_{ij} / N_i = a + b\,C_i$$

Between the price (cost) of visits actually made from each zone and the price at which the visitor rate would fall to 0 (that is the vertical intercept of the demand curve at point P in the figure)

$$C.S. = {}_{Ch=B}^{\;P}\,(a + bCh)dCh$$

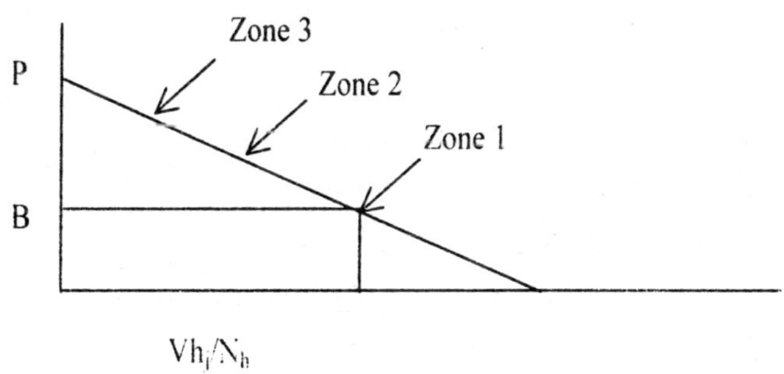

Annual total consumer surplus for the whole recreation experience can be estimated in each zone by first dividing total household consumer surplus by the zonal average number of visits made by each household to obtain the zonal average consumer surplus per household visit. Then the result can be multiplied by the zonal average number of visits per annum to obtain annual zone consumer surplus. Finally, cumulating zonal consumer surplus across all zones gives the estimate of total consumer surplus per annum for the whole recreational experience of visiting the site.

Basic Model of ITCM

The demand curve in this model relates individual's annual visits to the costs of those visits. That is:

$$V_{ij} = f(C_{ij}, X_i)$$

where:

V_{ij} = number of visits made by year by individual i to site j;

C_{ij} = visit cost faced by individual i to visit site j;

X_i = all other factors determining individual i's visits (income, time and other socio-economic characteristics)

The previous equation can be extended to allow for specification of a number of explanatory variable such as individual "i", estimate of the proportion of the day's enjoyment, which was contributed by the site visit to site j; individual i's estimate of the availability of substitute sites; size of individual's household; size of individual's party; whether individual i is a member of an environmental organization as well as other socio-economic data.

The demand curve for the site is defined by $\_V_{ij}/\_C_{ij}$. Integrating under this curve gives us the estimate of the consumer surplus per individual. The consumer surplus for the site is obtained by multiplying by the number of individuals visiting the site annually.

$$C.S. = Nj\ f(Cij, Xi).\ dCij$$

where:

Nj = number of individual visits to site j per year

(Cij, Xi) = defined as above

An example of a ZTCM is shown, which illustrates how the demand curve is derived and a consumer surplus (area under the demand curve) is calculated. In estimating consumer surplus for the on-site recreation experience, one assumes that people would react to increase in admission price in the same way as they would react to an increase in their travel costs and uses this information to estimate their response to an access fee. The second (Right) graph in the box uses the relationship derived from the left graph to estimate the reduction in the number of visitors to the site that would result from adding on access fee or price of \$5 or \$10. The relationship in the left graph gives the lower participation rate that could be expected from each zone when the price is added to the travel cost. That rate is then applied to the total population of the zone to yield the number of participants who could be expected at the price. These estimates are then used to plot the demand curve for the recreation site as in the right graph. The total Cs is $\frac{\$30 \times 29500}{2}$ or \$442,500

Options for Applying the Travel Cost Method

There are several ways to approach the problem, using variations of the travel cost method.

These include

1. A simple zonal travel cost approach, using mostly secondary data, with some simple data collected from visitors.
2. An individual travel cost approach, using a more detailed survey of visitors.
3. A random utility approach using survey and other data, and more complicated statistical techniques.

Application of the Zonal Travel Cost Approach

The zonal travel cost method is the simplest and least expensive approach. It will estimate a value for recreational services of the site as a whole. It cannot easily be used to value a change in quality of recreation for a site, and may not consider some of the factors that may be important determinants of value.

The zonal travel cost method is applied by collecting information on the number of visits to the site from different distances. Because the travel and time costs will increase with distance, this information allows the researcher to calculate the number of visits "purchased" at different "prices". This information is used to construct the demand function for the site, and estimate the consumer surplus, or economic benefits, for the recreational services of the site.

Step 1

The first step is to define a set of zones surrounding the site. These may be defined by concentric circles around the site, or by geographic divisions that make sense, such as metropolitan areas or counties surrounding the site at different distances.

Step 2

The second step is to collect information on the number of visitors from each zone, and the number of visits made in the last year. For this hypothetical example, assume that staff at the site keeps record of the number of visitors and their zip code, which can be used to calculate total visits per zone over the last year.

Step 3

The third step is to calculate the visitation rates per 1000 population in each zone. This is simply the total visits per year from the zone, divided by the zone's population in thousands.

Step 4

The fourth step is to calculate the average round-trip travel distance and travel distance and travel time to the site for each zone. Assume that people in Zone 0 have zero travel distance and time. Each other zone will have an increasing travel time and distance. Next, using average cost per mile and per hour of travel time, the researcher can calculate the travel cost per trip. A standard cost per mile for operating an automobile is readily available from AAA or other sources. Assume that this cost per mile is $30. The cost of time is more complicated. The simplest approach is to use the average hourly wage. Assume that it is $9/hour, or $ 15/minute, for all zones, although in practice it is likely to differ by zone.

Step 5

The fifth step is to estimate, using regression analysis, the equation that relates visits per capita to travel costs and other important variables. From this, the researcher can estimate the demand function for the average visitor. In this simple model, the analysis might include demographic variables, such as age, income, gender, and education levels, using the average values for each zone. To maintain the simplest possible model, calculating the equation with only the travel cost and visits /1000

$$\text{Visits} / 1000 = 330 - 7.755 * (\text{Travel Cost})$$

Step 6

The sixth step is to construct the demand function for visits to the site, using the results of the regression analysis. The first point on the demand curve is the total visitors to the site at current access costs (assuming there is no entry fee for the site), which in this example is 1600 visits per year. The other points are found by estimating the number of visitors with different hypothetical entrance fees (assuming that an entrance fee is viewed in the same way as travel costs).

Step 7

The final step is to estimate the total economic benefit of the site to visitors by calculating the consumer surplus, or the area under the demand curve. This results in a total estimate of economic benefits from recreational uses of the site of around $ 23,000 per year, or around $14.38 per visit ($23,000/ 1,600).

How do we Use the Results?

Remember that the agency staff's objective was to decide whether it is worthwhile to spend money on programs and actions to protect this site. If the actions cost less than $23,000 per year, the cost will be less than the benefits provided by the site. If the costs are greater than this, the staff will have to decide whether other factors make them worthwhile.

Application of the Individual Travel Cost Approach

The individual travel cost approach is similar to the zonal approach, but uses, survey data from individual visitors in the statistical analysis, rather than data from each zone. This method thus requires more data collection and slightly more complicated analysis, but will give more precise results.

For the hypothetical example of the recreational fishing site, rather than simply collecting information on number of visitors and their zipcodes, the researcher would conduct a survey of visitors. The survey might ask for the following information.

- Location of the visitor's home - how far they traveled to the site
- How many times they visited the site in the past year or season
- The length of the trip
- The amount of time spent at the site
- Travel expenses
- The person's income or other information on the value of their time

- Other socioeconomic characteristics of the visitor
- Other locations visited during the same trip, and amount of time spent at each.
- Other reasons for the trip (is the trip only to visit the site, or for several purposes)
- Fishing success at the site (how many fish caught on each trip)
- Perceptions of environmental quality or quality of fishing at the site
- Substitute sites that the person might visit instead of this site.

Using the survey data, the researcher can proceed in a similar way to the zonal model, by estimating, using regression analysis, the relationship between number of visits and travel costs and other relevant variables. This time, the researcher would use individual data, rather than data for each zone. The regression equation gives us the demand function for the "average" visitor to the site, and the area below this demand curve gives the average consumer surplus. This is multiplied by the total relevant population (the population in the region where visitors come from) to estimate the total consumer surplus for the site.

Because additional data about visitors, substitute sites, and quality of the site has been collected, the value estimates can be "fine tuned" by adding these other factors to the statistical model. Including information about the quality of the site allows the researcher to estimate the change in value of the site if its quality changes. To do so, two different demand curves would be estimated–one for each level of quality. The area between these two curves is the estimate of the change in consumer surplus when quality changes.

In the example, the researcher might recognize that development around the site is unlikely to totally destroy the fish population. However, it could diminish the population enough to adversely affect catch rates. By including catch rates in the model, the researcher can estimate the lost recreational benefits from reduced catch rates. However, the random utility model, described in the next section, is a more appropriate approach for this type of estimation.

Application of the Random Utility Approach

The random utility approach is the most complicated and expensive of the travel cost approaches. It is also the "state of the art" approach, because it allows for much more flexibility in calculating benefits. It is the best approach to use to estimate benefits for specific characteristics, or quality changes, of sites, rather than for the site as a whole. It is also the most appropriate approach when there is many substitute sites. In the example, the agency might want to value the economic losses from a decrease in fish populations, rather than from loss of the entire fish stock. The random utility approach would be the best way to do so, because it focuses on choices among alternative sites, which have different quality characteristics.

The random utility approach assumes that individuals will pick the site that they prefer, out of all possible fishing sites. Individuals make tradeoffs between site quality and the price of travel to the site. Hence, this model requires information on all possible sites that a visitor might select, their quality characteristics, and the travel costs to each site. For the example, the researcher might conduct a telephone survey of randomly selected residents of the state. The survey would ask them if they go fishing or not. If they do, it would then ask a series of questions about how many fishing trips they took over the last year (or season), where they went, the distance to each site, and other information similar to the information collected in our individual travel cost survey. The survey might also ask questions about fish species targeted on each trip, and how many fish were caught.

Using this information, the researcher can estimate a statistical model that can predict both the choice to go fishing or not, and the factors that determine which site is selected. If quality characteristics of sites are included, the model can easily estimates values for changes in site quality, for example the economic losses caused by a decreased in catch rates at the site.

Advantages of the Travel Cost Method

- The travel cost method closely mimics the more conventional empirical techniques used by economists to estimate economic values based on market prices.
- The method is based on actual behaviour–what people actually do rather than stated willingness to pay–what people say they would do in a hypothetical situation.
- The method is relatively inexpensive to apply.
- On-site surveys provide opportunities for large sample sizes, as visitors tend to be interested in participating.
- The results are relatively easy to interpret and explain.

Issues and Limitations of the Travel Cost Method

- The travel cost method assumes that people perceive and respond to changes in travel costs the same way that they would respond to changes in admission price.
- The most simple models assume that individuals take a trip for a single purpose–to visit a specific recreational site. Thus, if a triphas more than one purpose, the value of the site may be overestimated. It can be difficult to approach the travel costs among the various purposes.
- Defining and measuring the opportunity cost of time, or the value of time spend traveling, can be problematic. Because the time spent traveling could have been used in other ways, it has an "opportunity cost". This should be added to the travel cost, or the value of the site will be underestimated. However, there is no strong consensus on the appropriate measure–the person's wage rate, or some fraction of the wage rate–and the value chosen can have a large effect on benefit estimates. In addition, if people enjoy the travel itself, then travel time becomes a benefit, not a cost, and the value of the site will be overestimated.

- The availability of substitute sites will affect values. For example, if two people travel the same distance, they are assumed to have the same value. However, if one person has several substitutes available but travels to this site because it is preferred, this person's value is actually higher. Some of the more complicated models account for the availability of substitutes.
- Those who value certain sites may choose to live nearby. If this is the case, they will have low travel costs, but high values for the site that are not captured by the method.
- Interviewing visitors on site can introduce sampling biases to the analysis. Measuring recreational quality, and relating recreational quality to environmental quality can be difficult. Standard travel cost approaches provides information about current conditions. In order to estimate the demand function, there needs to be enough difference between distances travel costs to affect the number of trips made. Thus, it is not well suited for sites near major population centers where many visitors may be from "origin zones' that are quite close to one another.
- The travel cost method is limited in its scope of application because it requires user participation. It cannot be used to assign values to on-site environmental features and functions that users of the site do not find valuable. It cannot be used to value off-site values supported by the site. Most importantly, it cannot be used to measure nonuse values. Thus, sites that have unique qualities that are valued by non-users will be undervalued.
- As in all statistical methods, certain statistical problems can affect the results. These include choice of the functional from used to estimate the demand curve, choice of the estimating method, and choice of variables included in the model.

Background

This exercise draws on a actual study aimed at estimating the total use-value of forests in Linguria Region (Italy). Linguria

Region used to be an important producer of timber and chestnuts in the past. Over the years, the relative importance of these activities has dramatically declined, thus leading to forests abandonment and emigration of its population in the urban areas situated in the coastal zones. Consequently, coastal areas are now suffering of high and unsustainable population densities especially during the summertime, when significant flows of tourists add on the local population. As well, the internal forested areas progressively deteriorating due to lack of management and maintenance of both forests and historical sites within the forested areas.

From the financial point of view, revenues from forests of these areas are rather low compared to revenues from other uses of land. However, the forests provide other important non valued services such as soil-moisture conservation, open access recreation and hunting, as well as valuable secondary forest products such as berries, mushrooms and litter. Local authorities have therefore undertaken a study for appraising the net social benefits provided by the forested areas in a view of allocating financial resources for the rehabilitation and development of these areas.

This example estimates the benefits of the recreational use of these forests. In Liguria Region, recreation is a non-priced public good, therefore its value could not be assessed with conventional market techniques. The evaluation techniques used were the Individual Travel Cost Method and the Contingent Valuation Method.

Methodology

The procedure for carrying out a study with the ITCM method is similar, at least in the very first steps to the ZTCM method. The following 11 steps have been used:

Identification of the Site: a clear description of the site to be surveyed is essential for interviewers and interviewers to know which the object they are talking about is. It will also help identify the sites where interviews will be done. Maps and photographs of the site will provide a valuable contribution during the interviews. In the particular case of this study, sites have been identified on the basis of the following criteria: suitability of the area for tourism; actual flow of tourists; and the Plans of Liguria Region for creating regional parks.

Definition of the Environmental Good/service to be Valued: the good or service to be valued have been defined so that no misunderstandings occur. Goods or services of a forested area can be the whole area of the site, one of the particular services provided by the forest (i.e. aesthetics, cycling, fishing, etc), or the change in the supply of one attribute, both in quantity and in quality terms. In the Linguria case, the object was the whole site.

Questionnaire Design: the questionnaire is aimed at collecting information on the consumer's behaviour towards the particular environmental good/service to be valued. Information can be subdivided into compulsory and non-compulsory information. Compulsory are the origin of visitors, vehicle used for reaching the site, individual socio-economic features, other attributes (e.g., activities carried out in the site, member of environmental association, etc.); non compulsory are travel cost and opportunity cost of time (i.e., those variables that can be estimated outside of the questionnaire).

In this study, the questionnaire was constructed in order to be used both for the TCM method and the CVM. In particular it was subdivided in the following sections:

The first section was only for use of interviewers and aimed at providing some general information on the interview and interviewers (name of interviewer, location of interview, length of interview, etc). The second section was devoted to the collection of socio-economic data. It was assumed here that age, education, income, profession, number of family members, etc. are important determinants in visitor's behaviour towards recreational use or visits to forests. The third section attempted to identify the costs of travel costs faced by individuals to visit the site, by asking them their origin, the vehicle used to reach the site, the time employed to reach the area from their origin, the trip plan. The fourth section consisted of few questions aimed at investigating the environmental concern and awareness of visitors. It was assumed that the higher the awareness of environmental problems, the higher the perception of the value of forests recreational value.

Two further sections have been included for assessing the Willingness to pay with the CVM approach. In particular, the payment vehicle, the elicitation form and the market scenario for recreation have been defined. In this example only the ITCM is analysed. The information provide by the questionnaire will allow the analysts to derive the demand curve for the recreational site.

Survey Strategy: this step is mainly concerned with the organization and conduction of the survey. Before proceeding with the survey, the type of interview and the sample of interviewees must be defined. In the Liguria case is decided that interviews should be carried out on site. Therefore, for each single area, sites were identified for carrying out on-site face-to-face interviews. Interviewed visitors were selected randomly among the visitors throughout July, August and September 1993 on both weekdays and weekend days. The total number of interviews completed was 800.

The behaviour of people visiting the site during the summer times is most probably different than the behaviour of people visiting the same site in winter and spring (e.g., distance and time spent on site might be shorter). Thus, the total annual value of recreation services provided by the site might be overavalued if the observed behaviour in summer was assumed to be representative of the whole year. This possible bias was taken into account in the final interpretation of the results.

Elaboration of Calculated Variables and Statistical Description of the Sample: when the survey has been completed, a database is created like the one presented, using a simple spreadsheet. In order to check consistency of responses and reliability of data, some statistical analyses are carried out such as correlation analysis, contingency tables, Chi square analysis, mean median, variance, percentiles, etc.) All these analyses have been tested in the study carried out in Liguria.

Choice of Functional Form: a functional form relating the dependent variable (visits per year) and independent variables (travel cost, socio-economic variables) must be identified which better fits the particular database. The choice is among four functional forms : linear, log-dependent, log-independent, double log (also named log-log). In the particular case of Liguria, two functional forms were used : the double log form and the linear form.

Estimation of the Demand Function: once a functional form for the demand curve of the environmental services has been tentatively selected, its parameters are estimated using econometric models. Basic approaches requires models to be estimated by Ordinary Least Squares (OLS). More complex ones instead (e.g. truncated demand functions, simultaneous demand systems) require the use of Maximum Likelihood Estimation (MLE) methods, to be run by suitable econometric software packages.

Calculation of the Individual Consumer Surplus (CS). Once the most suitable functional form has been estimated, the individual CS can be calculated. It provides an analytical specification of the Consumer Surplus calculation for a linear demand function. The exercise below supplies the formula of the CS for a double log demand function.

Calculation of the annual individual average consumer surplus (averaging within the sample). When applying ITCM, the CS is calculated individual, i.e. for each observation of the sample. A summary measure of the samples CS can be obtained taking the sample mean of the CS. Several mean CS can however be calculated considering relevant specific socio-economic features of sample individuals, such as age, education, and gender or visit purposes.

Aggregation and further elaboration: if the objective of the study is the estimation of the total value of the site, the individual Willingness to pay resulting from the analysis has to be multiplied by the total visitors of the site during one time period (usually the year). Once the total value of the area has been computed, the value per hectare can be obtained by dividing the total value by the extension.

The travel cost method seeks to value recreational use of woodlands, wetlands, coastal zones and so on. The method assumes that travel costs (fares, cost, wear and tear, our of pocket expenses such as entry fees and value of time spent traveling and on site) are a proxy for the recreational value of visiting a given site. More recent applications of the TCM are referred to as count data models. The demand for the site is estimated by observing variation in the number of site visit according to variation in these costs. The methodology assumes there is an inverse relationship between the visit costs and the number of visits observed. For example, people living future away from the site incur higher transport costs and hence visit less often than those who live nearer the site. Because travel cost models are concerned with active participation they measure only the use value associated with any recreation site.

Evaluation of Travel Cost Method

Advantages

- Use of real market data

Disadvantages

- Can estimate use values alone
- May have substantial data requirements, if data are not readily available
- Requires estimates of value of travel / leisure time
- Problems arise with multi-purpose trips
- Cannot predict the changes in use values due to environmental changes without precedence.

Application

Recreational value (total) associated with national parks, reservoirs, woodlands, forest, wetlands, etc.

Random Utility Model

The random utility model is an extension of the travel cost method. It seeks to estimate recreational use values for individual features of a site. The random utility approach is concerned with explaining the choice between two or more goods with varying environmental attributes as a function of their characteristics. This can be useful where, for example, polluting activity causes damage to only some features of a recreational site. Travel costs and site attribute data are collected for a number of substitute sites in an area. The probability that an individual will visit site A rather than site B is then estimate depending on the costs of visiting each site and their physical characteristic relative to the characteristic of all sites in the individual's choice set. For example, for a forest, these could include species diversity and recreational facilities.

Evaluation of Random Utility Model

Advantages

- Estimates recreational use value of (i) changing environmental quality of site attributes and (ii) recreational use value of site in total
- Use of real market data.

Disadvantages

- can estimate use values alone
- may have substantial data requirements, if data are not readily available
- requires estimates of value of travel / leisure time
- problems arise with multi-purpose trips
- cannot predict the changes in use values due to environmental changes without precedence
- can be hard to handle participation decisions (i.e. whether to make the visit or not)

Application

- open access resources such as national parks, public woodlands, forests, wetlands, fisheries.

Hedonic Pricing Method

The hedonic pricing method seeks to estimate an implicit price for environmental attributes by observing actual markets in which those attributes are effectively traded. The market most frequently used is the housing market where property prices are determined not only by the structural characteristic (e.g. number of rooms, etc) and the access to public services (e.g. schools, etc) but also the environmental characteristic (e.g. clean air, peace and quiet, pleasant views, etc). By controlling for the non-environmental features and observing variations in house prices over the environmental attributes, the hedonic approach seeks to (i) estimate the implicit price that individuals are willing to pay for environmental characteristic associated with the house and (ii) infer how much people are willing to pay for an improvement in environmental quality, and hence estimate households demand for the environmental characteristic of concern.

Current evidence suggests that hedonic pricing is mostly used for estimating the implicit price is unique to the property market for which it is estimated, there is no justification for transferring them across property markets. Only studies that estimate a demand function (iii) above, can be transferred to other contexts.

Evaluation of Hedonic Pricing Method

Advantages

- Use of real market data

Disadvantages

- can estimate use values alone
- requires extensive house market data
- Cannot predict the changes in use values due to environmental changes without precedence.
- Current evidence suggest it is not suitable for use in benefits transfer

Application

- Amenities (such as proximity of woodlands), visibility (air pollution), noise; etc.

Avertive Expenditure / Avoided Cost Approach

The averting expenditure approach infers a monetary value for an environmental problem by observing the costs individuals are prepared to pay in order to avoid it. The increasing in averting expenditure such as purchase of water filters to ensure safe drinking water, or double glazing to reduce road/air traffic noise, is assumed to be a minimum measure of the welfare loss to the household of the decline in environmental quality. However problems of interpreting Willingness to pay estimates arise where the good used to trade-off against some environmental problem may have other benefits (e.g. double glazing provides multiple services such as energy conservation and anti-burglary), when averting expenditure is an imperfect substitute for the environmental loss, or when individuals engage in more than one form of averting behaviour.

Evaluation of the Avertive Expenditure Method

Advantages

- Modest data requirement
- Use of real market data

Disadvantages

- can estimate use values alone
- problems arise when : (i) individuals make multiple averting expenditure, (ii) there are secondary benefits of an averting expenditure and (iii) averting behaviour is not a continuous decision but a discrete one, i.e. double glazing is either purchased or not.
- Cannot predict the changes in use values due to environmental changes without precedence.

Application

- Water quality, noise nuisance, air pollution and random contamination.

Restoration Cost Method

The restoration cost approach values an environmental good according to the cost incurred in restoring it to its original state after it has been damaged. This approach is used extensively because it is relatively easy to find estimates of such costs. The approach also forms the basis of the Comprehensive Environmental Response, Compensation and Liability Act (CERCLA) 1980 in the USA. However, this method is not a proper valuation technique because there is no formal relationship between costs of restoration and Total Economic Value. The only link between restoration cost and Total Economic Value of a resources exist when there is a prior and unanimous social decision that the restoration must take place.

Evaluation of Restoration Cost Method

Advantages

- Costs easily obtained from direct observations of actual spending on restoring damage or from professional estimates.

Disadvantages

- Not considered a proper valuation technique.
- If restoration must take place, then cost become a minimum estimate of Total Economic Value.
- Potential to underestimate Total Economic Value because some damage may not be fully perceived, or may arise in the long term, or may not be fully restorable.
- Potential to underestimate Total Economic Value because there may be secondary benefits from restoration.

Application

- Use as an indication of the Total Economic Value of a resource only in exceptional circumstances such as :
- When restoration must take place due to quality standards
- As first approximations when there is an overall constraint not to let environmental quality decline.
- Useful for flood protection, water regulatory services supplied by forested watersheds, replacement of traditional medicines and the costs of crop insurance to replace natural insurance afforded by genetically diverse traditional cropping systems (Pearce and Moran 1994).

Stated Preference Techniques

Stated preference techniques (SP) use carefully structured questionnaires to elicit respondents' preferences for a given natural resources or environmental change. In principle, SP can be applied to any context. Thus, there exist numerous SP studies covering a wide range of non-marketed commodities. In addition SP are the only techniques that can estimate values for non-benefits of environmental resources. Non-use values have been shown to be a significant portion of Total Economic Value in the context of many natural resources, especially where the resources concerned is unique or the impact is irreversible. SP are also the most appropriate techniques to estimate indirect use values of ecological stability, biodiversity, watershed, protection and so on.

In the context of biodiversity, where respondents may not be entirely familiar with the resource in question especially with the concept of diversity as opposed to an individual biological resource, it is particularly important that surveys are designed in accordance with respected guidelines. Such a questionnaire will provide adequate information to permit respondents to give reliable estimates of Willingness to pay even though they are not directly familiar with the resource described.

There are two variants of SP contingent valuation and choice modeling. Both variants use similarly structured questionnaires but differ in the way they define the environmental resource of concern. Contingent valuation is concerned with the resource as the bundle of its attributes or characteristics, while choice modeling is concerned with the individual attributes of the resource.

Contingent Valuation

The aim of a CV study is to elicit individual's preferences, in monetary terms, for changes in quantity or quality of a non-market good or service. In CV studies, a contingent market defines the good itself, the institutional context in which it would be provided and the way it would be financed. The Willingness to pay / Willingness to accept question can be asked in a number of different ways, known as elicitation formats. These formats include open-ended (what are we willing to pay?), dichotomous choice (Are we willing to pay x?), bidding game (repetition of the dichotomous choice question with lower or higher amounts depending on the response to the initial question) and payment card (where the respondent chooses the Willingness to pay amount from a list of different amounts shown to him).

Econometric techniques are then applied to the survey results to derive the average Willingness to pay or Willingness to accept and to explain the variance in these results depending on the differences in the characteristic of the biodiversity resources and / or change of concern and the socio-economic characteristic of the respondents population, i.e. to define what is known as the bid function.

CV is likely to be most reliable for valuing environmental gains, particularly when familiar resources are considered. Using the standard CV approach to value complex and multidimensional scenarios raises some concerns, as complex and burdensome questionnaire designs as well as very large samples are required. In such situations, choice modeling may be preferred. There is also the issue of scope, which refers to the possibility that people may not be able to differentiate between different scales of the environmental impact of concern (especially very small changes). The academic debate on this suggests that estimated values may relate to some invariant 'moral satisfaction' or 'warm-glow' measure rather than the described impact. This issue affects both variants of stated preference techniques and is discussed.

Evaluation of Contingent Valuation

Advantages

- Can estimate both use and non-use values
- Suitable for valuing environmental changes irrespective of whether or not they have precedence
- Completed surveys give full profile of target population

Disadvantages

- Relative expensive
- Complex and multidimensional scenarios may be too much of a cognitive burden for respondents
- The concept of diversity may similarly be difficult to put across to the respondents.

Application

- Potentially can be applied to all biodiversity resources and changes, though the application is limited by the complexity of the questionnaire.
- Most suitable when all attributes of a natural resource are affected rather than individual attributes.

Valuing Public Goods: The Hypothetical (or Contingent) Market Experimental Techniques

The Contingent Valuation Method (CVM) has been used since the early 1960s. Robert Davis used questionnaire of 121 hunters and recreationists in the Maine area to estimate the benefits of outdoor recreation (Mitchell and Carson 1989). In the late 1960s, Ronald Ridker used the CVM in Philadelphia and Syracuse to estimate air pollution benefits. In the 1970s, CVM was used to value various recreational amenities.

Ciccheti and Smith used it to estimate willingess-to-pay to reduce congestion in a hiking area; Arthur Darling used the CVM to value amenities of three urban parks in California; Alan Randall used CVM to estimate air visibility benefits in the Four Corners area in the southwest; Acton applied the method to valuing programs which reduced the risk of dying from a heart attack (Mitchell and Carson 1989). By the late 1970s, CVM became a recommended method for determining project benefits for empirical resources such as environmental amenities. In 1979, the Water Resources Council recommended CVM in its Principles and Standards for Water and Related Land Resource Planning. In early 1980s, the U.S. Army Crops of Engineers began to use CVM to measure project benefits (Mitchell and Carson 1989) and CVM was also recognized under the Comprehensive Environmental Response, Compensation, and Liability Act of 1980 (Mitchell and Carson 1989).

CVM is a stated preference technique used to measure willingness to pay for certain nonmarket commodities. Since it may include nonuse values, it may be one's only alternative method for estimating use and nonuse values, especially for goods not traded in markets. Furthermore, the CVM is the only technique that measures Hicksian surplus directly, without requiring additional manipulation.

Under CVM a hypothetical scenario is created and individuals are surveyed through phone, mail or personal interview methods. To get a monetary measure of a scenario involving welfare change, individuals are asked the amount they would be willing to pay for the good (the welfare improvement) in question.

CVM assumes that individuals respond the same way to a hypothetical situation as they do to a real scenario. If so, their willingness to pay will be a realistic money measure of the individuals worth or utility gained or lost from changes in the availability or quality of non-marketed environmental goods. CVM is been widely used in environmental and natural resource economic and it is the most popular method for applied research (Hazilla 1997).

Contingent valuation approaches to projects benefits estimation necessarily involve surveys samples of the population of interests. If the sample is representative of the population, the sample mean of willingness to pay per capita (or per household) can simply be attributed to everyone in the beneficiary population of size N, so total project benefits are obtained as the product of N and per capita Willingness to pay.

There are three basic survey designs : (a) open-ended surveys where individuals are asked to state a willingness to pay; (b) closed-end referendums where individuals are presented with a bid and respond yes/no binary decision (single-response referendum); and (c) closed-end double referendums where individuals are presented with a sequence of two payments to obtain binary decisions (double-response referendum).

In the early years of CV, the method of payment elicitation was open ended. People were asked to reveal the specific monetary amount they would be willing to sacrifice for the provision of a non-marketed goods such as an improvement in ambient environmental quality. Obtaining measure of central tendency from this kind of data was a simple as calculating the mean or median of the willingness to pay values provided by the survey respondents. The econometric analysis involved was minimal, usually being confined to plausibility checks undertaken by split sample comparisons or by regressing the payment amounts on income and other socioeconomic variables to see if the signs on the parameter estimates in the relationship were consistent with prior expectations (e.g. Willingness to pay increasing with income).

Issues with Referendum Models

All of this changed with the advent of the referendum format, which only asks if the respondent would or not be willing to pay a specific pre-selected amount. Under this format it is not possible to know the true Willingness to pay of any individual directly. Because those who answer in the affirmative might actually be willing to pay even more, and those who answer in the negative might be willing to pay something less, econometric techniques have to be brought to bear to somehow interpolate and infer an expected value or other central tendency measure from the dichotomous choice information. Simplicity of data analysis was sacrificed in the referendum method in order to construct what many felt was a more realistic choice game.

In consequence, the notion that contingent valuation experiments of the referendum type can reveal a unique number which accurately and unambiguously represents individual willingness to pay for water quality improvement is unrealistic. Rather, there are several possible numbers, each dependent upon the way the initial survey was designed and administered and the way the resulting raw data was passed through the summarizing econometric sieve and reconstituted in the form of a central tendency measure. In short, such estimates are always uncertain when we acknowledge the existence of many routes that potentially can be taken to get at them and the several decision alternatives present at each step along the way. This is not a counsel of doom, or a suggestions that CB analysis based on referendum CV not be undertaken. But it is a fact that any benefit estimate to a greater or lesser degree is always a product of the analyst's protocol and judgment, something respectable analyst recognize and communicate to the users of their results.

The potential for a negative estimate of the expected value of willingness to pay is only a special example of a more general issue with referendum CV, which is that willingness to pay value extracted from the data can be heavily influenced by the methodological approach taken.

There are basically two routes to analyzing referendum data. The one most frequently pursued by project economists involves several steps, beginning with the specification and statistical estimation of one or more probability models of individual choice, employing prior assumptions about the form of the inverse distribution, and the covariates belonging in the distribution which serve to change its location and shape across respondents. This is followed by the evaluation of conditional mean or median formulas derived from the choice model, which depend on its estimated parameters. After calculating individual–specific means or medians, averages are taken over the entire sample to produce global central tendency measures. A less frequently traveled but much easier route ignores covariates and does not specify any particular inverse distribution. Instead it uses all the data in pooled form (i.e. the marginal distribution) to produce non-parametric measures of central tendency. Given its prominence, the next two sections concentrate on the parametric route, followed by a discussion of non-parametric options.

IE Contingent Valuation Method (Direct)

The contingent valuation method utilizes surveys to determine how much individuals are willing to pay to achieve improvements in environmental quality. Such methods are known as "Direct" non-market valuation methods, because they use individuals stated preferences to determine the value of individual characteristics. Contingent valuation methods can be applied to a wide variety of issues. The values captured in travel cost studies are known as revealed preference methods because they rely on actual behavior. Contingent valuation studies, however, can be applied to situations where there is not enough data to utilize revealed preferences. For examples, while we suspect that individuals in the US are supportive of saving endangered species, we don't observe people visiting regions where there are bears to see them. Most people try to avoid bears anyhow! Further, making a special trip to see bears is only one component of value. There are many people who value the existence of bears who will never actually go to see them. Many economists have argued that capturing these existence, option, or non-use values, are important components of environmental policy. If these values are important, they argue, then survey mechanisms such as contingent valuation methods can be used to determine the economic value of existence or the option to visit at some point in the future.

In order to better understand how contingent valuation studies can be used in practice, let's consider some of the important steps that should be undertaken to run a successful study.

Define the environmental problem to be addressed: for example, does our problem include a loss of environmental quality, does it include some again, or does it include the possibility of avoiding some potential loss. Good surveys state questions explicitly and the researchers can only get this information across if she has a good understanding of the problem herself. However, the fact that the contingent valuation method is based on asking people questions, as opposed to observing their actual behaviour, is the source of enormous controversy. The conceptual, empirical, and practical problems associated with developing dollar estimates of economic value on the basis of how people respond to hypothetical questions about hypothetical market situations are debated constantly in the economics literature. CV researchers are attempting to address these problems, but they are far from finished. Meanwhile, many economists, as well as many psychologists and sociologists, for many different reasons, do not believe the dollar estimates that result from CV are valid. More importantly, many jurists and policy-makers will not accept the results of CV. Because of its controversial nature, users must be extremely cautious about spending money on CV studies and about using the results of CV studies. This section continues with some example applications of the contingent valuation method, followed by a more complete technical description of the method and its advantages and limitations.

Hypothetical Scenario

A remote site on public land provides important habitat for several species of wildlife. The management agency in charge of the are must decide whether to issue a lease for mining at the site. Thus, they must weigh the value of the mining lease against the wildlife habitat benefits that may be lost if the site is developed. Because the area is remote, few people actually visit it, or view the animals that rely on it for habitat. Therefore, non use values are the largest component of the value for preserving the site.

Why Use the Contingent Valuation Method?

The contingent valuation method was selected in this case because of the importance of non-use values, and their potentially significant levels.

Alternative Approaches

Since non-use values are significant, and few people actually visit the site, other methods, such as the travel cost method, will underestimate the benefits of preserving the site. In this case, contingent choice methods might also be used, depending on the questions that must be answered, and whether contingent choice question formats are more effective than standard contingent valuation questions. This would be decided in the survey development stage of the application.

What to Value: the purposes of the survey and the object of the valuation exercise must be identified and stated clearly. What is the environmental good or service we want to place a value on? Are we valuing the whole environmental good, one of its attributes, a specific change in the quality / quantity of the good / service?

Type of the Value and Measure Unit: Is the analyst eliciting the marginal value or the average value to the individual of the good / service ? What is the measure unit?

Time Span of the Valuation: The analyst must decide whether to collect monthly, annual, multi-period Willingness to pay/Willingness to accept or lump sum Willingness to pay/ Willingness to accept.

Who Should be Interviewed: The relevant economic agents have to be defined, i.e., who is affected by the change in the provision of the environmental good or service (individuals, households, production units)

The second step concerns the design of the questionnaire. This activity is of fundamental importance. Indeed, a well-conceived questionnaire is the basis for the success.

Introduction: The interviewer explains some likely reasons (not necessarily all of them) of the survey to the interviewee, to make him involved.

Socio-economic Information: To analyse the answers and to interpret them in the socio-economic context of the respondent, data about the interviewee, his household and his social environment are normally collected (e.g. age, education, marital status, number of members of the household, annual income of the household etc.)

Scenario Design: The scenario description is of course different for each specific study. The scenario usually provides a clear and careful description of the environmental good / service object of the valuation, its changes under given conditions, the impacts of the change on the users/consumers, possibly the type of policies envisaged to secure (prevent) the change and who will pay for them. The Willingness to pay/Willingness to accept question must be phrased to present a clear, readily understood and plausible scenario. One important dimension of the scenario design is describing the availability of substitutes for the commodity valued. The value of a malaria vaccine, for instance, should depend on the existence of other methods for preventing the disease. Such methods should be clearly spelled out to respondents in the course of the survey; otherwise people may make different assumptions about the commodity being valued.

It is necessary also to include some final questions to check if the scenario has been communicated successfully (debriefing questions). Failure to communicate a realistic scenario to the respondent is likely to result in the respondent not taking the scenario seriously or replacing it with something of his own invention, either of which may bias Willingness to pay.

Elicitation Format: The elicitation procedure establishes the way the question for obtaining the value judgment is posed. Whatever the elicitation format however, it is important that before asking the individual what he would pay for the commodity valued, he should be reminded that this expenditures on other goods must be reduced by this amount; i.e., he should be remained of his budget constraint. It is also desirable to follow questions that elicit Willingness to pay with questions to determine whether a respondent who says he will pay nothing for the good is stating his true valuation or a protest bid. A respondent may say that he will pay nothing for a good that he in fact values if he believes that he should not have to pay for it or otherwise disagrees with the scenario presented in the survey.

Payment Vehicle: The choice of the payment vehicle is of utmost importance in the design of a CVM questionnaire. One procedure for determining the most appropriate PV is to carry out a pilot study in which several alternatives are tested. The analysis of the results will allow identifying the payment method the respondents are most familiar with and the most preferred by them. Possible payment vehicles are entrance fees (e.g., National Parks), taxes (e.g., pollution), contribution to funds (e.g. existence values such as protection of endangered species), charges (e.g., water use of agricultural, industrial, or domestic purpose). In rural areas in Africa, the choice of payment vehicles is a major challenge since many people have low incomes in a subsistence society.

The *third step* concerns the management of the survey. Sample selection and choices about how, where, and when to interview are presented.

Sample Selection: After determining the population who's Willingness to pay is to be measured, which, depends on the nature and location of the commodity to be valued, the researcher ideally uses probability sampling procedures to select a sample who will receive the contingent valuation questionnaire. This involves constructing a sampling frame and selecting from the sampling frame in such a way that the probability of each person being selected is known. Cost considerations have sometimes led to the use of convenience samples, i.e. interviewees selected at shopping malls or through newspaper advertisements. Such samples are usually adequate when the goal is testing how certain treatments (e.g. changes in scenario) affect responses, but do not allow Willingness to pay values to be generalized for the population.

How, When and Where to Run the Survey: The researcher must also determine the method of administering the questionnaire. Willingness to pay may be elicited through in-person interviews, or in mail or telephone surveys. The choice of administration method depends in part on the nature of (e.g., questions whose wording depends on the response to previous questions) are difficult to administer in mail surveys. Which method is chosen will affect both the cost of conducting the survey and the response rate. Response rates are generally lower for mail surveys and may lead to non-response or sample selection bias. This occurs when people who refuse to answer the survey (e.g., due to lack of interest in the topic) have Willingness to pay values that are systematically different from (e.g., lower than) those of respondents. The fourth step (data creation and analysis) is similar to most socio-economic surveys and will not be discussed here in more detail. Issue concerning the fifth step (Willingness to pay estimation) are presented in the following section.

Analysis of Contingent Valuation Responses

One goal of a contingent valuation survey is to estimate the population distribution of Willingness to pay for the environmental good in questions. This is the weighted frequency distribution of responses when the Willingness to pay question is open-ended. In the case of a closed-ended question, one can compute the percent of the sample that is willing to pay at least a given value, which provides an estimate of 1 minute the cumulative distribution of Willingness to pay.

To compute the total benefits of a policy, one typically computes the mean of the maximum Willingness to pay distribution and then multiples mean maximum Willingness to pay by the size of the population (N) affected by the policy. Issues arise concerning the definition of the population (N), the sampling choice, and the time period of the environmental benefits. The careful definition of target population is important and can obviously affect the size of society's aggregate maximum Willingness to pay/minimum Willingness to accept. For national and global issues with many stakeholders, the choice is difficult. The objective is to choose all those people whose utility will be significantly affected by the action.

As a alternative to mean maximum Willingness to pay, researchers frequently estimate median maximum Willingness to pay, which provides a robust lower bound for mean maximum Willingness to pay and is less sensitive to skewed distributions and the presence of gross outliers. It is also possible to compute ? – trimmed means to reduce the influence of extremely large and extremely small reported max. Willingness to pay amounts. The ? – trimmed mean is computed after deleting the ?100 percent smallest and ?100 percent largest maximum Willingness to pay observations. In addition to estimating the mean of the distribution of maximum Willingness to pay, it is of interest to see how mean maximum Willingness to pay varies with characteristics of the respondent (income, education). This provides a test of the internal validity of responses and is useful if one wishes to transfer Willingness to pay estimates from one population to another. The expression of the maximum Willingness to pay as difference of expenditure functions, as illustrated above, imply that maximum Willingness to pay can be written as a function of income (Y), prices, other non-market goods, taste variables (T), and the original and final levels of the good in question.

Applying the Contingent Valuation Method

Applying the contingent valuation method is generally a complicated, lengthy, and expensive process. In order to collect useful data and provide meaningful results, the contingent valuation survey must be properly designed, pre-tested, and implemented. Contingent valuation survey questions must focus on specific environmental service(s) and a specific context that is clearly defined and understood by survey respondents. In other words, a CV survey to assess the dollar value of the results of an environmental improvement cannot be based on the environmental improvement itself, but on increases in specific environmental services that the improvement is expected to provide.

The results of contingent valuation surveys are often highly sensitive to what people believe they are being asked to value, as well as the context that is described in the survey. Thus, it is essential for CV researchers to clearly define the services and the context, and to demonstrate that respondents are actually stating their values for these services when they answer the valuation questions.

A good CV study will consider the following in its application:

- Before designing the survey, learn as much as possible about how people think about the good or service in question. Consider people's familiarity with the good or services, as well as the importance of such factors as quality, quantity, accessibility, the availability of substitutes, and the reversibility of the change.
- Determine the extent of the affected populations or markets for the good or service in question, and choose the survey sample based on the appropriate population.
- The choice scenario must provide an accurate and clear description of the change in environmental services associated with the event, program, investment, or policy choice under consideration. If possible, convey this information using photographs, videos, or other multimedia techniques, as well as written and verbal descriptions.
- Unlike ordinary survey questions, which sometimes ask respondents whether they are willing to pay x dollars to improve "air quality", the nature of the good and the changes to be valued must be specified in detail in a CV survey. It is also important to make sure that respondents do not inadvertently assume that one or more related improvements are included. For example, if people are asked to value only air visibility, it would be important to make sure that they do not include their value for health-related improvements in their stated willingness to pay amount. Similarly, if people have a tendency to think

of environmental improvements in general, even when asked about water quality alone, it would be necessary to point out specifically that environmental quality, other than water quality, would remain the same.

- Questions can be asked in a variety of ways, using both open-ended and closed-ended formats. In the open-ended format, respondents are asked to state their maximum willingness to pay for the environmental improvement. With the closed-ended format, also referred to as discrete choice, respondents are asked whether or not they would be willing to pay a particular amount for the environmental improvement, or whether they would vote yes or no for a specific policy at a given cost. The discrete choice format is generally accepted as the preferred method.
- In addition to the hypothetical question that asks for willingness to pay, the survey must specify the mechanism by which the payment will be made, for example through increased taxes. In order for the question to be effective, the respondent must believe that if the money was paid, whoever was collecting it could effect the specified environmental change.
- Respondents should be reminded to consider their budget constraints.
- Specify whether comparable services are available from other sources, when the good is going to be provided, and whether the losses or gains are temporary or permanent.
- Respondents should understand the frequency of payments required, for example monthly or annually, and whether or not the payments will be required over a long period of time in order to maintain the quantity or quality change. They should also understand who would have access to the good and who else will pay for it, if it is provided.

- In the case of collectively held goods, respondents should understand that they are currently paying for a given level of supply. The scenario should clearly indicate whether the levels being valued are improvements over the status quo, or potential declines in the absence of sufficient payments.
- If the household is the unit of analysis, the reference income should be the household's rather than the respondent's, income.
- Thoroughly pre-test the valuation questionnaire for potential biases. Pre-testing includes testing different ways of sensitive to changes in the description of the good or resource being valued, and conducting post-survey interviews to determine whether respondents are stating their values as expected.
- Include validation questions in the survey, to verify comprehension and acceptance of the scenario, and to elicit socio-economic and attitudinal characteristics of respondents, in order to better interpret variation in responses across respondents.
- CVM can be conducted as in-person interviews, telephone interviews or mail surveys. The in-person interview is the most expensive survey administration format, but is generally considered to be the best approach, especially if visual materials are to be presented.
- Interview a large, clearly defined, representative sample of the affected population.
- Achieve a high response rate and a mix of respondents that represents the population.
- Whatever survey instruments and survey designs are used, and whatever response rate is achieved, make sure that survey results are analyzed and interpreted by professionals before making any claims about the resulting dollar values.

Advantages of the Contingent Valuation Method

- Contingent valuation is enormously flexible in that it can be used to estimate the economic value of virtually anything. However, it is best able to estimate values for goods and services that are easily identified and understood by users and that are consumed in discrete units (e.g. user days of recreation), even if there is no observable behavior available to deduce values through other means.
- CV is the most widely accepted method for estimating total economic value, including all types of non-use, or "passive use", values CV can estimates use values, as well as existence values, option values, and bequest values.
- Though the techniques require competent survey analysts to achieve defensible estimates, the nature of CV studies and the results of CV studies are not difficult to analyze and describe. Dollar values can be presented in terms of a mean or median value per capita or per household, or as a 'aggregate value for the affected population.
- CV has been widely used, and a great deal of research is being conducted to improve the methodology, make results more valid and reliable, and better understand its strengths and limitations.

Issues and Limitations of the Contingent Valuation Method

- Although the contingent valuation method has been widely used for the past two decades, there is considerable controversy over whether it adequately measures people's willingness to pay for environmental quality.
- People have practice making choices with market goods, so their purchasing decisions in markets are likely to reflect their true willingness to pay. CV assumes that people understand the good in question and will reveal their preferences in the contingent market just as they would in a real market. However, most people are unfamiliar with placing dollar values on environmental

goods and services. Therefore, they may not have and adequate basis for stating their true value.

- The expressed answers to a willingness to pay question in a contingent valuation format may be biased because the respondent is actually answering a different question than the surveyor had intended. Rather than expressing value for the good, the respondent might actually be expressing their feelings about the scenario or the valuation exercise itself. For example, respondents may express a positive willingness to pay because they feel good about the act of giving for a social good (referred to as the "warm glow" effect), although they believe that the good itself is unimportant. Respondents place importance on improved environmental quality in general. Alternatively, some respondents may value the good, but state that they are not willing to pay for it, because they are protesting some aspect of the scenario, such as increased taxes or the means of providing the good.
- Respondents may make associations among environmental goods that the researcher had intended. For example, if asked for willingness to pay for improved visibility (through reduced pollution), the respondents may actually answer based on the health risks that the or she associates with dirty air.
- Some researchers argue that there is a fundamental difference in the way that the researcher had not intended. For example, if asked for willingness to pay for improved visibility (through reduced pollution), the respondent may actually answer based on the health risks that he or she associates with dirty air.
- Some researchers argue that there is a fundamental difference in the way that people make hypothetical decisions relative to the way they make actual decisions. For example, respondents may fail to take questions seriously because they will not actually be required to pay the stated amount. Responses may be unrealistically high if respondents believe they will not have to pay for

the good or service and that their answer may influence the resulting supply of the good. Conversely, responses may be unrealistically low if respondents believe they will have to pay.

- The payment question can either be phrased as the conventional What are we willing to pay (Willingness to pay) to receive this environmental asset?, or in the less usual form. What are we willing to accept (Willingness to accept) in compensation for giving up this environmental asset?' In theory, the results should be very close. However, when the two formats have been compared, Willingness to accept very significantly exceeds Willingness to pay. Critics have claimed that this result invalidates the CVM approach, showing responses to be expressions of what individuals would like to have happen rather than true valuations.

If people are first asked for their willingness to pay for something has been found to depend on where it is placed on a list of things being valued. This is referred to as the "embedding effect".

In some cases, people expressed willingness to pay for something has been found to depend on where it is placed on a list of things being valued. This is referred to as "ordering problem". Respondents may give different willingness to pay amounts, depending on the specific payment vehicle chosen. For example, some payment vehicles, such as taxes, may lead to protest responses from people who do not want increased taxes. Others, such as a contribution or donation, may lead people to answer in terms of how much they think their "fair share" contribution is, rather than expressing their actual value for the good.

Many early studies attempted to prompt respondents by suggesting a starting bid and then increasing or decreasing this bid based upon whether the respondent agreed or refused to pay a such sum. However, it has been shown that the choice of starting bid affects respondents final willingness to pay response.

Strategic bias arises when the respondent provides a biased answer in order to influence a particular outcome. If a decision to preserve a stretch of river for fishing, for example, depends on whether or not the survey produces a sufficiently large value for fishing, the respondents who enjoy fishing may be tempted to provide an answer that ensures a high value, rather than a lower value that reflects their true valuation. Information bias may arise whenever respondents are forced to value attributes with which they have little or no experience. In such cases, the amount and type of information presented to respondents may affect their answers.

Non-response bias is a concern when sampling respondents, since individuals who do not respond are likely to have, on average, different values than individuals who do respond. Estimates of nonuse values are difficult to validate externally. When conducted to the exacting standards of the profession, contingent valuation methods can be very expensive and time-consuming, because of the extensive pre-testing and survey work. Many people, including jurist's policy-makers, economists, and others, do not believe the results of CV.

The Validity of the Contingent Valuation Method

Some economists have expressed skepticism of the method because contingent valuation questions are hypothetical and may therefore yield hypothetical answers. Criticism of the method has been also generated by the implausible answers observed in some contingent valuation studies (Hausman 1993). Contingent valuation studies sometimes produce many zero Willingness to pay values and/or a few implausibly large values. Zero values may be evidence of strategic behaviour, of attempts by the respondent to misrepresent Willingness to pay in order to influence the outcome of the survey. Or, zero values may represent protest bids, indicating that the respondent does not think he should have to pay for the good, even thought he has a positive Willingness to pay for it. Willingness to pay values that are large relative to the respondent's income and/or the availability of substitute commodities may indicate strategic behaviour, or that the respondents has failed to consider carefully his budget constraint.

In some studies Willingness to pay fails to increase very much with the quantity of the good valued. The value of reduction in risk of death may not increase with the size of the risk reduction (Jones Lee et al. 1985) or the Willingness to pay to save 1,000 birds may not differ significantly from the Willingness to pay to save 100,000 (Desvousges et al. 1993). This suggests either that the good that is valued is not well understood or that it is a category of programmes that is being valued; i.e., that the value attached to a large class of goods is embedded in the good valued (Kahnemann and Knetsch 1992).

To test whether contingent valuation provides accurate estimates of value, researchers have compared contingent valuation estimates with the results of actual or simulated market transactions. These tests assume that actual Willingness to pay measure true compensating variation, and is therefore an appropriate criterion to which to compare the contingent valuation method. This is perhaps reasonable for private goods, but is problematic for public goods, where actual Willingness to pay may understate compensating variation due to the free rider problem. The results of such comparisons are mixed. In the case of private goods, several experiments comparing hypothetical and actual Willingness to pay for hunting permits have found no statistically significant difference between mean values of hypothetical and actual willingness to pay (Mitchell and Carson 1989). Similar results were obtained by Dickie et al. (1987) in the case of strawberries. Comparisons of stated v. actual Willingness to pay for public goods have generally found that the former overestimates the latter (Kealy et al. 1987, Arrow et al. 1993).

A second test of validity compares contingent valuation estimates of Willingness to pay with estimates produced by indirect methods, such as hedonic pricing or the travel cost approach, that infer Willingness to pay from actual behaviour. To the extent that the two are similar, it is unlikely that people are overstating their Willingness to pay in a contingent valuation survey simply because they do not have to pay for the good. A meta-analysis of 83 studies that estimate Willingness to pay using both direct and indirect methods found that the mean ratio of the contingent valuation estimate to the estimate produced by indirect methods was 0.9 (Carson et al,1996). The mean ratio of the two estimates was not significantly different from 1.0 at the 0.05 level.

Comparing contingent valuation to other methods of valuing a good provides important evidence of the reliability of the contingent valuation method, but clearly cannot be applied in all cases. Tests of internal validity attempt to determine whether responses to a given contingent valuation survey are internally consistent. Tests of content validity ask whether the questionnaire describes the commodity to be valued in a meaningful way and avoids biases introduced by failing to mention substitute commodities or encouraging the respondent to consider his budget constraint.

CVM Guiding Principles

Litigation over natural resources damages has led to an upsurge of interest in the CVM method. Following the Exxon-Valdez oil spill. Exxon commissioned a number of CVM studies criticizing the reliability and validity of CVM as an approach to value environmental impacts. As a response the US government set a blue-ribbon panel under the joint chairmanship of Kenneth. Arrow and Robert Solow to advice on the use of CVM in natural resource damage for oil spills. The panel concluded that CVM studies can provide estimates reliable enough to be a starting point for a judicial or administrative determination of natural resources damages. To be acceptable to this purpose, though, such studies should adhere closely to the guidelines described in this report. The following main guidelines, not in ranking order, have been suggested by the panel (Arrow et al. 1993).

- For a single dichotomous question (yes-no type) format, a total sample size of at least 1000 respondents is required. Clustering and stratification issue should be accounted for and random sub-sampling will be required to obtain a bid curve and to test for interviewer and wording biases; involved, one must carefully weigh the pros and cons before embarking on CVM study and the depth of analysis required. It may be that a simple transfer of values from other similar studies could be as useful. This module presented a number of guidelines developed in the USA to improve the validity of CVM. When adhering to the guidelines, CVM studies can provide estimates reliable enough to be a starting point for a judicial or administrative determination of natural resources damages. This can eventually help assign CVM an important place among valuation methods.

Choice Modelling

The term choice modeling (CM) represents a range of SP techniques which take a similar approach to valuing non-market goods including:

– Choice experiments
– Contingent ranking
– Contingent rating, and
– Paired comparisons

Choice modelling were originally designed by marketing practitioners to isolate the value of individual product characteristic typically supplied in combination with one another. The techniques are based around the idea that any good can be described in terms of its attributes, and the levels that these take. For example, a forest can be described in terms of its species diversity, age structure and recreational facilities and a river can be described in terms of its chemical water quality, ecological quality and appearance. Changing attribute levels will essentially result in a different 'good' being produced and it is on the value of such changes in attributes that Choice modelling techniques focus.

Thus, Choice modelling approaches provide a direct route to the valuation of the attributes of a good and of marginal changes in these characteristic, rather than the value of the good in toto. This is of interest of those instances where damage affects only certain aspects of the natural resource of concern. Contingent valuation could also be used to value such changes, but the number of scenarios that can be considered in any one study is limited. Thus, it is generally assumed that choice modeling approaches are preferred over contingent valuation in contexts where it is important to value individual attributes. However, not all of the Choice modelling techniques are founded in the theory of welfare economics. In fact, only the choice experiment approach definitely fits the theory, while contingent ranking may do so. For this reason, this chapter considers choice experiments and contingent ranking only.

In a choice experiment respondents are presented with a series of alternatives and asked to choose their most preferred option. Each alternative is characterized by a number of attributes, of which one will be monetary (that is a price ticket), offered at different levels across options. Analysts can then see how respondents choices change as the attributes and monetary amounts are varied and from this information infer the value placed upon each attribute. A baseline status quo alternative is usually included in each choice set in order to ensure estimates are consistent with the theory of welfare economics.

Choice experiments give welfare consistent estimates for four reasons. First, they force the respondents to trade off changes in attribute levels against the cost of making these changes. Second, the respondents can opt for the status, quo, that is no increase in environmental quality at no extra cost to them. Third, because we can represent the econometric techniques used in a way which is exactly parallel to the theory of rational, probabilistic choice. Fourth, because we can derive estimates of compensating and equivalent surplus from the output of the techniques.

In a contingent ranking experiment respondents are asked to rank a number of alternative options according to their preferences. Each alternative is made up of a number of attributes and prices, which are set at varying levels across options. From the ordinal ranking (or choices), the Willingness to pay associated with each attribute can be indirectly calculated. As such, CR provides not only a monetary measure for each scenario but also uncovers the monetary values of each individual attribute within the scenario. In addition, CR avoids the need for an explicit elicitation of respondent willingness to pay relying instead on the ranking of (or choices between) a series of alternative scenarios or packages of attributes. It is important that one of the options must always be status quo for estimates to be consistent with standard welfare economics. Summaries the main advantages, disadvantages and the applicability of choice modeling in the current context.

Evaluation of Choice Modeling

Advantages

- Can estimate both use and non-use values
- Suitable for valuing environmental changes irrespective of whether or not they have precedence.
- Completed surveys give full profile of target population.

Disadvantages

- Not yet as widely tested as CV
- Some techniques are not based on economic theory
- The concept of diversity may be difficult to put across to the respondents.

Application

- Potentially can be applied for all biodiversity resources and changes, though the application is limited by the complexity of the questionnaire

- Appropriate method when damage affects only certain attributes of a natural resource.

Choice modeling can also be used to elicit respondents' opinion about the incident, damage and the characteristics of the restoration option without a monetary attribute. The choices respondents make provide information about their preferences about the different aspects of damage and restoration option and hence can be an input to designing the restoration options.

Note that choice experiments can be used to generate non-monetary estimates of damage to biodiversity or benefit of restoration and can also inform the choice of restoration options. This requires the design of the questionnaire according to the general rules but not including a cost or price attribute. This way, respondents can express their preferences for or against the damage and restoration without expressing their Willingness to pay/Willingness to accept in monetary terms.

Whether a SP study is successful or not depends on the existence of potential biases and how these are dealt with. The term 'bias' refers to the difference between the true Willingness to pay/Willingness to accept of a respondents and his/her stated Willingness to pay/Willingness to accept where such difference causes validity problems for the results of the study. Note that differences between value types do not always constitute a validity problem. Information bias (Willingness to pay/Willingness to accept sensitive to the amount and quality of information provided), for example, is defined as a reliability problem by some experts but as an expected and normal outcome by others

Benefits Transfer

Benefits Transfer (BT), i.e. the process of taking information about benefits from one context (the study site) and applying it to another context (the policy site) is the subject of a rapidly growing literature. The reason is obvious: if Benefits Transfer is a valid procedure, then the need for original (or primary) studies could be vastly reduced. In an ideal world, values would be taken off the shelf and applied to new contexts,

Less the ideally, this is how many cost-benefit studies of other contexts in which economic valuation is used proceed, and have proceeded for some considerable time. Less than ideally since various requirements for a valid benefits transfer exercise (as listed below) are rarely met in practice.

This section first outlines the general requirements for benefits transfer, and the different procedures, which may be used. This is followed by a discussion of the validity of benefits transfer and methodological considerations. Finally, the section is concluded with a review of the currently available electronic databases of valuation studies.

Requirements and Procedures for Benefits Transfer

It should be noted that certain conditions have to be met for a valid transfer of value to take place. These are widely recognized to be the following:

- The studies included in the analysis must themselves be sound
- The studies should contain Willingness to pay bid functions, i.e. regressions showing how Willingness to pay varies with explanatory variables such as those factors listed.
- The study and policy sites must be similar in terms of population and site characteristics, or differences in characteristics must be adjusted for;
- The change in the provision of the good being valued at the two sites should be similar, and Willingness to pay measures cannot be changed into Willingness to accept measures and vice versa and
- Property rights should be the same across the sites.

The process of benefits transfer is clearly less than ideal. Even if all the above criteria are met, benefits transfer are only as accurate as the original valuation study or studies. Thus, the quality of the data, and the methodology of the original study or studies need to be examined.

In implementing benefits transfer, the three most common procedure are to (i) transfer an average Willingness to pay estimate from one primary study, (ii) transfer Willingness to pay estimates from meta-analyses, and (iii) transfer a Willingness to pay function. These are discussed in turn overleaf. The analysis applies equally to Willingness to accept but Willingness to pay is used throughout this section as a short hand.

Transferring average Willingness to pay from a single study to another site which has no study

One elementary procedure is to borrow an estimate of Willingness to pay in context i (the study site) and apply it to context the policy site. The estimate may be left unadjusted, or it may be adjusted in some way. Transferring unadjusted estimates is clearly hazardous, although it is widely practiced. Reasons for differences in average Willingness to pay include differences in the:

- Socio-economic characteristic of the relevant populations;
- Physical characteristic of the study and policy site;
- Proposed change in provision between the sites of the good to be valued; and
- Market conditions applying to the sites for example variation in the availability of substitutes.

As a general rune there is little evidence that the conditions for accepting unadjusted value transfer hold in practice. Effectively, those conditions amount to saying that the various conditions listed above all do not hold, i.e. sites are effectively identical in all these characteristics. An alternative is therefore to adjust the Willingness to pay estimates in some way.

A widely used formula for adjusted transfer is

Willingness to pay_j = Willingness to pay_i $(Y_j/Y_i)^e$

Where Y is income per capita, Willingness to pay is willingness to pay and 'e' is the income elasticity of Willingness to pay, i.e. an estimate of how the Willingness to pay for the environmental attribute in question varies with changes in income. The typical practice in benefits transfers between countries has been to use the ratio of income in the two countries, as income is known to be one of the most important (if not the most important) factors resulting in changes in Willingness to pay. However, it is also possible to make a similar adjustment for, say, differences in age structure between the two sites, differences in population density, and so on. Making multiple changes of kind amounts to transferring benefit functions. Another approach to adjusted transfer involves the selection of a sub-sample from the original study site and then transferring the Willingness to pay estimate for that sub-sample to the policy site on the grounds that the policy site is more like the sub-sample than the complete sample. This may work is some contexts but sub-dividing samples may render the transferred values less reliable due to small sample size.

Transferring Benefit Functions

A more sophisticated approach is to transfer the benefit function (or bid function) from i and apply it to j. Thus if it is known that Willingness to pay_i = f(A,B,C,Y) where A,B,C and Y are factors affecting Willingness to pay at site i, them Willingness to pay_j can be estimated using the coefficients from this equation but using the values of A,B,C, Y at site j. Given that the characteristics of the population to which the estimate will be transferred is likely to differ from those of the study population, it is hoped that benefits estimates can be improved by using the transfer equation to modify the estimate of average Willingness to pay to account for these differences.

This approach relies on the availability of an appropriate valuation function in the original study, which may be used for Benefits Transfer purposes. When benefits transfer is the objective, the analysis of valuation results according to the best fit model, which is usually the only model reported, will usually be inadequate for Benefits Transfer purposes. A Benefits Transfer model will contain only a limited number of covariates: those that can easily be gathered for the transfer population. Usually this amounts to basic socio-economic details, such as the respondent's income, age and sex, since details of these characteristic in the transfer population can be easily collected from census returns. For resources that have a spatial dimension, however, it is crucial to include a variable that measures the household's distance from the site of provision.

Although valuation practitioners are becoming more aware of potential uses of their results for Benefits Transfer, it is not yet standard practice to present the results of such a model in a report on the analysis of data from any valuation study.

Transferring Benefits Functions: Meta Analysis

An alternative procedure is to use meta-analysis to take the results from a number of studies and analyse the in such a way that the variations in Willingness to pay found in those studies can be explained. This should enable better transfer of values since we can find out what Willingness to pay depends on. In the meta-analysis case, whole functions are transferred rather than average values, and the functions do not come from a single study, but from a collection of studies.

Interest in the application of meta-analysis to the field of economic valuation has expanded rapidly in recent years, and many are of relevant to the current context of natural resources valuation. Studies have taken place in respect of, among others, outdoor recreation, the ecological functions of wetlands, and the local income generation effects of tourism.

Validity of Benefits Transfer and Methodological Considerations

The previous section illustrated the various ways in which Benefits Transfer might be practiced. A further consideration is whether Benefits Transfer is affected in any particular way by the nature of the valuation studies that are included. For example, is there any reason to suppose that stated preference studies will perform better or worse than revealed preference studies? This is of concern in the current context since stated preference techniques are the only methods capable of estimating non-use values, which are likely to be important in the valuation of damage to natural resources.

On the basis of the requirements for Benefits Transfer outlined above, there is no particular reason to suppose that stated preference studies would fare any worse than any other form of study, although care will need to be taken to ensure that views about who has effective property rights are fully accounted for. Otherwise, the tests are very much a matter of carefully scrutinizing the accuracy and professionalism of the original studies.

More formal tests might therefore, involve seeing how far the transferred values are accurate. If transferred value from stated preference studies have more error than transferred values from revealed preference studies, then there may be a legitimate concern about the validity of stated preference transfers. However, it is only in the last decade that the question has been properly raised as to whether Benefits Transfer is valid, and formal testing of this type has taken place. Brouwer (1998) observes that there are comparatively few studies that test for the validity of Benefits Transfer, on whatever basis the original estimates were obtained.

There are three broad procedures for the validation of BT results, which have been undertaken to date although these procedures cannot be required from each BT exercise due to time and budget limitations.

(i) Transfer a value and then carry out a primary study at the policy sites as well. Ideally, the transferred value and the primary estimate should be similar. If this exercise is repeated until a significant sample of studies exists in which primary and transferred values are calculated for policy sites, then there would be a justification for assuming that transferred values could be used in the future without the need to validate them with primary studies.

(ii) Conduct a meta-analysis on existing studies to explain why studies result in different mean (or median) estimates of Willingness to pay. At its simplest, a mean analysis might simply take an average of existing estimates of Willingness to pay, provided the dispersion about the average is not found to be substantial, and use that average in policy site studies. Or, average values might be weighted by the dispersion about the mean, the wider the dispersion the lower the weight the estimate would receive. A more sophisticated approach takes a set of n primary studies and uses n-1 of the studies to estimate the value at the nth site; and

(iii) Benefits transfer may be tested by estimating Willingness to pay before an actual change occurs and then revisiting the area later when the change is complete to see if people behaved according to their stated Willingness to pay.

- That transferring benefit functions is more accurate than transferring average values ;
- Stated preferences studied appear to perform no worse than revealed preference studies in terms of transfer error;
- But transfer error can be quite large, 1 to 75% if outliers, i.e. responses which are unrealistically low or high are ignored, but up to 450% if they included:
- Individuals attitudes are often important determines of Willingness to pay in SP studies, yet most BT makes little effort to test for variability in attitudes across sites. This suggests the BT would have to be supplemented by social surveys at the policy site.
- Meta-analysis of contingent valuation studies can explain a reasonable proportion of the variation in the original studies, but the original studies do not include sufficient information to test whether more information would have increased the explanatory power of the meta-analysis and

– The missing information may well be of the motivational type, i.e. why people adopt the value stances they do. This conclusion fits will the current focus in economics on the analysis of motives for preferences.

The new focus on motive suggests that more attention should be paid to motives in SP studies so that there is a better chance of explaining Willingness to pay variation between studies. Study context may also be important (e.g. the historical factors affecting a particular site, or the cases of the problem that is being valued). Efforts to conduct SP studies in a context - free environment should help to resolve this issue, although care must be taken not to render the questions meaningless to respondents.

Cost Benefit Method

Cost benefit analysis (CBA) is a more encompassing framework. By comparing both the costs and benefits of different restoration options, it can assist with determining whether a restoration target is desirable as well as choosing the best restoration option. The direct comparison of costs and benefits requires both identities to be expressed in the same unit, i.e. money. However, this may not always be possible. Thus, Cost benefit analysis framework is usually expanded to take both monetary and non-monetary expressions of costs and benefits into account. Multi-criteria analysis decision analysis (MCDA) is developed to compare costs and benefits when they are expressed in different units. This annex covers MCDA as well as Cost-Effectiveness Analysis and Cost benefit analysis.

This section discusses a number of issues, which apply to all of the above analyses, namely, the definition of costs and benefits, definition of the 'affected' population (also relevant to the valuation techniques discussed) and discounting. It present guidance on how to implement Cost-Effectiveness Analysis, while Cost benefit analysis framework is discussed. Some relevant aspects of MCDA are discussed. Risk and uncertainty associated with costs and benefits and how to deal with these are discussed. A set of criteria for selecting the relevant analyses and information requirements for each is given.

Definition of Benefits

A benefit is defined as any addition to human wellbeing. Wellbieng sounds vague but it has a strong linkage to what people want i.e. to their preferences. Essentially, if an individual prefers to be in situation A rather than situation B, he or she can be said to have a higher wellbeign in A than in B. In the context of restoration of biodiversity damage, benefits of a restoration option is equal to the avoided damage to biodiversity (both initial damage and interims losses). Just as damage is assessed in relation to the baseline condition of the damaged resource, benefits should be measured in relation to the same baseline. This is referred to as with/without principle of cost-benefit analysis.

Economic approach is based on the assumption that individuals are the best judges of their preferences and uses the valuation techniques presented in Annex 1 to estimate the value placed on the (damage to) resources in monetary terms. For this value to be estimated, however, the link between the damaged resource, the services it provides and the value placed on those services and hence the resource needs to be established. Given this resource service value link, benefits can be expressed in either resource, service or value units. Note that the first two are generally based on expert judgment, while the latter is held by individuals but elicited by experts.

For example, suppose that 100 ha of a wetland is damaged due to an oil spill and a restoration option devised to recreate the same area of wetland in an adjacent site. Benefits of this restoration measure (which is equal to the cost of damage) can be expressed in resource units such as 100 ha of wetland recreated. Benefits can also be expressed in service units. Suppose that the wetland was used for recreation and prior to damage there were 120,000 visits per year. The unit for benefits can be 1,20,000 visits to the affected wetland. Finally (land, strictly, what is preferred for the economic approach), benefits can be expressed in value units i.e., value of the services provided by the damaged resources expressed in monetary units. Supposing that a visit to the damage wetland was worth Euro 10, the value of damage (and benefit of restoration) would using benefits transfer if the relevant literature is available or by travel cost or stated preference techniques if an original valuation study is to be conducted. Note that the only unit in which non-use values can be expressed is money. Also note, that, in Cost-Effectiveness Analysis, where the restoration target is predetermined, baseline or damage is assessed in order to set the restoration target but benefits of restoration do not need to be measure.

Definition of Costs

A decrease in human wellbeing is defined as a cost. In the current context, there are two aspects to cost : (1) cost of damage to biodiversity (2) cost of restoring this damage. Cost of damage to biodiversity is dealt with in definition of benefits since cost of damage avoided by restoration is assumed to be equal to the benefit of that restoration. This section deals with the second aspect : the cost of (both primary and compensatory) restoration.

Costs of restoration are expressed in monetary units and include items such as cost of undertaking the damage assessment and preparing restoration and monitoring strategy; cost of species population restoration, habit restoration and cleaning; cost of implementing the monitoring and surveillance strategy.

It is important to note the distinctions between how a conventional financial analysis and how Cost benefit analysis and Cost-Effectiveness Analysis conducted from the point of view of the society as a whole differ in the way costs are defined and measured. The later uses a concept called opportunity cost. Given the limited availability of financial (and indeed all other) resources, funds that are used for one purpose, biodiversity damage restoration in this case, cannot be used for other purposes. Therefore, decisions whether or not undertake restoration should include considerations of not only benefit of restoration but also what else could have been done with, or the opportunity cost of, the funds spent for restoration.

It is usually assumed that market prices used to measure the cost items reflect the opportunity cost. To ensure this is the case, sometimes it is necessary to undertake shadow pricing or to use real prices. At its simplest, this involves taking account of taxes and subsidies that maybe included in the market prices of goods and services used for restoration. Taxes and subsidies are known as 'transfer payments' since they are transferred from one group to another (from the consumers to the government in the cases of taxes and from the government to the recipients in the case of subsidies) but have no net effect on the society's resources. Thus, shadow pricing involves using market prices net of any taxes and subsidies. This is valid both for Cost benefit analysis and Cost-Effectiveness Analysis.

Cost Effectiveness Analysis

The cost-effectiveness analysis (CEA) is a truncated form of cost-benefit analysis. It draws inspiration and guidance only from the cost side–or alternatively, only from the benefit side of cost-benefit analysis. As mentioned above, the cost side is the appropriate use of Cost-Effectiveness Analysis in the current context. Thus, to implement Cost-Effectiveness Analysis, we need to undertake the following tasks :

- Damage assessment and significance analysis in order to identify the baseline and the restoration target.
- Determine primary restoration options, and identify and measure the costs of primary restoration options.
- Determine compensatory restoration options, and identify and measure the costs compensatory restoration options.
- Calculate the present value of the costs of primary and or compensatory restoration.

Cost-Effectiveness Analysis compares the primary or compensatory restoration options that are identified as technically capable of meeting the restoration target in terms of their costs. Costs should be expressed in present value terms as shown. The option that achieves the restoration target at the lowest (present value of) cost is chosen as the optimum option according to Cost-Effectiveness Analysis.

Cost-Effectiveness Analysis is sufficient when there is an agreement on the restoration target. It is also sufficient so long as the cost of the choose (primary or compensatory) restoration option is not deemed to be excessive. If the cost is deemed to be excessive, the only way to test whether this is the case is to compare the cost of the option with its benefit, in other words, cost-benefit analysis, which is presented in the next section.

Cost Benefit Analysis : Procedure

Cost-benefit analysis (CBA) is a framework for comparing the present value of the cost and the present value of the benefits of an action. In the current context, Cost Benefit Analysis is needed only when the cost of the chosen primary and/or compensatory restoration option is deemed to be excessive Cost Benefit Analysis builds on the tasks of cost-effectiveness analysis. Thus, the following steps need to be taken to implement Cost Benefit Analysis.

- Damage assessment and significance analysis in order to identify the baseline and the restoration target.
- Determine primary restoration options and identify and measure the costs of primary restoration options.
- Determine compensatory restoration options, and identify and measure the costs of compensatory restoration options.
- Calculate the present value of the costs of (primary and/or compensatory) restoration options.
- Calculate the present value of the benefits of (primary and/or compensatory) restoration options.

Cost Benefit Analysis uses two decision-making criteria, net present value (NPV) and benefit-cost ratio (BCR). NPV is calculated as follows :

$$NPV = PV(B) - PV(C)$$

Where NPV is net present value, PV(B) is the present value of benefits, PV(C) is the present value of costs, B_i is the benefit which occurs in time period t, C_i is the cost which occurs in time period t, t_i is the period when the incident causing the damage occurs. T is the last period in the assessment, and d_t is the weight used to convert the past and future costs and benefits to present values.

NPV is useful for determining whether a given restoration project should go ahead :

- Negative NPV implies that the present value of the costs of a restoration option is greater than the present value of its benefit. On the basis of NPV alone, such options should be rejected.
- Zero NPV implies that the present value of the cost of a restoration option is equal to the present value of its benefits. Conventionally, if this is the result, Cost Benefit Analysis would be indifferent whether the option should be rejected or accepted since the main purpose of Cost Benefit Analysis is to identify options that generate net increase in social wellbeing. However, in the current context, the main purpose of Cost Benefit Analysis is to demonstrate that meeting the restoration target does not entail excessive costs. Thus, so long as the present value of costs is equal to the present value of benefits, it cannot be argued that the costs are excessive. This is why, the guidance in the USA (DARP, 1999) recommend that the options with zero NPV should be accepted.
- Positive NPV, implies that the present value of the costs of a restoration options is less than the present value of its benefits. On the basis of NPV alone, such options should be accepted.

Note that if there is no option with zero NPV and that there are more than one options with positive NPV, the option with the highest NPV should be chosen. if there are more than one option with zero or positive NPV, either the option with the lowest cost or the option that generates the highest benefits (the highest positive NPV) should be chosen. The choice depends on whether the purpose is just to demonstrate that the restoration target is met and the cost is not excessive (zero NPV) or to generate net benefit (the highest positive NPV).

The benefit-cost ration is another way of comparing the discounted (present) value of costs and benefits.

$$\text{Benefit / Cost} = \frac{PV\,(B_i)}{PV\,(C_i)}\ \frac{PV\,(B_i)}{PV\,(C_i)}$$

This ratio enables a comparison of options on the basis of value per Euro spent. Therefore, it should be used if the restoration budget is predetermined and fixed. The decision rules are the same as those for NPV :

- Benefit-cost ratio of less than one implies that the present value of costs of a restoration option, is greater than the present value of its benefits. On the basis of this ratio alone, such options should be rejected.
- Benefit-cost ratio of exactly one implies that the present value of the costs of a restoration option is equal to the present value of its benefits. Conventionally, if this is the result, Cost Benefit Analysis would be indifferent whether the option should be rejected or accepted since the main purpose of Cost Benefit Analysis is to identify options that generate net increase in social wellbeing (ratio greater than one). However, as for NPV, in the current context, the main purpose of Cost Benefit Analysis is to demonstrate the meeting the restoration target does not entail 'excessive' costs. Thus, so long as the present value of costs is equal to the present value of benefits, it cannot be argued that the costs are excessive.

– Benefits-cost ratio of greater than one implies that the present value of the cost of a restoration option is less than the present value of its benefits. On the basis of this ratio alone, such options should be accepted.

In the case of fixed restoration budgets, maximum benefits are obtained by first implementing the option with the highest benefit-cost ratio, then the second highest and so on. Once the restoration budget limit is reached the remaining projects cannot be implemented even if they have benefit-cost ratios greater than unity.

As discussed above, costs of restoration are expressed in monetary units. Therefore, the benefits should also be expressed in monetary units. This allows for the comparison of like with like and Cost Benefit Analysis can directly suggest options for implementation.

However, due to data, time and resource restrictions, this may not be possible and hence benefits may be partially expressed in monetary units and partially in non-monetary units. Although conventionally not incorporated in Cost Benefit Analysis, non-monetary expressions of benefits can still be incorporated in the Cost Benefit Analysis alone cannot directly suggest options for implementation. Professional judgment would be required to choose the option.

In this case, the procedure should be to compare all monetised costs and benefits as above, and list the non-monetised effects. Note that a non-monetised negative indicator constitutes a cost and a non-mentioned positive indicator constitutes a benefit. There are them four ways of dealing with mixed outcomes.

1. If monetised benefits exceed monetised costs and the non-monetised indicators are judged mainly to be positive, then proceed since benefits more than outweigh the costs.
2. If monetised benefits exceed mentioned costs and the non-mentioned indicators are judged mainly to be negative, then compare net mentioned benefits with the non-monetised costs. Using professional

judgment ask if the non-monetised costs are likely to be greater than the net monetised benefits. If they are, the option is not worthwhile. If they are not, then the option is potentially wroth pursing.

3. If monetised costs exceed monetised benefits and non-monetised indicators are judge mainly to be positive, the compare net monetised cost with the non-monetised benefits. Using professional judgment, ask if the non-monetised benefits are likely to be greater than the net monetised costs. If they are, the option is potentially worth pursuing. If they are not then the option is not worthwhile.
4. If monetised costs exceed monetised benefits and the non-monetised indicators are judged mainly to the negative, then the scheme is not worth pursuing.

Risk and Uncertainty

Discussion so far is based on the implicit assumption that the costs and benefits of restoration can be identified and measured with certainty. In reality, this may not always be the case. There are a number of different aspects to risk and uncertainty surrounding the restoration of biodiversity damage including, but not limited to :

Pre-incident resource status may not be known with certainty if there are no data collected prior to the incident.

The extent of the damage may not be possible to measure with certainty because of (i) lack of data about the pre-incident resource, and (ii) lack of understanding of the process through which an incident causes damage ;

- Restoration options cannot be said to be certain to succeed because of : (i) lack of data about the time period required to meet the restoration target, and (ii) uncertainties about the future conditions.
- The costs of restoration options may not be known with certainty since more funds can be required if unforeseen factors threatening the success of restoration need to be compensated;

- The benefits of restoration options may not be know with certainty because (i) the extent of damage and the success with restoration option are not known with certainty, (ii) affected population may not be accurately identified. (iii) future uses of damaged resource may not be predicted, (iv) measurement of the preferences may not be accurate especially if benefits transfer is not used appropriately, and (v) if non-monetary measure of benefits is used, expert judgments may be disputable and
- Choice of restoration option can be significantly affected by the choice of discount (and compound) rate applied.

The different aspects listed above and others can be dealt with using different approaches. First of all, it is crucial to distinguish whether the above lead to risk or uncertainty since their treatment can be significantly different. Risk is defined as some known combination of the probability of an event occurring, and the scale of the event. Uncertainty arises when this probability distribution is not known and the scale of the event, if it occurs, may be known accurately or only imperfectly. The distinction between risk and uncertainty is important because the means of dealing with them are different. The rest of this section presents an outline of these different approaches to risk and uncertainty.

Incorporating Risk and Uncertainty into Costs and Benefits

For situations involving risk there are three options depending on the level of available information about risk.

If the probability distributions of the costs and benefits are known, stochastic simulation methods (e.g. Monte Carlo method) can be useful for developing insights into these probability distributions. Although sophisticated and hence desirable, the information requirements of such simulation exercise can be too high. It can be applied only if the distribution of probabilities (across time and space) attached to costs and benefits are known, which is unlikely to be the case for most natural resource damage cases.

The second option involves estimating expected values for the outcomes of concern and can be chose when we have only point estimates for the risk, i.e. we only have one absolute value for the outcome and one estimate of the probability of that happening.

For example, if an event with a cost valued at Euro 100 occurs with a probability of 0.1 one approach might be to multiply the two numbers so that risk equals Euro 10. This is an example of an expected value approach to representing risk. Equation below shows how the net present value estimate would change, when expected rather than absolute values for costs and benefits are used.

$$NPV = PV(B) - PV(C)$$

Where p_b is the probability of benefits occurring, p_i is the probability of costs occurring and the other variables are as above. The value these probabilities are likely to take depends on the characteristics of the resource in questions, the restoration project alternative appraised and the background conditions at the site. Therefore, it is not possible to make a general statement about what these values should be other than to recommend that relevant experts in the field are consulted to identify these risks, and their likely effect on the future costs and benefits of various restoration options.

Possibly the most sophisticated approach to incorporating risk and uncertainty into the values of costs and benefits is by estimating a risk premium. This involves measuring the certain benefits that affected individuals would accept in lieu of the uncertain benefits. The difference between the certain benefit and the uncertain benefits is referred to as the risk premium (Wilson, 1982). Risk premia may be estimated through the application of stated preference techniques to measure the certain benefit that is equivalent to an uncertain stream of benefits. For example, a stated preference study and an uncertain stream of natural resource services and an uncertain stream of natural resources services. The choices thus elicited would yield information on individual risk premia (DARP, 1999). If a stated preference study is already implemented to estimate the benefits of restoration, the issue of risk premium can be incorporated into the design of the questionnaire. However, implementing a questionnaire with the sole purpose of estimating risk premium is not likely to be cost-effective. This is recognized by the USA guidance on the issue which states that the risk-adjusted measures of value (all three approaches mentioned here, namely, Monte Carlo simulation, expected values and risk premium) are rarely fully implemented due to their extensive requirements for generally unavailable information.

Incorporating Risk and Uncertainty into the Discount Rate

An alternative, less preferred, approach presented in the USA guidance (DARP, 1999) involves selecting discount rates that reflects the level of systematic, i.e. non-diversifiable, risk associated with the restoration option. A range of after-tax interest rates that reflect varying levels of risk are available for this purpose. For the USA, the lower end of the range of rates representing a risk-adjusted time preference is the riskless average real after-tax return on the market portfolio of stocks. Based on the historical data, the lower end is identified as about 1%, while the upper end is identified as about 7%. It is suggested that sensitivity analysis over the range of risk-adjusted rates to generate a range of outcomes. We do not have historical data about Europe to recommend risk-adjusted values for Europe but the USA suggestions can be tested in the European context. However, the USA guidance notes that adjusting discount rates to accommodate risk and uncertainty is rarely fully implemented due to the extensive requirements for information.

Institutional Mechanisms for Accommodating Risk and Uncertainty

The USA guidance (DARP, 1999) states that the most common method for addressing uncertainty in restoration implementation is through institutional mechanisms. There are three approaches to institutional mechanisms: (i) performance standards, (ii) design standards, and (iii) contingency factors.

With a performance standard, the liable party agrees to undertake restoration and is held to a, performance standard for the outcomes of primary and compensatory restoration projects. The performance standard shifts the uncertainty associated with restoration to the liable party rather than the public authority. Should the liable party fail to comply with the performance standard, such as restoring the damage resource to its pre-incident condition within a given time period, the liable party is subjected to a pre-determined penalty.

In some circumstances, there may be substantial uncertain about project outcomes due to factors external to project implementation. In these cases, the liable party may not agree to any performance since performance is, to a large degree, out of its control. Design standards offer a mechanism for sharing uncertainty between the liable party and the public authority. With a design standard, the liable party is subject to standards on project actions and may incur penalties for non-compliance. The advice given by the English Nature in the UK (State nature conservation agency) can be likened to what is referred to as the design standard in the USA.

Guidance on CBA Methods

What is the Market Failure? Where are the Response Options?

A basic tenet of the economic approach to regulatory assessment is that a successful regulation improves on the allocation of resources that would otherwise arise. To be able to describe the benefits of regulation, therefore, the first step is characterizing the market failure. Sometimes, as in the case of pollution spillovers, the nature of the failure is clear (although its magnitude might be uncertain). Other cases might be less clear. In the case of consumer safety, is the problem the production of unsafe products or the failure of the market to offer adequate information?

Given that the existence of a market failure is established, the set of potential response options must be identified. These can include nonregulatory, as well as regulatory, options, e.g., information provision, voluntary standards, use of fees, or tax breaks. Regulation also can be applied at different levels of government, with different levels of stringency, and to different populations. These options need to be considered in justifying a new federal regulation.

What is Baseline for Assessing the Effects of Regulation?

The answer to this question would seem straight forward in most cases: assume business as usual in the absence of the regulation. Some nontrivial judgment calls do arise, however : what level of baseline compliance with current regulations should be assumed? What future shifts in market conditions or other regulations should be considered? More troublesome is the issue of deciding what is in the baseline and what is varying in the policy analysis when multiple regulatory actions are under consideration. Here it is necessary to ensure consistency in assumptions that affect benefits and costs and to avoid self-serving hypotheses. To illustrate, suppose that a particular pollution standard would require a mixture of lower-cost and higher-cost responses for compliance. If the higher-cost options are assumed to occur in the baseline, then the cost of generating incremental environmental improvement will seem low. If the low-cost options are assumed to be in the baseline, because emitters are supposed to seek them out to meet existing environmental requirements, a different picture will emerge.

How Should the Analysis Address Risk and Uncertainty?

Uncertainty and risk are ubiquitous. CBA seeks to identify outcomes with maximal net benefits. Distributional issues are discussed below. To implement this evaluation measure under uncertainty, it is necessary for the economic analyst to get as much information as is reasonably possible about the probability distributions underlying important components of benefit and cost e.g., reductions in health threats, improvements in resource productivity, costs of compliance, and potential for innovation. Of particular importance is information about central tendencies (mean or median) of benefits and costs for the population as a whole, so that aggregate expected net benefits can be evaluated. Information about the variability of possible outcomes also is relevant for judging the "certainty equivalents" – the adjusted net benefits after taking into account the burden of risk-bearing on risk-averse parties. If the scientific assessment of risk provides only information on upper bounds of hazards, the analysis will either overstate the net benefits to the general population or have relevance only to the tail of the distribution whose experience is represented by such outcomes. A similar points applies to worst-case characterizations of compliance cost.

Scientific and economic information about risks and uncertainties often provides little insight into the nature of the probability distribution, and even under more favourable circumstances the distribution might be measured with considerable error. These points underscore the need for considering the sensitivity of the assessment to changes in key assumptions, including the properties of risks and the estimates of costs. Risk assessments and evaluation also need to be updated as the state of science (e.g., the assessment of health risks associated with different air pollutants) grows. Information about the variance of benefits and costs also is important, both to indicate the level of analytical uncertainty and because people can be presumed to be concerned about uncertainty themselves. The Monte Carlo analysis described is an example of an appropriate treatment of uncertainty.

Aside from information about the probability distribution of aggregate costs and benefits, one report puts additional emphasis on "distributional" - i.e., equity – issues, including effects on especially sensitive subpopulations or other groups (such as the historically disadvantaged). Subgroup benefits estimates might be more scant or uncertain than estimates for the overall population. Nevertheless, the analyst should seek to provide information about any important distribution effects. Explicit consideration of these effects is preferred to assumed conservatism in the risk assessment to protect the most sensitive individuals.

What Costs and Benefits Should Be Considered?

The economics paradigm seeks the widest practicable range of benefits and costs for inclusion. In all cases, however, it is necessary to ensure that the costs included are truly opportunity costs rather than just transfer payments. That poses analytic challenges when there are market distortions (the exercise of market power or distorting tax and subsidy provisions). Similarly, the focus in benefits assessment should in principle be on aspects for which there is Willingness to pay. Where willingness to pay is highly uncertain (e.g., in assessing the social value of some ecosystem services), other information needs to be provided.

A particularly nettlesome issue in CBA is the treatment of impacts on employment and local economic activity. Many analyses of regulations have purported to calculate how many jobs a regulation "creates". Closer scrutiny of those studies reveals that they often treat the supply of labour services as unlimited, so increases in employment as a consequence of a regulation have no opportunity cost in terms of diversion of productive activity from other parts of the economy. Economists generally view the job-creating potential of regulation with considerable skepticism, although a regulation can increase the use of unemployed or underemployed workers. In the absence of evidence that such conditions prevail, employment effects should be note but not counted as net benefits of regulation.

A related issue is the potential effects of regulation on technical innovation, including that embodied in increased worker productivity. Generally, analyses of the future consequences of regulation should attempt to account for potential changes in technology and productivity. Such changes could include regulation-caused innovation, slowing of technical progress, and deflection to technical progress into regulatory response in other parts of the economy. Given the uncertainties surrounding technical change, assumptions about alternative scenarios and their likelihood need to be clearly specified.

How Should Key Market Failures (or Nonmarket Goods) Be Assessed?

Different methods will be more or less applicable under different circumstances. A key general point in the OMB guidelines is the need for the use of methods that reflect the current state of the art and of consistency in the use of analytic tools undergoing considerable continuing refinement, such as stated-preference methods for valuation, that imposes a burden on agencies to maintain consistency with the state of the art. In other cases, consistency is needed in the valuation method itself, such as the valuation measures for assessing the benefits of reduced mortality risk.

In some cases, economic measures will be limited or even nonexistent for some important effect of regulation. A complete assessment should include a description (and quantification in noneconomic units to the extent possible) of such effects. Information on the incidence of benefits and costs should also be provided, to the extent possible.

How Should Benefits and Costs Be Compared Over Time?

The economic approach to this issue is discounting, in which future benefits and costs are deflated to reflect the basic idea that people are impatient and prefer benefits sooner rather than later (and mind later costs less than current costs). Modern economic theory points to a preferred approach, the shadow-price-of-capital model (Lind, 1990). This approach can be used to assess how either government outlays or mandated private expenditures displace private consumption or investment and, for the investment that is displaced, to assess the ultimate cost to the economy in reduced future consumption possibilities. With all regulatory impacts converted to consumption equivalents, analysts can discount streams of benefits and cost at a rate that reflects intertemporal consumption tradeoffs. The appropriate rate of such tradeoffs in the United States is generally taken to be around a 3% real, riskless rate. That rate is lower than the official OMB rate for the opportunity cost of capita, currently set at 7%. Note that the rates measure different things (the opportunity cost of capital and the consumer rate of time preference generally will differ, given taxation of corporate income and potential limitations on risk markets, including the difficulties in hedging inflation risks).

Whatever the theoretical appeal of the shadow-price-of-capital approach, it must be admitted that there is little practical experience with it and that it can be very sensitive to the choice of means for calculating displaced consumption associated with reduce investment. The OMB rate reflects the overall average return to capital in the economy (including a risk premium for undiversifiable risk). That rate is easier to apply than the shadow-price-of-capital approach and might be a reasonable approximation if regulations primarily affect domestic investment rates. This discussion illustrates the need for sensitivity analysis to gauge how benefits and costs change with alternative discount-rate hypotheses and for careful assessment of the extent to which the burdens of regulation fall on consumption or investment.

How Should Agencies Gauge the Appropriate Level of Effort for Assessing the Economic Effects of Regulation?

Requirements for full assessments apply only to significant regulatory actions, in particular those expected by the principal decision-makers to have impacts on the economy in excess of @100 million per year. Within that limit, it is still possible to scale the depth of analysis to the need for it. When benefits are compelling and costs seem limited, less analysis might be warranted than when the reverse is true. Even when the stakes of good decision-making are high, gross uncertainties can limit the scope of useful analysis (a fact that should also affect decisions). Statutory limits also constrain the set of regulatory options (although some analysis of the opportunity costs of these constraints would be valuable, as the OMB guidelines note).

In striking the necessary balance, regulatory guidance has tended to emphasize the establishment of general concepts and principles, rather than specific quantitative rules. For example, there has not been consistent numerical guidance for the valuation of mortality-risk reduction. But the establishment of clearer numerical guidelines runs the risk of over prescribing valuations that are not comparable with a situation under consideration (e.g., the valuation of avoiding cancer death risk through a measure that reflects Willingness to pay for lower risk of an occupational-accident death). Ultimately, the process works only if there is a commitment from agency heads on down to follow the principles and to justify principal valuation assumptions, including their consistency with existing empirical information; an equally strong commitment on the part of all concerned to follow good scientific principles in issue farming, analysis, and review, and OMB is able to enforce quality control.

Interpretation of Statutory Requirements

Agencies often interpret basic statutory requirements for protecting the environment, health, and safety very differently. They also interpret constraints on the use of economic information is setting regulatory requirements differently. That is at least partly due to the vagueness of statutory language, difference in interpretation of judicial decisions, and differences in regulatory philosophy. One consequence is substantial differences in the thoroughness and quality of regulatory assessments.

Although basic uncertainties and disagreements about statutory requirements must be addressed at the policy level or through judicial actions, differences in agency philosophy or interpretations regarding the importance of CBA can and should be addressed as part of the implementation of the regulatory review. The new OMB guidelines lay out a set of requirements that, although flexible enough to be tailored to individual circumstances, provide a norm against which all assessments can be judged. Moreover, even where statutory language limits the scope of use of CBA information in rulemaking (explicitly or through judicial interpretations), some assessment adequate to indicate the benefits and costs should accompany any proposed significant regulatory action. That information is important for informing legislators, policy-makers, and the general public about the consequences of regulation, for building support for regulation that serves the public interest, and for highlighting opportunities for legislative or regulatory improvements. These issues are closely linked to the broader question of what information has standing in the regulatory process, which we discussed.

Damage versus Benefit

A reasonable conjecture is that people would be willing to pay more to avoid worsening conditions than to obtain improved conditions. To the extent that, that is true, reliance on a damage estimate when a benefit estimate is called for leads to an overestimate of Willingness to pay and *vice versa.* It is important to use the appropriate type of estimate.

The distinction between damage and benefits does not affect only the valuation step. The presence of thresholds or important, continuous nonlinearities in does-response functions could also lead to errors in transferring a benefit study to a damage context. Damage to lakes as a results of acid rain is a good example. Lakes have thresholds beyond which input of acid surpasses the buffering capacity. Estimating damages from an increment of acidity in a group of lakes, of which some have passed this threshold, some are close to it, and some have much buffering capacity left is difficult in the best of circumstances. Bur reversing a sign on the results of a study of the benefits of reducing acid deposition to the same group of lakes could not possibly provide a reasonable damage estimate.

Marginal versus Average Damages

The appropriate estimate of damages is one taken at the margin. In the context of the debate over "adders" to value spillovers from electricity generations, for example, that means the value of impacts with respect to an addition to existing generation capacity. The literature of damages and benefits rarely in casts in marginal terms; rather one usually sees estimates of the total damage from a pollutant divided by the number of tons of that pollutant emitted to obtain an estimate of damages per ton of pollutant. Where all components of the impact pathway are linearly related (i.e., from emissions to valuation), marginal damage and average damage are equal, and this approach is acceptable. But this condition is exceedingly stringent, inasmuch a few pollutant dispersion models are linear and many does-response functions and some valuation functions in the literature are nonlinear.

One of the most important uses of the travel cost method is that it can be used to value recreational sites. Because many of these sites, such as public beaches, forest, parks etc., are usually free (or very low cost to enter), there is no obvious price information available to determine the value of the site. However, economists have adopted the use of indirect measures, such as the cost of traveling to the site, as a proxy for the price of obtaining a visit to the site.

The travel cost model therefore suggests that there is a relationship between the number of trips that each person will take to a recreational site and the distance they have to travel to get there :

$$\text{Trips} = F(\text{Distance})$$

Individuals who live close have low price of obtaining the resource, and therefore are more likely to demand more trips. Individuals living further away have a higher price and will take fewer trips to a site. The relationship between trips and distance defines a demand curve for trips, with travel costs as the measure of price.

When the number of trips taken is 'T', the travel costs for each trip are Willingness to pay. Over all trips, the total travel cost is 'T' times Willingness to pay or the area noted as travel costs.

For this same number pf trips, the value of the recreation site to a visitor who travels there 'T' times per year is the area of consumer surplus, or the area underneath the Willingness to pay, or demand, function.

Factors that Influence Site Visitation

Environmental Quality

Environmental quality can encompass many different aspects of a site, ranging from scenic beauty to infrastructural development like parking lots, hiking paths, or bathrooms. These different factors can influence the demand function for sites. A more precise way to define the travel cost demand function would account for differences in environmental quality across sites:

Trips = F (Travel cost, Environmental quality),

Unique Features

These are more specific form of environmental quality. The Geysers at Yellowstone National Park, or Half-dome at Yosemite are two unique features that attract individuals from very long distances to visit these parks. Unique features may alter the shape of the demand function, making parks with the feature more valuable than other similar parks without the feature.

The essential point about unique features and environmental quality is that they determine the amount of substitution that exists with particular parks. Parks like Yellowstone and Yosemite have few or no substitutes. However, local parks or recreational areas have many substitutes, unless they have particular environmental amenities that are desired by society. The travel cost model will reflect this.

The difference between the two types of parks is quite distinct. Unique Parks will attract visitors from long distances, who take few trips there every year (or may be only 1 trip in a lifetime). The demand curve for unique parks crosses the Y axis at a high point, but slopes downward very quickly to reflect this. Such a park has few substitutes. The local park, however, does not attract people from long distances to it, so it never achieves a high willingness to pay. However, visitors to this park may make many trips to it simply because it is close and it has some amenities that they desire. The exact specification of the travel cost demand function will depend on the park chosen for analysis and the sapling techniques used to estimate the demand function.

Other Activities

With many sites, one must be careful to assess only the value directly attribute to the use of the site. For example, individuals visiting beaches in Florida may be on a family vacation, and they may have other things that they plan to do in Florida, like visit family, visit Disney World, or visit the Everglades. Therefore, we cannot assign the entire cost of the trip to Florida to beaches alone. This can affect trips to unique or local sites equally, and must be considered carefully in travel costs studies.

Uncertainty

Benefits cost analyses are fraught with uncertainties. The list below includes some of them.

- The nature, strength and timing of the hazard.
- The relationship between the hazard and the mitigation
- The outcomes and effectiveness of mitigation.
- The technical, economic, and social environment of the future.
- Whether impacts are benefits or costs.
- The future value of presently known benefits and costs.
- The length of the project's effectiveness.

Uncertainties arise from limitations of data, or our understanding of the relationships between the natural environment, technology and human behaviour, or failure to model all relevant relationships in the BCA calculations. Good practice requires that the analyst identify as many sources of uncertainty as possible and an attempt made to account for them rather than convey the impression that all benefit and cost values are fixed and guaranteed. The simplest way of including uncertainty into the calculation of net present value is to use the expected value of uncertain impacts. This requires that the analyst knows all possible values for the benefit or cost, and specifies the associated probability distribution over those values. More often than not we know little about these two components of the expected value calculation, so we must settle for a best estimate of the mean, or median value. If the analyst can specify the probability distribution of net benefits is should be presented in the BCA to provide not only the mean, or average net benefit, but also the range and variance of net benefits. While decision makers have little experience dealing with analyses that provide more than one single-valued conclusion, providing some statistical measure of the variability of the calculation to uncertainty is at least honest and consistent with the data and the modeling.

According to the OMB guidelines for performing BCA, it is inappropriate to use variations in the discount rate to adjust the calculation for particular projects risk. But many analyses can capture uncertainty by adding a factor to the discount rate to compensate for the added risk associated with uncertainty (Zerebe and Dively 1994 p328). According to standard portfolio investments analyses where variation in return is used to measure risk, a premium could be added to the discount rate to account for the uncertainty in benefits or costs be added to the discount rate to account for the uncertainty in benefits or costs increasing the variance of the estimate of net benefits. If the uncertainties arise from benefits and costs that fall on individuals, these risks should be discounted at a private rate that individuals would choose. Alternatively, to the extent that public projects are a way of spreading the risks across the population through the funding mechanism (taxes usually) government investment decisions should be discounted at a rate that ignores uncertainty (Graham 1981; Arrow and Lind, 1970, reprinted 1994, p.163). Unfortunately, no definitive prescription for handling uncertainty in BCA has emerged.

Sensitivity analysis can provide the analyst with a means of communicating the uncertainties in the analysis. It also allows the analyst to indicate the effect of the assumption made regarding the data, the relationship between elements of the project under review, and any modeling used to obtain values. Zerbe and Dively (1994, p.372) list two general formats for sensitivity analysis : the variable-by-variable approach and the scenario approach. In the first approach, alternative values for certain benefits and costs are inserted in the calculations and the new values of net benefits recorded. The analyst does this for each value, or category of impacts, with high and low values the most common choices for alternatives. Rather than indicate the sensitivity of the net benefit measures in such a piece-metal fashion, the second method creates scenarios of values for variables together to produce something like a "best-case" and "worst-case" scenario. Sensitivity analysis can also show how the net benefit measure varies with the scale and scope of the project.

When using sensitivity analysis to reveal the effect of assumptions, the analyst should pay most attention to assumptions regarding the models used to estimate benefits and costs. In the case of hazard mitigation projects, the technical and physical links between the hazard, the mitigation action, and the natural and built landscape are key elements in determining the effectiveness of mitigation. Often the analyst will employ sophisticated statistical methods such as Monte Carlo to generate the empirical distribution of estimates. Since value estimates and net benefit calculations rest on many parameters not directly measured or observed for the particular project, such as the discount rate, the inflation area if real values are calculated, the duration of the project and any terminal net benefits, these become the primary candidates for sensitivity analysis. Because a sensitivity analysis can produce a large number of different net benefit values, often the best way to present the findings is to produce interval estimates of important values as a function of the values of selected parameter (Boardman, et al 2001 p.171).

Other Special Considerations for Hazard Mitigation BCA

Doing benefit cost analysis for a particular hazard mitigation project properly means comparing the well-being of citizens in two states or the world: after a disaster with mitigation and after a disaster without mitigation. As mentioned earlier, because the comparison involves a counter factual, such an analysis cannot be performed based on observation alone. This problem leads to the more frequent approach of comparing a before-mitigation world to an after-mitigation state. However, the comparison likely includes the impacts of more than just the mitigation project since many things other than mitigation activity could have, and probably did, change between the two states. The challenge to BCA for hazard mitigation is to creatively describe and hence measure the world in the two properly comparable states–disaster with mitigation, and disaster without mitigation. Analysts can combine data from past disasters and mitigation projects with models of new mitigation projects and disasters as they occur to create more realistic frameworks with which to identify impacts and value them. Doing BCA for hazard mitigation requires special cooperation between the people "on the ground"–emergency response teams, risk management people, local bureaucrats, researchers such

Selection Criteria for Valuation Techniques

There are four options for expressing biodiversity damage, which are discussed. Below these options are summarized in the order from the option with the least resource requirement to the option with the most resource requirement. The first option of 'scoring' is possibly the easiest, option requiring the least time and resources. However, it is sufficient only for cost-effectiveness analysis (CEA), even though due to difficulties with estimating the monetary expression of damage they can also be used for cost-benefit analysis (CBA). The time and resource requirements of the options increase as we go down the list but so does the coverage of impacts and suitability of the option to the current context. Therefore, the choice between these options depends largely on balancing the time and resource requirements of an option with its usefulness in expressing biodiversity damage and choosing the restoration options. Some criteria that can help with this choice are presented after the options.

Available Valuation Techniques

Scoring and weighting generally uses expert judgment to identify the scale of the different aspects of biodiversity damage and the effectiveness of different restoration options in achieving the predetermined restoration targets. The scales can also be informed by the stakeholders to identify the relative importance of the aspects of use and non-use values. The stakeholder participation can be through the implementation of a choice modeling survey without the use of money as the unit of measure. Clearly, the scales are specific to the environmental issue, thus direct comparisons of environmental significance between different issues cannot be made. Such comparison would require weights to be assigned based on local characteristics.

This is the most practical option since it generally only requires expert opinion to be collated based on their experience and the results of the damage assessment process. Thus, it can potentially be used for assessing all aspects of biodiversity damage.

Its applicability is limited to being an input to CEA (e.g. score per Euro of restoration cost) of primary and compensatory restoration options and the service-to-service approach for determining the scale of the compensatory restoration options. Since it cannot express biodiversity damage in monetary units, it should not be used in CBA, even though limitations in monetary damage estimates may necessitate its use in CBA.

Benefits transfer is the process of borrowing the Willingness to pay/Willingness to accept estimates from one study (study site) and applying it to the context in hand (policy site). The transfer usually involves some adjustment, which is generally based on the income differences between the study site and policy site.

The attraction of benefits transfer is clear it can generate monetary expressions of biodiversity damage at a fraction of time and resources requirements of undertaking an original valuation study. All that is required for benefits transfer is a thorough literature review, which is becoming easier with the provision of on-line databases and an understanding of the adjustment processes. On the other hand, there can be significant limitations to benefits transfer. First of all, it is unlikely that there is currently sufficient literature covering all aspects of biodiversity damage. Secondly, some evidence in the literature show that using estimates from other studies can lead to results that are significantly different from those achieved if the estimates were obtained from site-specific valuation studies. This difference is larger than more site-specific the damage is and the more unique the resource is.

Since it can generate monetary expressions of biodiversity damage, benefits transfer can be an input to CBA. However, it is likely that not all impacts can be covered by benefits transfer and that the remaining has to be expressed in monetary terms using revealed or stated preference techniques or in non-monetary terms using scoring and weighting techniques. Moreover, due to the limitation mentioned above, its use is recommended to be limited to small impacts on non-unique resources. It is also useful in that it generates ballpark figures for the magnitude of the different aspects of biodiversity damage. For example, in the case of damage to a wetland, if the valuation literature can tell us which functions of wetlands are valued more, we can make an informed judgment about the severity of the damage based on which functions are affected most. This can inform both the design of the restoration options (especially of the compensatory restoration) and the design of the monetary valuation exercise, if it is to take place. Brouwer et al (1999) present just such a summary of the valuation literature about wetland.

Revealed preference techniques analysis consumers behaviour in actual market for goods and services, which have environmental attributes, in order to derive estimates of Willingness to pay for the environmental attribute. For hedonic property pricing, the marketed goods is private property, the price of which (and hence the demand for it) is affected by environmental attributes such as landscape surrounding the property, peace and quite etc. For travel cost and random utility models, the marketed goods is the recreational site. The cost of traveling to and from a site and the money spent while at the site are indications of people's preferences for visiting that site based on its attributes, some of which are environmental. It is likely that for most biodiversity damage cases, travel cost method and random utility models will be more relevant than hedonic property pricing.

If the necessary data are available for these techniques to be implemented, the analysis takes a relatively short time. However, if such data need to be collected, it could take up to a year, which is as long as a full-scale stated preferences study. In fact, in the case of hedonic pricing technique, which requires panel data (time series and cross-sectional), such data would not be possible to collect within the time frame of a damage assessment exercise. There are two other limitations to revealed preference techniques. Firstly, only those resources or habitats that are used for recreation, or possibly have an effect on property prices, can be covered. Secondly, only use value attached to a resource can be measured since the data analysed are collected from users behaviour alone.

Since revealed preference techniques can generate monetary expressions of biodiversity damage, they can be an input to CBA. However, they are relevant for impacts on landscape and biodiversity especially when such impacts are experienced in areas used for recreation. Since a revealed preference study can only assess use value, if non-use values are thought to be affected, then revealed preference results need to be supplemented by a stated preference study. Consideration of non-use values is likely to be relevant where any of the following apply: (i) the magnitude of the damage is high, or irreversible: (ii) the damaged resource is considered to be of critical importance, or (iii) compensatory restoration with a resource of the same type, same quality and comparable value is infeasible. Where any of these critical apply, especially in combination, it will be difficult to be reasonably sure that restoration really does provide an appropriate level of compensation without a thorough investigation of preference, including non-use values. Stated preference techniques such as contingent valuation or choice modeling would provide the most accurate information for this purpose.

Once a stated preference study is needed, however, it would be more cost-effective to use stated preference techniques alone.

Stated preference techniques make use of carefully structured questionnaire to elicit people's preferences for or against an environmental change like damage to biodiversity. Surveys generate data on people's opinions and attitudes as well as their Willingness to pay/Willingness to accept. The Willingness to pay/Willingness to accept questions provide respondents with information about the change of concern (biodiversity damage), restoration options (if available), institutional context in which the restoration will take place and mechanisms through which people can make or demand a payment. Contingent valuation version of stated preference techniques concentrates on people's preferences for the whole of an biodiversity resources and asks Willingness to pay/ Willingness to accept questions directly. Choice modeling version, on the other hand, concentrates on individual attributes of a resource and asks respondents to choose between or rank the scenarios they are presented with, where the scenarios show the different levels of different attributes of the same environmental resource. One of the attributes is usually the cost of providing the scenario, such as the cost of the restoration option.

If the whole of the environmental resource is damaged, contingent valuation is the more appropriate method, while if only some of the attributes of a resource are damaged (e.g. only some of the functions of the wetland), then choice modeling would be more appropriate. Moreover, choice modeling can be used to derive scores or weights if the choice questions are asked without a monetary attribute.

Finally, both versions of the stated preference techniques have the advantage of being able to estimate non-use values associated with environmental resources, which can be significant especially if the damaged resource is unique (either locally, nationally or globally). The main drawback of stated preference techniques is the time and resource required for their implementation. Other drawbacks such as potential biases can largely be dealt by careful design. While valuation estimates obtained through stated preference techniques may depend on whether the exercise is conducted ex ante or ex post, this should not be an issue of concern. In some cases, the latter is known to produce lower estimates than the former, once the damage is observed to be not as bas as expected. However, this is largely due to differences in information and knowledge at the time the surveys take place. There is no rule that this will always be the case. In the current context, since valuation studies for estimating the relevant compensation amount are likely to take place after the damage occurs. It is possible that, with poor design, survey design and appropriate piloting, it should be possible to obtain accurate results. For example, use of a Willingness to pay question (rather than Willingness to accept) may be found to be more suitable, or indeed the use of choice modeling techniques which elicit preferences for trade-offs between resources directly without the use of monetary measures.

Since stated preference techniques can generate monetary expressions of biodiversity damage, they can be an input to CBA. In fact, the extensive coverage of the techniques both in terms of different aspects of damage and in terms of both use and non-use values, they are the only option that a lead (at least close) to a full monetary implementation of CBA.

Before going into the discussion about the criteria for selecting between these options, it is crucial to note that any one of the above options can be only as good as the information provided by damage assessment. If damage assessment cannot provide information, which are in an easily understandable format, it will be difficult for the experts to make the links between damage and the loss of economic value regardless of which of the above options is chosen.

Choice of Valuation Techniques

The following are the criteria recommended for choosing between the above options that can generate monetary expressions of biodiversity damage. Since scores and weights are much more straightforward in their application and do not generate monetary estimates, they are not included in these criteria. The skills and resources of the analysis undertaking the damage assessment are not included as a criteria since it is assumed that the assessment should be undertaken by an interdisciplinary team consisting at least of experts in ecology and related technical issues, (environmental) economics, lawyers and where necessary social consultation experts. Although there is a growing guidance literature about undertaking economic valuation and appraisal ranging from text chapters to official guidance documents, to an non-expert, they can only provide guidance in assessing the quality of the process and the results not is actual implementation.

Likely magnitude of the damage: The more severe the magnitude of the damage to a natural resource, the more important it is that the valuation of damage is carried out thoroughly to ensure full compensation. Moreover, in the case of severe damage the required assessment is likely to be more complex, and it may well be the case that non-use values are affected. Benefits transfer techniques are unlikely to be appropriate in this case, as they are unlikely to be able to provide estimates which capture values for the full range of impacts, nor provide sufficient information regarding the relevant population who might hold non-use values for the damaged resource. Original studies, in particular stated preference techniques, are therefore likely to be the most appropriate.

Critical importance of the environmental resource impacted, the significance of the impact and the type of value to be measured the more important the resource and the more significant the impact, the greater the need for as comprehensive an analysis as possible. Benefits transfer is limited by the limited coverage of all aspects of biodiversity damage in the literature and its insensitivity to site-specific characteristics. Revealed preference techniques are limited to the use values associated with those resources that are reflected in actual markets. The scope of stated preference techniques is the largest in this context given that potentially all impacts can be covered and both use and non-use values can be estimated.

Feasibility of compensatory restoration with resources of the same type, same quality and of comparable value. The less similar and the more distant the resources identified for compensatory restoration, the harder it will be to be reasonably sure that restoration really does provide an appropriate level of compensation without conducting valuation. This criterion should be considered in conjunction with the two criteria above, namely the magnitude of damage and the critical importance of the damaged resource. Where the magnitude of damage is relatively minor, and the resource is not of critical importance, benefits transfer techniques may provide acceptable estimates of the relatively severe and the resource concerned is unique or of critical importance, there may be a strong case for a thorough investigation of preferences to provide some assurance that the scale of restoration is appropriate to provide full compensation. Stated preference techniques such as contingent valuation or choice modeling are likely to provide the most accurate information for this purpose.

Applicability: the purpose for which the above options are implemented for determine which option should be chosen. If service-to-service approach is sufficient for the design of the compensatory restoration option and CEA is sufficient for the choice of the restoration option, then scoring/weighting system are sufficient. As discussed above, choice modeling techniques can be used to elicit the public's view of the relative importance of the different aspects of damage and restoration. However, these do not need to include monetary expressions of damage. If the value of damage needs to be measured for monetary compensation or value-to-value scaling of compensatory measures, or CBA is needed to choose the restoration option, them monetary expressions of biodiversity damage (or benefit of restoration) would be necessary. If there is sufficient literature available, benefits transfer could be used for small damages and interim losses. If not, original valuation exercise would be required.

Time and data available for analysis availability of data about the physical measures of environmental impacts is a concern for all valuation options. The availability of economic valuation data is typically not a concern for stated preferences studies since the necessary data are generated by the study itself. However, for this reason, stated preference studies can take between six months to a year depending on the complexity of the resources and damage to it. Therefore, the decision to implement stated preference applications are completed. Typically, damage assessment and compensation negotiations proceed on parallel tracks, so that information generated during the damage assessment may play a role in negotiations, and settlements may be reached before the damage assessment is completed. In the Exxon Valdez and Blackbird Mine cases, for example, settlements were reached after completion of stated preference pilot studies.

Provided that the necessary data are available, reveled preferences studies can take about six months to implement. But if such data do not exist, a revealed preferences study can take as long as a stated preference one. Data for travel cost method and random-utility models can be collected using surveys. However, since such surveys are likely to take similar time and effort as stated preference surveys, it would be more efficient to design a questionnaire that would enable the application of stated preference and travel cost and or random utility models simultaneously. On the other hand, in the absence of data, hedonic pricing is unlikely to be appropriate since the necessary (especially time series) data cannot be collected within the time scale of a damage assessment. Benefits transfer can be applied in a matter of weeks provided that the appropriate literature exists.

The cost of the valuation exercise depends on the complexity of the damage and restoration options which affect the complexity of the questionnaire design, the size of the sample and the complexity of the data analysis. Valuation of damage is relatively inexpensive for small injuries. Various versions of benefits transfer can successfully be used for small incidents. Benefits transfer is relatively inexpensive given that there is little need for data collection and significant less time input requirement. Stated preference studies can cost between 50,000 and 200,000 (although some studies have cost millions of dollars in the U.S.A.), while a comprehensive revealed preference study could also cost towards the higher end of this range if necessary data are not readily available and towards the lower end of this range (or may be even lower) if such data are ready. However, the crucial criterion is not the absolute cost of a valuation exercise but its incremental cost in terms of additional information it provides and the increased accuracy and reliability of the results produced at the end of the assessment process. The cost of the valuation exercise should also be seen in the light of the cost of the total restoration activity. Even the most expensive option of a stated preference study is a small amount relative to the total cost or restoration option that can cost millions of pounds. It is usually more cost-effective to spend more at the beginning of an assessment process than to spend to mitigate the outcome of a wrong decision afterwards.

Whether the results of a valuation exercise are legally defensible: this depends on how strongly a valuation approach is grounded in theory and how well it is implemented in the particular study of concern. For example, benefits transfer is largely based on assumptions about the suitability of Willingness to pay/Willingness to accept estimated borrowed. Further, assumptions need to be made regarding the boundaries of the population affected. Such assumptions can be easy to contest and difficult to defend. Again, scoring and weighting techniques are generally based on expert opinion, which can be contested. For stated preference studies, approaches such as pairwise comparisons etc. that are not founded in economic theory, should be avoided. Moreover, the design specifications and assumptions behind econometric analysis of the responses to the stated preferences surveys should be clearly stated and defensible.

Timing of the valuation exercise: it can be difficult for a state preference study to elicit people's preferences for the environmental impacts of concern and distinguish these from strategic choices once the incident causing damage becomes a political issue. However, this is an empirical issue and one which can be dealt with adjusting the questionnaire design. For example, the contingent valuation study undertaken for the Exxon Valdez oil spill asks the respondents how much they are willing to pay to avoid a similar incident in the future rather than asking for Willingness to pay for restoration or Willingness to accept for damage (Carson et al, 1992). This puts the valuation question in a more neutral to answer truthfully. This can be achieved if people care about the damage and believe that their responses will be taken care.

The hedonic pricing method is used to estimate economic values for ecosystem or environmental services that directly affected market prices. It is most commonly applied to variations in housing prices that reflect the value of local environmental attributes.

It can be used to estimate economic benefits or costs associated with :

- Environmental quality, including air pollution, water pollution, or noise
- Environmental amenities, such as aesthetic views or proximity to recreational sites.

The basic premises of the hedonic pricing method is that the price of a marketed good is related to its characteristics, or the services it provides. For example, the price of a car reflects the characteristics of that car – transportation, comfort, style, luxury, fuel economy, etc. Therefore, we can value the individual characteristics of a car or other good by looking at how the price people are willing to pay for it changes when the characteristics change. The hedonic pricing method is most often used to value environmental amenities that affected the price of residential properties.

This section continues with an example application of the hedonic pricing method, followed by a more complete technical description of the method and its advantage and limitations.

Hypothetical Situation

Agency staff wants to measure the benefit of an open space preservation program in a region where open land is rapidly being developed.

Hedonic Methods (Indirect)

Why do we purchase a particular shirt, car, or house? There are many different shirts available for we to purchase, however when we are shopping for a shirt, we have to settle for one or two that attract we the most. What qualities does that shirt have that attracts we to it : the colour, the pattern, the materials, the buttons, the collar, the logo on the front, the brand-name, the fact that it was made in America? There are many different components to shirts, and we all have certain aspects that are most important to us.

The same thing is true of automobiles and houses. The fact is that most goods we purchase have many different characteristics. When we purchase goods, we are most concerned about purchasing a mix of characteristics rather than the good itself. Economists have developed methods that allow us to separate the values for the characteristics from value of the entire good. For instance, we can determine the value an extra bedroom provides to a house that already has 3 bedrooms and a mix of other rooms. The methods we use to estimate the values for characteristics are called hedonic methods.

Hedonic methods are used in a variety of ways in environmental valuation. For instance, if we consider the same exact house design, but put one house near the beach and one house in a residential community away from the beach, we are likely to get two different values for the house. By comparing these two houses, we can determine the additional value that having the house next to the beach provides.

Hedonic Housing Studies

In reality, there are many different house designs, and it is unlikely that we will find any two houses that are exactly the same as described above. However, using statistical analysis, we can compare the values of many different houses, with and without the environmental amenity (or externality), to determine the effect of the amenity or externality on house values. By considering many different houses, we can determine the "marginal" effect of the environmental externality. The exact size of the effect depends on the specific region where the study is done.

Hedonic housing studies requires substantial data on house sales and house amenities. In order to separate the effect of the externality from the effects of different house sizes, room numbers, bathrooms, garages, etc., one must consider a large sample and run a regression.

A price function of housing is of the following type: $P_a = f(S,N,E)$

where:

a = unit of housing

S = structural characteristics

N = neighbourhood characteristics

E = environmental characteristics

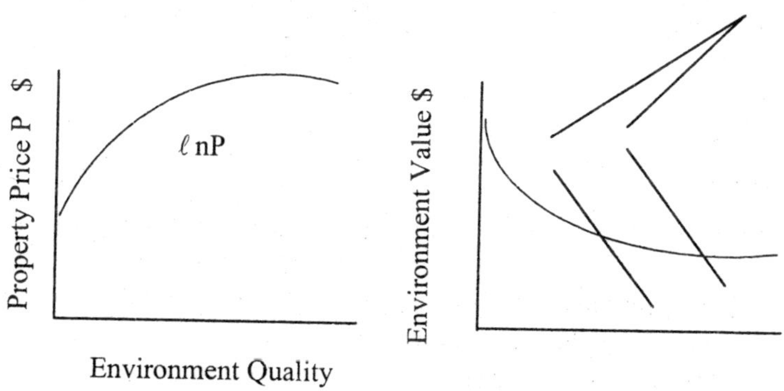

Households Marginal WTP
(increase demand) for
Ea and b are observed price paid for
marginal increments for E by household a and b

The equation can be specified as a logarithmic form :

$$\text{In Pa} = \text{a InSa} + \text{b InNa} + \text{c InEa}$$

Where: c tells us by how much property prices vary if we alter the values of the environmental variable. Figures below show the relationship between environmental changes and property prices.

Property prices increase with environment improvement at a diminishing marginal utility. Differentiating the equation above with respect to Em gives the implicit marginal purchase price of the environmental good: $f\text{P} / f\text{E} = \text{c} \times \text{P/Em}$. Figure (b) plots the slope of the implicit marginal purchase price of the environmental good or the amount by which property values would increase if landscape (environment) were to be improved assuming that the only explanatory variables is given by the environmental attributes, the hedonic price function would be In Pa = c.

In Em. Using the integral log-log rule : $1/\text{x dx} = \text{In x} + \text{d}$, we can rewrite the function : $f\text{P/P} = \text{c} f\text{E/E}$ or $f\text{P}/f\text{E} = \text{cP/E}$. However, the marginal purchase price function does not necessarily correspond to the household demand curve for E. That is, it is unlikely that it corresponds to the household marginal willingness to pay for E. This non-correspondence can be demonstrated intuitively by considering an individual's decision about whether or not to buy an additional unit of E. At a low level of E the individual / household may be more than willing to pay the marginal purchase price for an additional unit of E (MWTP>implicit marginal price). At high levels of E that same household may not be willing to purchase an additional unit (MWTP < marginal implict purchase price). Yet, there will be an intermediate level of E where the household is just willing to pay for the incremental Em unit (MWTP = marginal implicit price). Figure (b) illustrates the relationship between households WMTP and implicit marginal purchase price for two households.

As shown in figure (b), each household will choose a location where its marginal willingness to pay for one more unit of E equals the marginal implicit price of E. But as it can be observed, this amount is only a single point on the household's inverse demand function for the environmental good. What this demonstrates is that the slope of the estimated *l* nP is a locus of point on the demand curves of many households. Yet, the question is if the curve *f* P/*f* E of figure (b) can be interpreted as an inverse demand function for E. If all individuals were identical, the demand curve for landscape would correspond to *f* P / *f* E insofar as all households would have identical demand functions. But since individuals do not have the same preferences for landscape quality and differ in income, the HPM should be complemented by estimations of how the marginal Willingness to pay varies with income and other individuals characteristics, and should take into account the supply side of housing market. Once the demand function has been estimated, the area under the demand curve between two values of the landscape quality determines the change in consumer surplus caused by the change in "quantity" of this attribute. By aggregating all individuals' CSs we obtain the overall value of landscape improvement.

Let's consider a purely hypothetical example. Suppose we are interested in determining whether or not beaches enhance the value of houses. We would be interested in comparing houses both near and far from a beach. We might hypothesize the following relationship:

House value = $\square_0$ + $\square_1$ (sq.feet) + $\square_2$(bedrooms) + $\square_3$(garage) $\square_4$(neighborhood) + $\square_5$(age of house) + $\square_6$ (distance from beach) + $\square$

The values for $\square_0$,$\square\square_1$,$\square\square_2$,$\square\square_3$,$\square\square_4$,$\square\square_5$,$\square\square_6$$\square$describe how each of the variables influence the total value of the house. A positive value indicates that they increase the value of a house, and a negative value indicates that they reduce the value of a house. An environmental economist may be most interested in how housing values relate to distance from the beach, but without including the first five variables (square feet, bedrooms, garage, neighborhood and the age of the house), the model will not be specified correctly. Including the first five variables allows us to separate the impact of distance from the beach from the other influences we know to exist.

One can estimate the □ parameters by using data from a mix of houses both near the beach and far from the beach. Say we have 500 observations of house sales near and far from a beach, but still within the same housing market. We could run a regression analysis to estimate the □ values. These are prices for the individual characteristics. Suppose we get the following results:

House value = 2000 + 50* (sq.feet) + 1000*(bedrooms) + 5000*(garage) 10000*(neighborhood) – 1500*(age of house) – 1300* (distance from beach)

In this example, each additional square foot in a house add $ 50 to the value of houses. Thus, $50 is the marginal benefit, or price, of an additional square foot of house. All of the attribute variables add value to the house, except the age and distance from the beach. In the case of age, this makes sense (expect in rare cases) because older houses generally have more maintenance problems, and will be worth less than newer houses.

As for the environmental variable, distance from the beach, we find greater distances from the beach have a negative influence on housing values. Thus, houses closer to the beach are valued higher than houses farther from the beach. This results from the positive externality associated with being near the beach. If the distance from the beach is measured in 1/10ths of a mile, each tenth of a mile reduces the value of a house by $1,300.

Note that the hedonic method is an indirect way to determine the impact externalities have on goods commonly purchased in society. Here are two additional examples that allow we to see how these methods work:

Application of the Hedonic Pricing Method

Step 1

The first step is to collect data on residential property sales in the region for a specimen time period (usually one year).

The required data include:

- Selling prices and locations of residential properties
- Property characteristics that affect selling prices, such as lot size, number and size of rooms, and number of bathrooms
- Neighborhood characteristics that affect selling prices, such as property taxes, crime rates, and quality of schools
- Accessibility characteristics that affect prices, such as distances to work and shopping centers, and availability of public transportation
- Environmental characteristics that affect prices.

In this case, the environmental characteristics of concern are the proximity to open space. The researcher might collect data on the amount and type of open space within a given radius of each property, and might also note whether a property is directly adjacent to open space. Often, this type of data may be obtained from computer-based GIS (geographical information systems) maps. Data on housing prices and characteristics are available from municipal offices, multiple listing services, and other sources.

Step 2

Once the data are collected and compiled, the next step is to statistically estimate a function that relates property values to the property characteristics, including the distance to open space. The resulting function measures the portion of the property price that is attributable to each characteristic. Thus, the researcher can estimate the value of preserving open space by looking at how the value of the average home changes when the amount of open space nearby changes.

How do we Use the Results ?

The results can be used to evaluate agency investments in open space preservation. For example, specific parcels may be under consideration for protection. The hedonic value function can be used to determine the benefits of preserving each parcel, which can then be compared to the cost.

Applying the Hedonic Pricing Method Using Housing Prices

In general, the price of a house is related to the characteristics of the house and property itself, the characteristics of the neighborhood and community, and environmental characteristics. Thus, if non-environmental factors are controlled for, then any remaining differences in price can be attributed to differences in environmental quality. For example, if all characteristics of houses and neighborhoods throughout an area were the same, except for the level of air pollution, then houses with better air quality would cost more. This higher price reflects the value of cleaner air to people who purchase houses in the area.

To apply the hedonic pricing method, the following information must be collected :

- A measure or index of the environmental amenity of interest.
- Cross-section and /or time-series data on property values and property and household characteristics for a well-defined market area that includes homes with different levels of environmental quality, or different distances to environmental amenity, such as open space or the coastline.

The data are analyzed using regression analysis, which relates the price of the property to its characteristics and the environmental characteristic(s) of interest. Thus, the effects of different characteristics on price can be estimated. The regression results indicate how much property value will change for a small change in each characteristic, holding other characteristics constant.

The analysis may be complicated by a number of factors. For example, the relationship between price and characteristics of the property may not be linear-prices may increase at an increasing or decreasing rate when characteristics change. In addition, many of the variables are likely to be correlated, so that their values change in similar ways. This can lead to understanding the significance of some variables in the analysis. Thus, different functional forms and model specifications for the analysis must be considered.

- The method's main strength is that it can be used to estimate values based on actual choices.
- Property markets are relatively efficient in responding to information, so can be good indications of value.
- Property records are typically very reliable.
- Data on property sales and characteristics are readily available through many sources, and can be related to other secondary data sources to obtain descriptive variables for the analysis.
- The method is versatile, and can be adapted to consider several possible interactions between market goods and environmental quality.

Issues and Limitations

- The scope of environmental benefits that can be measured is limited to things that are related to housing prices.
- The method will only capture people's willingness to pay for perceived differences in environmental attributes, and their direct consequences. Thus, if people aren't aware of the linkages between the environmental attribute and benefits to them or their property, the value will be reflected in home prices.
- The method assumes that people have the opportunity to select the combination of features they prefer, given their income. However, the housing market may be affected by outside influences, like taxes, interest rates, or other factors.
- The method is relatively complex to implement and interpret, requiring a high degree of statistical expertise.
- The results depend heavily on model specification.
- Large amounts of data must be gathered and manipulated.

– The time and expense to carry out an application depends on the availability and accessibility of data.

The wage differential is similar to HPM in that the benefits of environmental improvement (decreasing mortality and morbidity) to human health can be deduced from variations in wages across different work settings and localities. It is assumed that an individual can make trade-offs between income and health, thus revealing the value of health and the work place. Both the supply side and the demand side of the labour market define a job and the corresponding wage rate (hedonic wage rate) by a bundle of characteristics or attributes such as, age, skill level, education, and so forth, including exposure to environmental or occupational risk. There is a relationship between wage level and environmental/risk characteristics. The analysis of these relationships is called wage risk study.

Methods

Most of the applications of this method have dealt with the valuation of risks of death or injury due to accidents on the job. However, the same approach can be used to value adverse environmental effects on morbidity and mortality.

A wage risk study involves four steps :

- To estimate earning function, W = f(J,S,R) where the wage (W) depends on job (J) related characteristics (occupation, union, working conditions, number of weeks/year, location...), the individual socio-economic characteristics (S) (income, age, education, experience..) and the job risk (R) of accident (mortality or morbidity);
- Collect cross-sectional data ;
- Derive the coefficient on risk or the implicit value of the risk of an accident ?W/?R using multiple regression analysis; and
- Derive the value of the benefits of reducing the risk of damage to human health in terms of probability of death or injuries i.e. for extra 1 in 100,000 homes, the

Value of Life = 100,000x ?W/?R, or if logarithm function is used to VoL = 100,000 x average wage xW/R.

Strong Assumptions for this technique to work are that

- Labour markets functions freely. In developing countries, labour markets are highly imperfect and are characterized by important surplus of labour and uneducated workers who do not have a good perception of job risk. This leads to situations where low wages may coexist with poor environmental conditions;
- Labour is mobile;
- It is possible to isolate the exclusive impact of risk on wages;
- Perfect comparability between different types of risks; and
- Good quality of information on risks.

10

Biodiversity Measurement

Ecosystem-a system formed by the interaction of a *community of organisms* with their *physical environment*. So, natural ecosystems are made up of a biotic factors such as air, water, rocks, energy and biotic factors such as plants, animals, and microorganisms. Areas with similar biotic and abiotic factors form ecosystems, such as grassland, woodland, mires etc. Ecosystem ecology studies the flows of matter like water, nutrients etc. and energy in the overall system composed of biological organisms and abiotic environment.

Community

Community is the biotic component of ecosystem. Community ecology studies the effect of climatic and a biotic conditions, disturbance, biotic interactions on the species assemblage. Components of a community are species composition and species abundance. For a concrete study it is reasonable to make further distinction and specify if diversity of plants, animals, microorganisms, fungi, lichens, etc has been considered. Heywood and Iriondo (2003) noted that plant communities are usually defined by the dominant species which influence and create the main characteristics and structure of the community. Plant communities are in fact basic components of the landscape and their extent and arrangement has consequences both for species survival and for ecosystem processes.

Based on the origin and development of the ecosystems it is important to make the following distinction between ecosystems: there exist natural, semi-natural and artificial ecosystems.

In *natural communities* plant composition and their abundances have developed naturally over long temporal scales through natural succession. Structure and species composition are rather stable in these communities. Example: active raised bogs in North-hemisphere, broadleaved forests in Atlantic Europe.

Semi-natural Communities are plant communities where long-term management grazing or mowing has biased the natural succession of vegetation towards climax ecosystem and where herb layer is the dominating layer of the community. In practical definition semi-natural communities are traditional agricultural landscapes like wooded meadows, alvars, floodplain meadows, coastal meadows, but also other meadows and pastures, where grazing animals and mowing have been the only human activities. In temperate region semi-natural communities represent the very species rich ecosystems, especially those on those calcareous soils in Western and Northern Estonia and Southern Sweden. If the traditional human management of these communities stops, the bushes and trees succeed to establish and the herb dominated community is firstly replaced by brushwood and later by forest.

Artificial communities are purely human created plant assemblages like forest plantations, cultural grasslands, parks, gardens. As these habitats often aim to have economical use such as forestry and timber production, cultivation of grassland for a greater biomass production and for increasing the value as forage, the habitats are usually rather species poor and the interest is concentrated to one or few species. On the other hand, some of the artificial communities like the old parks might significantly contribute to biodiversity presenting habitats for a plenty of species: bats, owls, beetles. However, that kind of importance is often coupled with the high proportion of alien plant species in the habitats e.g. decorative herbs, bushes, trees.

Change in Time–Succession

Plant communities are dynamic systems. Odum (1971): "Succession is an orderly process of community development; it is reasonably directional and, therefore, predictable ... it results from modification of the physical environment by the community; that is, succession is community-controlled. Species replacement occurs because populations tend to modify the physical environment, making conditions favourable for other populations until an equilibrium between biotic and abiotic is achieved." During the succession different ecosystems and their plant communities change and replace each other.

Annual plants are replaced by perennial herbs, later shrubs enter and finally trees form the forest layer. Primary succession occurs when plants become established on land completely un vegetated and soil lacking land. A plant community may be disturbed causing some plants to be destroyed, as from a fire or from human logging or cultivation. If the disturbance stops, the community will begin a secondary succession, changes in the vegetation that will lead to the final stage of succession-a climax community. The climax community is the one with rather stable composition and structure. It depends on the theoretical concept, if semi-natural communities present climax or not. In the conditions of ongoing management grasslands are also rather stable and do not change to forest. Changes in plant communities drive changes in related animal community.

Species Diversity: Measures and Scales

How to describe the biodiversity of communities? When it is needed at all? There might be many reasons why communities should be compared for the matter of diversity. The nature conservational aims are among the most important ones. The diversity needs to be measured somehow, to be able to assess the local sites for the nature conservational purposes. Therefore, some quantitative measures must characterize the biodiversity. Subdividing the concept of biodiversity into various components facilitates its measurement. The species richness of any region or ecosystem is a consequence of two factors: the richness of each of the smaller areas that compose it, and the turnover in species composition among them. For this reason the following types of diversity are used: point, alpha, and gamma-diversity. In addition to these, beta-diversity measures the change in species composition among different communities in a landscape or along an environmental gradient.

Measurement of Biodiversity

Biodiversity is a measure that combines richness and evenness across species. It is often measured because high biodiversity is perceived a synonymous with ecosystem health. In general diverse communities are believed to have increased stability, increased productivity, and resistance to invasion and other disturbances.

Diverse habitats with a variety of plants can have benefits such as:

- Provides forage for a variety of insect and vertebrate species.
- Stability resulting from plants in the community that are able to survive drought, insect plagues, and disease outbreaks so that the site will have some soil protection/forage/etc. in those years.
- Plants contains a variety of genetic material that may be useful in long-term survival and stability of the community.
- The community benefits from a mixture of plants: soils improve with nitrogen fixers, deep rooted plants bring nutrients up from soil layers below other plants roots some species work together so that both can survive called commensalism and therefore, diverse communities can be more stable
- Healthy diverse plant communities generally have all niches filled and are theoretically less likely to be invaded by noxious or opportunistic introduced species.

Species richness is the number of different species in a given area. It is represented in form as Species richness is the fundamental unit in which to assess the homogeneity of an environment. Typically, species richness is used in studies to determine the sensitivity of ecosystems and their resident species. The actual number of species calculated alone is largely an arbitrary number. These studies, therefore, often develop a rubric or measure for valuing the species richness numbers or adopt one from previous studies on similar ecosystems.

Species Richness

This is the simplest of all the measures of subspecies diversity. All we do is count of the number of subspecies found in a community e.g., the number of the subspecies found in the Meadow Fields zone.

However, this does not indicate how the diversity of the population is distributed or organized among those particular subspecies. For example, if there were 4 different subspecies observed in Zone 1 and Zone 2 the richness would be equal. This does not indicate what percentage of the abundance there were of each subspecies. In Zone 1, 80% of the abundance could have been Blue Jay while at Zone 2 there could have been an even 25% of each subspecies.

Factors Affecting Species Richness

There is a strong inverse correlation in many groups between species richness and latitude: the farther from the equator, the fewer species can be found, even when compensating for the reduced surface area in higher latitudes due to the spherical geometry of the earth. Equally, as altitude increases, species richness decreases, indicating an effect of area, available energy, isolation and zonation. It could be noted that intermediate elevations can receive species from higher and lower.

Latitude

Latitudinal Gradient

- The species richness increase from high latitudes to the low latitudes.
- The peak of the species richness is not at Equator. It is deduced that the peak is between 20-30°N. The tropics fall within this range which is about 24.3 degrees north and south and this attests to the fact that species richness and biodiversity is highest here.
- gradient of species richness is asymmetrical about the equator. The level of species richness increase rapidly from the north region but decrease slowly from the equator to southern region.

Area Effect

The latitudinal gradients of the species richness may result from the effect of area. The area at lower latitudes is larger than that at higher latitudes, leading to higher species richness at lower latitudes than normal.

Productivity

The latitudinal gradients of species richness may be result from the energy available to the ecosystems. At lower latitudes, there are higher amounts of energy available because of more solar radiation, more resources for example, minerals and water; as a result, even higher levels of species richness can be allowed at lower latitudes. However, there have been relevant studies showing that species richness and primary productivity are actually negatively correlated.

The Millennium Ecosystem Assessment, an international ecological effort initiated by the United Nations, states: "In most ecosystems, changes in the number of species are the consequences of changes in major abiotic and disturbance factors, so that the ecosystem effects of species richness per se is expected to be both comparatively small and very difficult to isolate. For example, variation in primary productivity depends strongly on temperature and precipitation at the global scale and on soil resources and disturbance regime at the region-to-landscape-to-local scales. Factors that increase productivity, such as nutrient addition, often lead to lower species richness because more productive species outcompete less productive ones. In nature, therefore, high species diversity and high productivity are often not positively correlated."

Area

The species-area relationship is commonly approximated as following equation:

$$S = cA^2 \text{ or } \log(S) = \log(c) + z \log(A)$$

where S is the number of species, reflecting the species richness sometimes also called species diversity, A is the area given in hectares, and c and z are constants. c is the species richness factor, usually between 20 and 2000; z is the species accumulation factor, usually between 0.2 and 0.5. This equation was first described by Arrhenius in 1921 and explains the variation of species richness among different areas.

Sampling Measurements of simple species richness are related to area size, but are also affected by sampling intensity. When relatively small numbers of samples are taken from a given area, the numbers of species identified will be lower than if greater numbers of samples are collected. More samples are collected, the greater the number of species will be found. Thus in some cases, measurements of species richness can be considered a statistical artifact.

Habitat Diversity

It is possible that larger area contain more habitats as it is said that larger area is more topographically and environmentally diverse. Therefore, there are more opportunities for more species to set up their populations due to higher habitat diversity.

Relationship between Endemism and Species Richness

The levels of endemism and that of species richness are frequently positively correlated. However, on some oceanic islands, there are high levels of endemism but the levels of species richness are quite low.

Adjusting the Species Richness

The most common formula for working out Species Diversity is the *Simpson's diversity index,* which uses the following formula:

$$D = \frac{\Sigma n(n-1)}{N(N-1)}$$

Where:

D = diversity index

N = Total number of organisms of all species found

n = number of individuals of a particular species

A high D value suggests a stable and ancient site, while a low D value could suggest a polluted site, recent colonisation or agricultural management. Usually used in studies of vegetation but can also be applied to animals. In order to account for the probability of missing some of the actual total number of species present in any count based on a sample population, the Jackknife estimate may be employed:

$$S = n + \left(\frac{n-1}{n}\right)^h$$

where

S = species richness

n = total number of species present in sample population

k = number of "unique" species of which only one organism was found in sample population

Similarly the equation may also be noted as:

$$S = E + k\left(\frac{n-1}{n}\right)$$

where

E = the summation of number of species in each sample

k = number of rare/unique species

n = number of sample

As well, when looking at local diversity the appropriate formula to use is:

$$S = c\,(A)^2$$

where

c = a specific number for each taxa

A = the area of study

= the slope perimeter

Other measures of biodiversity may also take into account the rarity of the taxa, and the amount of evolutionary novelty they embody.

Weakness

Species richness fails to take into consideration species evenness. Other measures of biodiversity, such as the Simpson index, the Shannon index, and the fundamental biodiversity parameter θ of the unified neutral theory of biodiversity take species evenness into consideration.

More About Measuring Biodiversity

When measuring diversity it is good to remember that what we are trying to describe is the relationship of individuals of varying subspecies within a zone. In our research, we use the number of individuals of each subspecies observed e.g., found in each zone. There are some underlying assumptions that all the measures of biodiversity have in common:

The Categories are Well Known

In most cases, such as with our study, this assumption is not violated since the classification of subspecies that we use is accepted world-wide. However, this may be difficult if everyone does not use the same classification system e.g., some people may put similar individuals into several subspecies categories while others use one category.

All Categories are Equally Different

Categories are equally different from each other category. This is not always true since two subspecies from the same genera are treated the same as two subspecies from different families.

Use a Measure of Species Importance

Usually, one uses the number of individuals, percent coverage, relative density or biomass. The choice usually depends on the ease of measurement.

The Community under Study is Well Defined

The relative importance of an individual category will vary greatly depending on the definition of the extent and makeup of the community.

The most common measures of biodiversity are Species richness, Simpson's index and Shannon-Wiener index.

Simpson's Index

A measure that accounts for both richness and proportion of each species is the Simpson's diversity index. It has been a useful tool to terrestrial and aquatic ecologists for many years and will help us understand the profile of biodiversity across our Zones.

The index, first developed by Simpson in 1949, has been defined three different ways in published ecological research. The first step for all three is to calculate P_i, which is the abundance of a given subspecies in a zone divided by the total number of subspecies observed in that zone.

Simpson's Index: D = sum (P_i^2)

The probability that two randomly selected individuals in the zone belong to the same subspecies.

Simpson's Index of Diversity: 1 - D

The probability that two randomly selected individuals in a zone belongs to different subspecies.

Simpson's Reciprocal Index: 1/D

The number of equally common subspecies that will produce the observed Simpson's index.. D is influenced by two parameters - the equitability of percent of each species present and richness. For a given species richness, D will decrease as the percent of the species becomes more equitable. The researcher must observe the species patterns carefully to interpret the values effectively.

Shannon-Wiener Index

This diversity measure came from information theory and measures the order or disorder observed within a particular system. In ecological studies, this order is characterized by the number of individuals observed for each subspecies in the sample plot. It has also been called the Shannon index and the Shannon-Weaver index.

Similar to the Simpson index, the first step is to calculate P_i for each category subspecies. We then multiply this number by the log of the number. While we may use any base, the natural log is commonly used. The index is computed from the negative sum of these numbers. In other words, the Shannon-Wiener index is defined as:

Using species richness (S) and the Shannon-Wiener index (H), we can also compute a measure of evenness:

Evenness (E) is a measure of how similar the abundances of different species are. When there are similar proportions of all subspecies then evenness is one, but when the abundances are very dissimilar e.g. some rare and some common species, then the value increases.

Best Estimates of Biodiversity Value : Using Genealogy to Predict Genetic or Character Richness

It should be based on principles, phylogenetic and taxonomic measures and lack of phylogenetic information.

Principles

Biologists needing to estimate richness in a particularly valued currency of different genes or characters are usually unable to measure this directly, or at best are only able to study small samples of genes or characters. However, because genes and characters are inherited, biologists have been able to respond by proposing *phylogenetic* or *taxonomic measures* of diversity. These measures predict the biodiversity value of different biotas, using knowledge of the genealogical or hierarchical relationships among organisms in combination with models of gene or character evolution.

To illustrate how classifications may be used to predict diversity value by estimating richness at the level of the genes or characters, the example below shows a classification for some African species of milkweed butterflies (family Nymphalidae). The branching pattern of the classification is derived by analysis of 217 morphological and chemical characters, and the branch lengths are scaled by the number of character changes found within this sample, shown as vertical ticks. Considering all combinations of three species, the most diverse set of three species is *niavius, echeria* and *damocles,* because these three have the longest total branch lengths with the largest numbers of character differences between them is shown in black *below*:

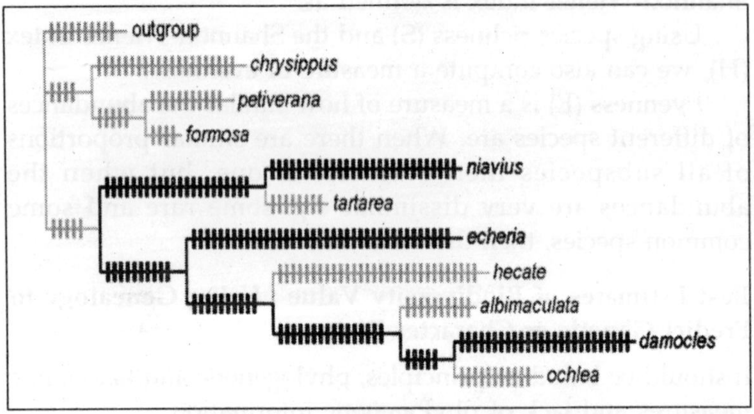

Figure 1

To illustrate how using this approach can effect the relative values of different faunas or floras, consider first one of the most popular measures of diversity, species richness. The example below shows counts of the numbers of species of *sibiricus*-group bumble bees among equal-area grid cells. For these bumble bees, a large representative sample of character-difference data is not yet available, only a genealogical classification. Nonetheless, a simple evolutionary model can still be used to estimate the way changes in characters for these bees are likely to have accumulated, given this classification tree. When combined with a measure of tree length to count the relative expected number of character differences within each fauna, the next example shows an improved prediction of relative diversity value for the same bee faunas, to take into account the expected numbers of gene or character differences:

Phylogenetic and Taxonomic Measures

In the absence of complete knowledge of the valued genetic or character differences among organisms, the most direct approach to estimating relative value is to use phylogenetic measures. This is used to scale the branch lengths on the tree for use by the measure in the third step. It can use sample data for gene or character variation, or it can be based on just a taxonomic classification, if that is all that is available and if it is believed to represent genealogical relationships. These require three components:

Cladograms

The phylogenetic approach is based on the general evolutionary model of descent with modification: that genes and characters are inherited, with rare alterations or changes. Some of these changes are reversals, so that data often conflict with any one tree *homoplasy*, although such conflicts are minimised in the construction of cladograms. If the resulting trees are good estimates of genealogical relationships, then they should be more reliably predictive of the unsampled genetic or character variation, which always remains the great majority.

Models of Gene/Character Evolution

Next, an explicit choice has to be made of a special evolutionary model for linking the branching pattern of the tree with the way genes or characters change along the tree. Three extreme options for this model are envisaged assumes that changes occur at random and are subject to little constraint by selection. Consequently, in effect, changes accumulate more or less in proportion to the time elapsed along the branches assumes that the distribution of changes in a small sample of variation is representative of the great majority of unsampled variation. The pattern of changes is usually expected to be intermediate between those from the other two models assumes that most changes are associated with speciation or divergence events, for example if strong selection constraints are relaxed at these times. Although the numbers of changes associated with each branching event may differ, in effect changes accumulate more or less in proportion to the number of branching events including those to extinct branches.

Measure of Gene/Character richness

Third, a measure of relative richness in different genes or characters is needed, which sums the relative degree of change along the tree using the branch lengths scaled by the chosen special evolutionary model. There is now broad agreement on the form of this measure, as a measure of the length of the subtree that spans any given set of species of interest on the tree. A simple case of this measure is illustrated graphically, for calculating the increase in diversity or complementary diversity when a species C is added to a tree for two species A and B from the pair wise differences between species, AB AC and CB:

For the example of the butterfly cladogram shown at the top of the page, branch lengths can be scaled on the horizontal axis, the vertical components should be ignored using the three evolutionary models described above. For each model, the set of three species that is estimated to have the greatest genetic or character richness is sought. For this tree, only with the sample model is there a unique solution. With the other models, there are at least two equivalent choices for each species, as shown by the vertical tie bars:

Most disagreement is now centred on the choice of special evolutionary model. The sample model is the easiest to use, and its consequences in this example show that it may given less equivocal answers concerning which species represent the greatest relative diversity. However, the apparent advantages of the sample model are only real if the sample truly represents the overall pattern of variation. For example, if a sample of unconstrained genetic data were available that behaved as though following the clock model, whereas the value being sought for expressed characters was under strong stabilising selection and distributed as though it followed the saltatory model, then using the sample model could introduce a severe bias into the measure. Consequently, any apparent increase in resolution arising from using information from the sample in this way could actually be misleading.

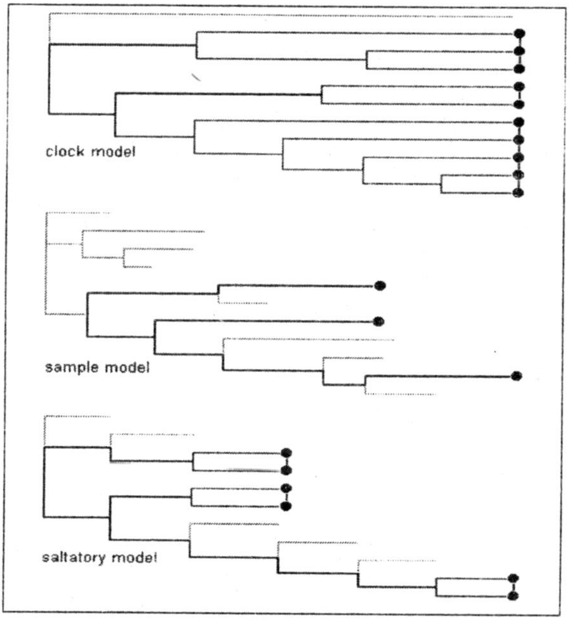

Figure 2

Lack of Phylogenetic Information

In practice, detailed and reliable phylogenetic i.e. genealogical and difference information is often unavailable. Nevertheless, arguments for measuring biodiversity value as gene or character richness do at least provide a philosophically and economically defensible starting point, as one possible answer to the problem of what is valued in diversity. Accepting that phylogenetic or taxonomic diversity measures can use genealogical pattern as a predictor of value also provides a possible key to the problem of finding more practical measures

Diversity Measurement is based on Three Assumptions

1. *All Species are Equal:* this means that richness measurement makes no distinctions amongst species and threat the species that are exceptionally abundant in the same way as those that are extremely rare species. The relative abundance of species in an assemblage is the only factor that determines its importance in a diversity measure.

2. *All Individuals are Equal:* this means that there is no distinction between the largest and the smallest individual; in practice however the smallest animals can often escape for example by sampling with nets.

   Taxonomic and functional diversity measures, however, do not necessarily treat all species and individuals as equal.

3. *Species has been recorded in using Appropriate and Comparable Units.* It is clearly unwise to use different types of abundance measure, such as the number of individuals and the biomass, in the same investigation. Diversity estimates based on different units are not directly comparable.

Species richness S is the simplest measure of biodiversity and is simply a count of the number of different species in a given area. This measure is strongly dependent on sampling size and effort. Two species richness indices try to account for this problem:

Margalef's Diversity Index

$$D_{Mn} = \frac{S}{\sqrt{N}}$$

$$D_{Mg} = \frac{(S-1)}{In\,N}$$

Where N = the total number of individuals in the sample and S = the number of species recorded. Despite the attempt to correct for sample size, both measures remain strongly influenced by sampling effort. Nonetheless they are intuitively meaningful indices and can play a useful role in investigations of biological diversity.

Heterogeneity Measures

Heterogeneity measures are those that combine the richness and the evenness component of diversity. Heterogeneity measures fall into two categories: parametric indices, which are based on a parameter of a species abundance model, and nonparametric indices, that make no assumptions about the underlying distributions of species abundances.

Parametric Indices

The log series index is a parameter of the log series model. The parameter is independent of sample size. describes the way in which the individuals are divided among the species, which is a measure of diversity. The attractive properties of this diversity index are: it provides a good discrimination between sites, it is not very sensitive to density fluctuations and it is normally distributed, in this way confidence limits can be attached to.

The log series takes the form:

$$\alpha x, \frac{\alpha x^2}{2}, \frac{\alpha x^3}{3}, \ldots, \frac{\alpha x^n}{n}$$

αx is the number of species to have one individual, $\frac{\alpha x^2}{2}$ those with two individuals, and so on. Since $0 < x < 1$ and α and x are presumed to be constant, the expected number of species will be the highest in the first abundance class. is calculated interactively from:

$$\frac{S}{N} = \frac{(1-x)}{x} \cdot \ln \frac{1}{(1-x)}$$

And α can be calculated from the equation:

$$\alpha = \frac{N(1-x)}{x}$$

Non-Parametric Indices

The first two indices are based on information theory. These indices are based on the rationale that the diversity in a natural system can be measured in a similar way to the information contained in a code or message.

- The most widely used diversity index in the ecological literature is the *Shannon-Wiener diversity index.*

It assumed that individuals are randomly sampled from an infinitely large community, and that all species are represented in the sample. The Shannon index is calculated from the equation:

$$H' = -\Sigma_{pi} \ln_{pi}$$

p_i is the proportion of individuals found in the with species.

- Where the randomness cannot be guaranteed, for example when certain species are preferentially sampled, the *Brillouin index* is the appropriated form of the information index. It is calculated as follows:

$$H = \frac{1}{N} \cdot \ln \frac{N!}{\pi N_i}$$

In which $\pi N_i = N_1 . N_2 . N_3 ... N_i$ and N_i = the number of individuals in species i and N is the total number of individuals in the community.

$$\gamma = \sum_i p_i^2$$

One of the best known and earliest evenness measures is the $p_i$is the proportion of individuals found in the with.species This index is used for large, sampled communities. Simpson's index expresses the probability that any two individuals drawn at random from an infinitely large community belong to the same species.

- *The Hill numbers* show the relation between the species-richness indices and the evenness-indices. Hill defined a set of diversity number of different order. The diversity number of order a is defined as:

where p_i = the proportional abundance of species in the sample and a = the order in which the index is dependent of rare species.

The most known are

$$H_0 = S \quad H_\alpha = \left(\sum_i p_i^\alpha \right)^{\left(\frac{1}{1-\alpha}\right)}$$

$H_1 = expH'$ (Exponential of Shannon-Wiener diversity index)

$$H_2 = \frac{1}{\gamma} \text{ (the reciprocal of Simpson's } \gamma)$$

Taxonomic Indices

If two data-sets have identical numbers of species and equivalent patterns of species abundance, but differ in the diversity of taxa to which the species belong, it seems intuitively appropriate that the most taxonomically varied data-set is the more diverse. As long as the phylogeny of the data-set of interest is reasonably well resolved, measures of taxonomic diversity are possible.

- *Clarke and Warwick's taxonomic distinctness index* which describes the average taxonomic distance–simply the "path length" between two randomly chosen organisms through the phylogeny of all the species in a data-set–has different forms: taxonomic diversity and taxonomic distinctness.
- Taxonomic diversity (Δ) reflects the average taxonomic distance between any two organisms, chosen at random from a sample. The distance can be seen as the length of the path connecting these two organisms through a phylogenetic tree or a Linnean classification. This index includes aspects of taxonomic relatedness and evenness.

$$\Delta^* = \frac{\Sigma\Sigma_{i<j}\omega_{ij}x_i x_j}{\frac{N(N-1)}{2}}$$

$$\frac{\Sigma\Sigma_{i<j}\omega_{ij}x_i x_j}{\Sigma\Sigma_{i<j}x_i x_j}$$

Taxonomic distinctness (Δ*) is the average path length between two randomly chosen but taxonomically different organisms. This measure is measure of pure taxonomic relatedness.

When only presence/absence data is considered both Δ and Δ* converge to the same statistic Δ+, which can be seen as the average taxonomic path length between any two randomly chosen species.

$$\Delta^{+} = \frac{\Sigma\Sigma_{i<j}\omega_{ij}}{\frac{S(S-1)}{2}}$$

Functional Diversity

The positive relationship between ecosystem functioning and species richness is often attributed to the greater number of functional groups found in richer assemblages. Petchey and Gaston proposed a method for quantifying functional diversity. It is based on total branch length of a dendrogram, which is constructed from species trait values. One important consideration is that only those traits linked to the ecosystem process of interest are used. Thus a study focusing on bird-mediated seed dispersal would exclude traits such as plumage color that are not related to this function, but traits such as beak size and shape should be included With standard clustering algorithms a dendrogram is then constructed. The method makes sense. For example a community with five species with different traits will have a higher functional diversity than a community of equal richness but where the species are functional similar.

Nearly all diversity and evenness indices are based on the relative abundance of species, thus on estimates of pi in which:

$$p_i = \frac{N_i}{N}$$

with Ni the abundance of the i-th species in the sample and

$$N = \sum_{i=1}^{S} N_i$$

with S the total number of species in the sample. If one records the abundance of different species in a sample, it is invariably found that some species are rare, whereas others are more abundant. This feature of ecological communities is found independent of the taxonomic group or the area investigated. An important goal of ecology is to describe these consistent patterns in different communities, and explain them in terms of interactions with the biotic and a biotic environment. Different investigators have visualized the species-abundance distribution in different ways.

The rank/abundance plot is one of the best known and most informative method. In this species are ranked in sequence from most to least abundant along the horizontal (or x) axis. Their abundances are typically displayed in a log10 format on the y axis, so that species whose abundances span several orders of magnitude can be easily accommodated on the same graph. In addition proportional and or percentage abundances are often used.

The k-dominance plot shows the cumulative percentage i.e. the percentage of the k-th most dominant plus all more dominant species in relation to species (k) rank or log species (k) rank.

The Lorenzen curve is based on the k-dominance plot but the species rank k is transformed to (k/S) x 100 to facilitate comparison between communities with different numbers of species.

The collector's curve addresses a different problem. When one increases the sampling effort, and thus the number of the animals N caught, new species will appear in the collection. A collector's curve expresses the number of species as a function of the number of specimens caught. As more specimens are caught, a collector's curve can reach an asymptotic value but they often don't due to the vague boundaries of ecological communities: as sampling effort increases, also the number of different patches increases.

The species-abundance distribution plots the number of species that are represented by r = 0,1,2,... individuals against the abundance r. This can only be drawn if the collection is large and contains many species. More often than not the species are grouped in logarithmic densities classes. A diverse range of models has also been developed to describe species abundance data. The fitting of a model to field data is meaningful if the parameter estimates are to be used in further analysis.

The Niche Preemption Model or the Geometric Model

It assumes that a species preempts a fraction k of a limiting resource, a second species the same fraction k of the remainder and so on. If the abundances are proportional to their share of the resource, the ranked abundances list is given by geometric series:

$$k,\ k(1\text{-}k),\ \ldots,\ k(1\text{-}k)(S\text{-}2),\ k(1\text{-}k)(S\text{-}1)$$

where S is the number of the species in the community. The geometric model gives a straight line on a plot of log abundance against rank (species sequence). It is not very often found in nature, only in early successional stages or in species poor environments.

The Broken Stick Model or the Negative Exponential Distribution

In this model a limiting resource is compared with a stick, broken in S parts at S-1 randomly located points. The length of the parts is taken as representative for the density of the S species subdividing the limiting resource. If the species are ranked according to abundance, the expected abundance of species i, Ni is given by:

$$E(N_i) = \frac{1}{S}\sum_{x=i}^{S}\frac{1}{x}$$

The negative exponential distribution is not often found in nature. It describes a too even distribution of individuals over species to be a good representation of natural communities.

The Log-Series Distribution

Fisher's logarithmic series model see also the log series index) describes the relationship between the number of species and the number of individuals in those species.

The Log-Normal Distribution

Preston first suggested to use a log-normal distribution for the description of species-abundances distributions. The distribution is traditionally written in the form:

$$S(R) = S_0 \exp\left(-\alpha^2 R^2\right)$$

With S(R) = the number of the species in the Rth octave to the right, and to the left of the symmetric curve; S_0 = the number of the species in the modal octave; and $\alpha = 2\left(\sigma^2\right)^{(-1/2)}$ = the inverse width of the distribution

Existing Measures of Biodiversity

Various measures of biodiversity have been developed both by ecologists and economists. Whereas, the first group is more interested with the number of species, the second group try to take into account the contribution of species to biodiversity. In what follows, we present briefly these two approaches.

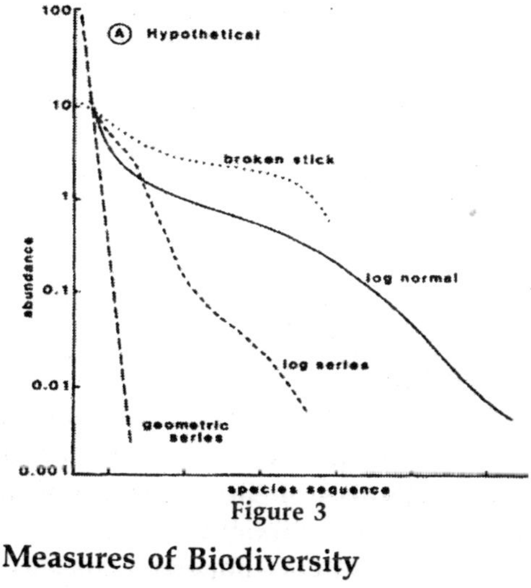

Figure 3

Economic Measures of Biodiversity

Different economic measures of biodiversity have been developed. Here, we present the most frequently cited one formulated by Weitzman (1993). His approach form the basis of most recent papers concerned with biodiversity.

The diversity measure, D(S), of a set S of species is based on a function d(i, j) that measures the distance or dissimilarity between two species i and j in the set S. Suppose we have a subset Q such as S and another subset Q0 which includes one species I such as and Weitzman compared the diversity of Q and that of Q0 by measuring the dissimilarity between species i and the set Q as follow

The biodiversity measure of the set S is then calculated, implicitly, by a recursive algorithm starting from an arbitrarily chosen start value, D0, of a set with only one species One problem with this approach is that the value of D depends on the sequence in which species are added to construct the full set S. Hence, the solution of equation is not unique. Nevertheless, Weitzman introduces a nice intuition when he focuses on the distance between species. But, by doing so, such elements of as the role of a species in helping conservation or allowing interaction are ignored. Indeed, according to this approach, it is better to have a set of fish and gorilla than a set of monkey and gorilla since the dissimilarity between the two components of the first set is larger than the dissimilarity between the two components of the second set.

We conclude that all measures seen above miss out an important element of biodiversity valuation that is a reference value of biodiversity relative to each ecosystem. Indeed, all ecosystems are not the same and their biodiversity must be different.

Gamma Diversity: the species diversity in the landscape; it is the diversity of a region; it concerns the diversities of the all communities in a certain landscape. Scale of >1000 km2; richness of a landscape that comprises different communities. Gamma-diversity or the diversity of a region is related to the abiotic and biotic features of the region like water-energy dynamics, climate, history, topography etc.

Beta Diversity: measures the change in species in terms of composition among local communities; e.g. between two forest types; it describes the species replacement. It is also called by term "differentiation" or named habitat patchiness. Thus, beta-diversity measures how similar or different sites, habitats or samples are. The respective spatial scale for beta-diversity is 100 km2. Beta diversity connects levels of richness at local and regional levels. Beta diversity constitutes the turnover between patches at the alpha scale. Species presence and absence data are used to measure the beta diversity. This measure is used to describe the species diversity along ecological gradients gradual change in values and transects. Beta diversity increases as the similarity in species composition decreases.

The more heterogeneous is the landscape, the higher is the beta-diversity. In other words, the more different habitat patches are present in the landscape, the more diverse is the landscape and the higher is the beta-diversity. Landscape consisting of different habitats with no overlap in species composition is more diverse than landscape of habitats sharing the same species. There exists a common pattern of increase of beta diversity from poles to tropical areas for number of organisms: mammals, birds, trees and herbaceous plants.

Alpha Diversity: species richness within a defined community or habitat; local diversity. Consists of several subsamples in the habitat or community. The diversity of a subsample is called *point diversity,* i.e. richness of a subset of a community. Alpha diversity can measure e.g. the number of all vascular plants found in an Estonian calcareous grassland site or all nesting bird species in the forest habitat site of Finland. The more different more various in the species composition are the subsamples or points of a community, the higher is the diversity of the community. Alpha, beta and gamma diversity of different organism groups are often measured at different scales taking into account their body size, mobility, habitat use and other relevant features. For example, for terrestrial plants 100-10 000 m2 is appropriate scale for alpha diversity and 106 – 108 m2 for gamma diversity. For phytoplankton 10 x 10 cm might be the appropriate scale to measure alpha diversity.

Measurement of Alpha Diversity. Indices of Heterogeneity

The fundamental measure for diversity is species number: the easiest way to assess the diversity is to count the number of species, i.e. to assess the *species richness* in the system. However, a difficulty of using alone the number of species in a system arises, because species richness is always dependent on the size of the sampled area. The larger the sample the greater the expected number of species, but see Example 1. Larger areas contain more individuals, more habitats, and more biomes or biogeographic provinces than do smaller areas. The greater the habitat variety, the greater the species diversity.

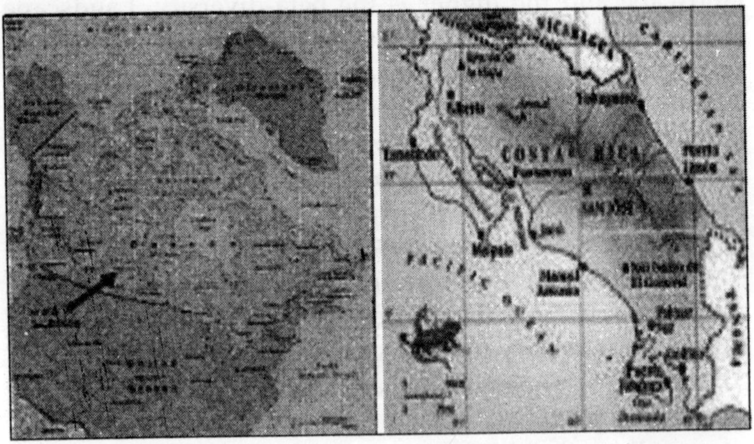

The area of Costa Rica is 100 times smaller than area of Canada. Following the general relationship of species richness–area, the biodiversity of Canada should be much higher than that of Costa Rica. But the actual species numbers are: Reptiles: Canada 32, Costa Rica 218 species Birds: Canada 434, Costa Rica 796 species Mammals: Canada 94, Costa Rica 203 species This example indicates the complex way how biodiversity is packed in the nature. Species richness area relationship does not explain why in some cases the areas of similar size or the area of smaller size can be more species rich. The reasons of species diversity are not related only to the area of described site, but to the evolutionary events, speciation, location of the site, variety of climatic conditions etc.

Species-area relationships for *Lumbricidae* (*Annelida, Oligochaeta*) in Europe. From Gaston and Spicer (1998) and Gaston (2000). Relationship between species number and

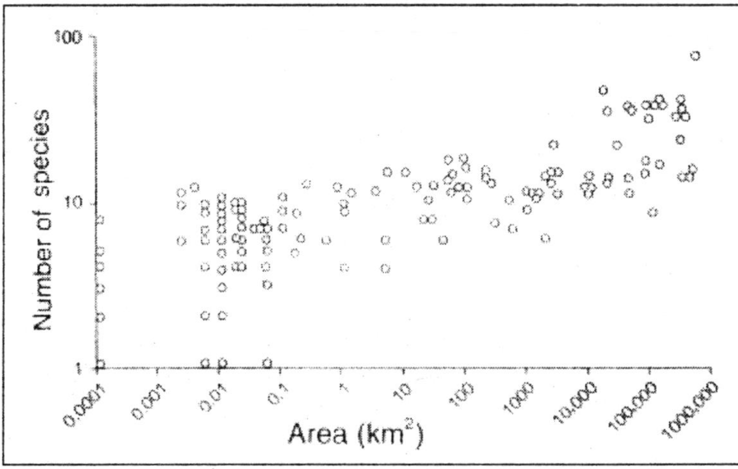

sample area is one of the general laws in ecology and can be described by the simple equation: where **S** - number of species, **c** - a constant measuring the number of species per unit area, **A** - area of the habitat, **z** - a constant measuring the slope of the line describing the relationship between S and A. For example, on islands, **z** is around 0.3. By taking logarithms, it makes easier to describe the relation between species number and area, as it is characterized by linear function: **log S = (log c) + z (logA).** That's why very often the logarithmic values of both species number and sample area are used. But comparison of species numbers requires equal sample sizes. In this reason, it is desirable to have a richness index independent of sample size. To eliminate the difference in diversity caused by the differences in sizes of sampled areas following indices are used: 1) species density–number of species per area or unit. In addition, the number of species per fixed number of individuals or biomass may be used instead of area.

Species density gives ıdea both about how many species there are per unit area or cubic measure, it is - how tightly species are packed or how evenly they are distributed in the community. Calculation of species density allows to elliminate the difference in species number in sites caused by the difference in areas of the sites.

Species density, e.g. the number of species per metre squared, is favoured especially in the botanical studies. The extraordinary high species densities in the world are described in grasslands of temperate region on the calcareous base - 76 vascular plant species per 1 square metre, Estonia tropical rainforest - 450 tree species per 1 hectare of forest, Mata Atlantica With the help of the analyses of the dominance-diversity curves it is possible to visualize the distribution pattern of species abundances in the community. Dominance diversity curves characterize both species richness and concentration of dominance. The longer the tail of the curve, the more species in the community. The abrupt the slope of the curve, the higher concentration of dominance exists in the community. The characteristics of the respective curves - shape and slope-are calculated and used as variables of

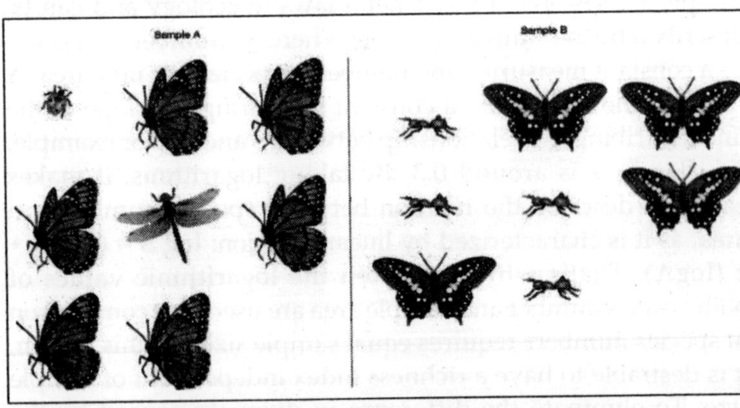

diversity of communities. Incorporating the relative abundances of species besides species richness makes the estimate of diversity more truthful. For instance, is the diversity of a community of 10 species, one of which comprises 90% of the individuals, equal to the diversity of another community sharing the same 10 species which all are equally abundant? This example shows light why two components of the biological diversity of a community (alpha diversity) are needed to be distinguished: the number of species (**species richness**) and the abundance of each species

Figure 2. Two insect communities. Which is more species rich? Which is more even? Which is more diverse? **Evenness** describes the relative abundance of species in a given system, taking into account the different statuses of the species in the community. This indirect characteristic of diversity shows how evenly are the resources between species distributed. Rarely are species equally abundant, some species are always more successful than others: some are better competitors or more fecund, resulting in higher abundance.

The more equal the species abundances are, the higher is the evenness of the community. On the contrary, if one or few species are dominating, the evenness of the community is low. Various measures of abundance like biomass, cover in case of plants, number of individuals per study unit can be used for assessing the contribution of each species. The high evenness indicates the missing dominant species. It reflects often a good health of the community. The dominance of one or few species in ecosystems results often from the negative human impacts. Cessation of mowing or grazing of the grassland ecosystems is followed by the dominance of higher growing species with higher nutrient demand. These strong competitors lead to the disappearance of many lower growing ones and to the decrease of species diversity.

But some ecosystems are naturally species poor and much less even than others. Taiga forests are less even and diverse than deciduous forests. Tundra ecosystem is naturally species poor, as the extreme climatic conditions limit the distribution of many species to these habitats. 3) Species richness and evenness can be used alone as measures of diversity, but it is more appropriate to use them together. Based on the two components the diversity statistics - **heterogeneity indices-** are produced. The indices differ in the mathematical method of relating the two features. The more often used heterogeneity indices are Simpson's Index of Diversity and Shannon-Wiener diversity index relative abundance of species where ni = number of individuals of species i, N = total number of individuals in all species Simpson's Index is the species dominance measure. It is sensitive to changes in the most abundant species. This index measures the probability that two individuals selected at random from a sample will belong to the same species. With this index, 0 represents infinite diversity and 1, no diversity. That is, the bigger the value of the index, the lower the diversity. This is neither intuitive nor logical, so to get over this problem, D is often subtracted from 1 to give:

The value of this index also ranges between 0 and 1, but now, the greater the value, the greater the sample diversity. This makes more sense. In this case, the index represents the probability that two individuals randomly selected from a sample will belong to different species. The value of this index starts with 1 as the lowest possible figure. This figure would represent a community containing only one species. The higher the value, the greater is the diversity. The maximum value is the number of species or other category being used in the sample. For example if there are five species in the sample, then the maximum value is The name 'Simpson's Diversity Index' is often very loosely applied and all three related indices described above Simpson's Index, Simpson's Index of Diversity and Simpson's Reciprocal Index have been quoted under this blanket term, depending on author. It is therefore important to ascertain which index has actually been used in any comparative studies of diversity, is the proportion of individuals biomass, abundance, cover etc natural logarithm the exponential of Shannon index, Exp (H') is widely used and called Shannon index.

It is sensitive to changes of the rare species in the sample. This index is difficult to interpretate. The higher the evenness, the higher the value of Shannon index. It equals to the number of species if all species have equal abundances. To calculate heterogeneity indices for a particular area, the area must first be sampled. The number of individuals of each species present in the samples must be noted. Several samples need to be taken to get statistically reliable data.

Measurement of Beta Diversity. Indices of Similarity

Beta diversity measures the diversity caused by variation environmental factors. Indices of beta-diversity are measure the extent of change in species composition along predefined gradients, i.e. species turnover. Therefore they are also called similarity indices and are based on the similarity tables. The selection of a particular index depends on characteristics of the focal samples, the domain or gradients over which they are evaluated, and the hypotheses of interest.

Originally beta-diversity was measured as the proportion by which the species richness of a region exceeds the average richness of a single locality within that region: where S= total number of species recorded in the system, á = the average sample diversity, where each sample is a standard size and diversity is measured as species richness. Range: 1 complete similarity - 2 no overlap in species composition.

Subtracting 1 changes the range of beta-diversity to 0-1, the scale what is used in many other measures of beta-diversity. The majority of beta diversity indices are based on consideration of species losses and gains along a gradient or transect, i.e. on presence/absence data of species. Similarity indices are based on the following 2 x 2 similarity table: The possible spatial distribution of a species across a pair of quadrates. (a) Component *a* comprises the total number of species that occur in both quadrates; (b) component *b* comprises the total number of species that occur in the neighbouring quadrate but not in the focal one; and (c) component *c* comprises the total number of species that occur in the focal quadrate but not in the neighbouring one. More widely used similarity indices:

Sörenson index Cs = 2 a/(2a+b+c) * Jaccard index: Cj = a/(a+b+c) where a is the total number of species present in both samples, b the number of species present in only in quadrate 1 and c the number of species present only in sample.

Important to know: 1) Sörensen measure will be always large if samples differ markedly in terms of species richness.

2) Similarity indices do not take into account the relative abundance of species, both dominant and rare species have equal impact to the measure. 3) Beta-diversity decreases as progressively larger areas are investigated.

		Sample A	
Sample B	**Species**	Present	Absent
	Present	a	B
	Absent	C	d

Figure 4. The possible spatial distribution of a species across a pair of quadrates. (a) Component *a* comprises the total number of species that occur in both quadrates; (b) component *b* comprises the total number of species that occur in the neighbouring quadrate but not in the focal one; and (c) component *c* comprises the total number of species that occur in the focal quadrate but not in the neighbouring one (Koleff *et al.* 2003).

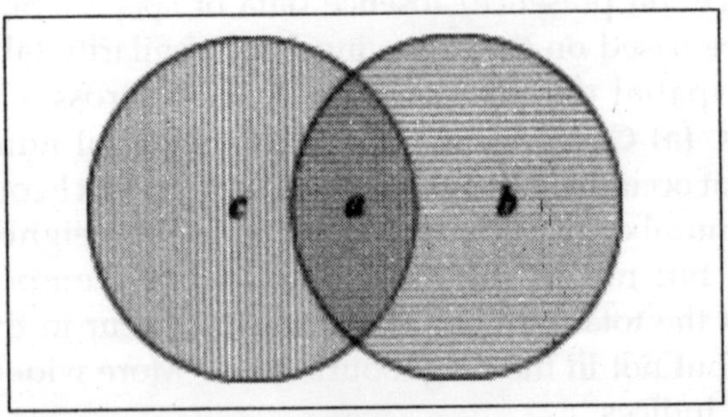

Figure 4

More Widely Used Similarity Indices

Where a is the total number of species present in both samples, b the number of species present in only in quadrate 1 and c the number of species present only in sample 2. Important to know: 1) Sörensen measure will be always large if samples differ markedly in terms of species richness. Similarity indices do not take into account the relative abundance of species, both dominant and rare species have equal impact to the measure. 3) Beta-diversity decreases as progressively larger areas are investigated.

Indicators often do not include Information on Biodiversity Itself

Many 'biodiversity indicators' actually give little information about overall biodiversity levels or status. Most monitoring, particularly in data poor countries, focuses on large species, usually mammals and birds or on land cover types. While these may give general information about the 'existence' of habitat, they often say comparatively little about its 'quality' including its ability to support other species. Forest elephants in both Africa and Asia, for instance, can continue to live in quite degraded forests as long as sufficient food plants remain and the same is true for many of the great apes such as gorilla (*Gorilla beringei*) and orang utan (*Pongo pygmaeus*). Forests can appear intact or only slightly degraded using remote sensing tools when fauna has been virtually destroyed.

The Need to Choose Indicators that Resonate with the Public and Decision-Makers

Although indicators of biodiversity condition should clearly be driven by good science, they also have an important political aspect, which needs to be acknowledged and addressed if they are to be successful in influencing management. Biodiversity indicators are not only important in providing scientists with a means of measuring the state of nature, but they also have to resonate with the policy-makers who respond to the information that they contain, to politicians, industry representatives, land managers, non-governmental organizations and perhaps most important of all, with the wider public to whom governments are responsible. Indicators in a politically driven process such as the CBD, therefore also have a role in telling a wider story about biodiversity and in setting and driving responses to the pressures that they illuminate. Indicators not only have to be technically coherent and defensible, but to provide information in a way that many people outside the fields of the biological sciences will be able to understand and to sympathize with. To be of political value, an indicator or suite of indicators may need to embody some kind of 'vision' for how the problems it is illuminating could be addressed. This implies a level of negotiation and trade-offs may be needed in reaching a final indicator list.

Agreement on specific indicators within an international agreement implies a certain level of commitment to it being applied and therefore to the message that they contain. Dudley and Jeanrenaud (1996) pointed out that good biodiversity indicators need to have more attributes than would be needed from a strictly scientific point of view, and in an ideal world would:

- be easy to identify and locate in the field without heavy reliance on specialists and often encountered when the management or uses are at an acceptable level;
- provide reliability but also a broad range of information within a single indicator;
- give good value for the resources invested i.e. avoiding data that is more complex or expensive to collect than necessary;
- be usable by many people and if possible not relying on a small number of specialists to identify or collect;
- resonate with public and policy makers and telling a clear and understandable story about what is happening within an ecosystem.

This also implies that all indicators will need accompanying explanation, with their limitations clearly articulated. Once indicators leave the confines of the scientific world and are used in wider policy-making and by many more stakeholders, they will also be shaped and sometimes constrained as a result of these wider interests and varying applications. Indicators can 'foster better management. However, there are confusions and tensions to reconcile between general and local applications, between the ideal and the pragmatic, and between the scientific and the democratic. To overcome this requires a sober appraisal of what can realistically be achieved in each location and how this can best be promoted. Good judgment remains the foundation of competent management.

11

Biodiversity Survey Techniques

Biodiversity Survey

A biodiversity survey is an inventory of the plant and animal life of an area including their habitats. There are many complex data collection methods used for biodiversity surveys. The data collection methods have been discussed on the basis that they are the most commonly used and have been effectively trailed by non-expert people. These methods are recognized and used by the government agencies responsible for the management of our natural environment. This section looks at two types of study, baseline surveys and comprehensive surveys. Each approach requires a different level of participation and advice from scientific experts.

Need for Biodiversity Survey

How will a biodiversity survey assist the conservation of our natural environment? The continued existence of native plants and animals on private land provides an indication that something right is being done. Many landholders maintain small native bush areas which provide a haven for our native species. In addition the knowledge and experience held by the community about our natural environment can contribute significantly to the conservation of biodiversity. However, it's true that we still lack some simple information about our natural environment. Without this knowledge it can be difficult to assess the impacts humans have on our biodiversity. Increasing our knowledge will help us to ensure biodiversity is protected for future generations.

Government planning and policy development at all levels has started moving towards integrating biodiversity considerations into the decision-making process. It is essential that environmental planning processes and reporting mechanisms have accurate and up-to-date information for them to be effective.

Ways to Make the Best Use of Survey Results

The methods used in this section will provide data to discover what plants and animals are present or absent. If particular plants and animals have not been found a survey area and it was thought they would be, this is important information as recommendations can be made to ensure that additional surveys are conducted in other seasons. Some plants and animals may only be present or obvious in particular seasons. It may mean that new sites within the area should be surveyed to establish if those plants and animals are there or not.

The reason why scholars are doing the survey and that will help to determine what should happen to the data we collect. The most important thing to remember is that the more people that have access to collected data so that it can be used for planning and management purposes, the more useful collected data will be.

In this context survey results could be used in the following way:

- Help policy makers to plan the management of land. By knowing what is there one can think about the most sustainable ways of managing his land.
- Produce a short report with the data attached for the local council so that it can be used to make management decisions.
- Share the knowledge with neighbours so that a bigger picture of what is in particular area can be established.

– Use the results to produce fact sheets and a short report in a format that can be used by various interested agencies.

When undertaking a biodiversity survey it is not possible to record the identity, location and condition of every individual plant or animal over a whole survey area. The cost would be prohibitive, the human effort huge, disruption to the environment would be great, and the amount of data collected daunting. So, for practical reasons, we take samples of the biodiversity of an area and drawing conclusions about the distribution of species within the whole survey area based on the information the samples give us.

The aim of a biodiversity survey is to sample the species living in the survey area across the full range of environmental variation so that we can be confident we have the clearest, most accurate understanding of the biodiversity of the survey area. How comprehensive a survey will be is determined directly by the strategy used to locate sample sites within a survey area. In other words, where samples are taken is critical to the quality and usefulness of the results.

The fact that most species are known to have environmental preferences means that we can use information on key environmental factors for example average rainfall, soil type and the vegetation type to devise a logical strategy to sample the species. We can use that fact to divide the survey area into environmental types that will guide the distribution of survey sites and ensure that the full range of environment has been sampled. In that way, we can be confident that we can explain the distribution of species within the survey area and relate it to significant environmental characteristics.

The Side Concept

A survey site is a fixed location on the face of the earth at which one will take a sample of the biodiversity presented by applying one or more survey methods. The mix of survey methods applied at a site will be determined by one aim and objectives, and by practical constraints of time and available resource. Selecting that mix of methods and adjusting the extend of a comprehensive survey to fit these constraints are dealt with in the Survey Coordinators Notes. For a comprehensive survey, there will be many survey sites. It is very important that each site be given a unique identifying code that appears on all field data sheets associated with that site.

It is important to accurately locate each survey site using the guidelines given in 'Recording site details'. An accurate location and a unique site identifying code to tag results with will enable the environmental conditions of that site for example average rainfall, soil type, together with the weather conditions at survey time to be associated with the results of the survey methods applied at the site. It will also enable one or others to relocate each site for future surveys.

Sampling Effort

Determining an acceptable amount of work done at each site to detect the presence of species or measures their abundance is a significant decision to be made during comprehensive survey design. The general term sampling effort is often used to describe how much work is done at each site to record species. Different units are used to measure sampling effort for different survey methods. For example, trapping for small mammals is measured in the number of trap-nights. This is calculated by multiplying the number of traps set out at a survey site by the number of nights that they are left open for small mammals to enter.

In the case of vegetation surveys, the size of the sampling quadrat is the measure of sampling effort at a site. In a few instances, sampling effort in vegetation surveys may be increased by placing more than one quadrat at a site. As sampling effort is increased there comes a point where extra effort yields no significant increase in the number of species detected. This point may vary according to the nature of habitat in which the survey is being conducted. For example, vegetation quadrat can vary between 10 meters x 10 metres in closed coastal heaths to 20 meters x 20 meters in typical eucalypt forests or woodlands.

Practical experience has shown that quadrats larger than this in these habitat types is wasted effort. It is best to seek expert advice on this. The greatest value comes from biodiversity surveys where sampling effort is constant across all sites. Assuming that the sampling effort at a site is adequate to give a reasonable list of species for a reasonable amount of work, when sampling effort is constant that is, doing exactly the same sampling work at each site this ensures that when results are being analysed we are comparing like with like.

Strategies for Locating Sites

The most effective way to ensure that the range of environment within the survey area has been defined is to conduct an environmental stratification. This involves selecting those environmental factors for example average rainfall, soil types thought to be significant factors of the distribution of species within the survey area and marking patches on a map of the survey area that can be characterized by combinations of these factors. Each defined combination of environmental factors is called a stratum.

Some environmental factors may vary continuously along environmental gradients. An environmental gradient occurs when an environmental factor varies smoothly from a low value to a high value. Environmental gradients can occur at a number of scales. Where environmental factors that are important for of species distribution exist as environmental gradient, it is usual to subdivide the gradient in ranges of values or bands.

Example: If average maximum temperature is seen as important in large scale survey, and it varies across one's survey area from a low of 20 degrees Celsius to a high of 25 degrees Celsius. Consider selecting survey sites in areas between say 20 degrees Celsius and 22 degrees Celsius. Then select a set of sites whose average annual temperature lay between 23 degrees Celsius and 25 degrees Celsius.

For smaller-scaled surveys, local gradients are often seen in things like soil moisture gradients. Usually, this is related to the position of a possible site down the slope from ridge top i.e. dry soil conditions to gully bottom moist soil conditions. In those situations, it is possible to roughly divide the moisture gradient into three bands corresponding to dry soil ridge tops, medium soil moisture mid-slope positions, and moist gullies.

Random Site Selection

It is best to place survey sites within each stratum at random. This will avoid any bias in the placement of sites within a stratum and ensure that any variation of environmental conditions within each stratum will be randomly sampled. Because, the residual environmental variation is randomly sampled.

Another important guideline is that it is best to adjust the number of sites in a stratum in proportion to the total area of the stratum within the survey area. However, a rough rule of thumb is that one should aim for a minimum of four sites per stratum.

Placing Sites within a Stratum

In an ideal stratified random sampling design, plots should be located at random within each stratum. However, there are a number of reasons for departing from the ideal design. The first reason relates to the need to avoid the effect of human disturbances that will confound an understanding of the patterns of species distribution in the survey area. The second reason is the need to avoid placing sites near the boundaries of strata. Sites placed here are likely to be in transition zones between the environmental conditions characterizing each stratum. Finally, the placement of quadrats or transects at each site should take account of local gradients and edge effects.

Example: A 20 meters by 20 metre i.e. 400 square metres vegetation sampling quadrat placed in a narrow strip of riparian vegetation many extend well beyond the riparian zone into adjacent vegetation. In those situations, the quadrat's shape should be adjusted to fit within the riparian vegetation stratum i.e. adjusted to say 10 metres by 40 metres, but still 400 square metres in area.

The final location of sites within strata, and the arrangement of quadrats or transects at a site, should be made according to the following criteria.

– Away from the boundaries of the stratum;
– Homogenous conditions considered to be representative of the stratum, and within the same vegetation growth-stage where possible;
– Away from local disturbances such as roads, mines, quarries, and eroded areas; and Animal welfare

Not only should team leaders and team members ensure their own safety, they should also consider the welfare of animals that they are surveying. Team leaders and the Survey Coordinator should ensure that all participants are aware of the general principles and the specific guidelines relevant to surveying animals, at their welfare.

There are a number of practical steps which one can take which becoming involved in animal survey work:

– Ensure that any study with which researcher is involved, and which involves capturing, handling or interfering with vertebrates, has approval from wild life conservation Act.
– Researcher should also ensure any work carried out complies to the Prevention of Cruelty to Animals Act.
– Familiarize us with the specific procedures, which are approved by the Animal Ethics Committee for use in our study, especially those designed to minimize distress to the captured animals. They may include the following commonly - used practices :
– *Check traps early in the morning,* and place insulating material such as newspaper or dry leaves into metalsided traps in cold weather, to minimize stress from heat or cold.
– *Keep animal handling times to the* minimum required to identify and if necessary measure the animal. This is particularly important with frogs, which have porous skins capable of rapid dehydration and absorption

of toxic substances from human hands e.g.sunscreen, mosquito repellent, nicotine.

- *Handle animals quietly* and speak in soft voices.
- *Ensure that animals are kept in quiet,* cool and well-ventilated conditions if it is necessary to keep them for any length of time which should not be more than one day. For example, it is sometimes necessary to hold insectivorous bats until release that night.
- *Release animals in the location where they were captured,* at a time appropriate to each species, and into shelter such as vegetation cover or leaf litter. This protects them from predators and minimizes stress due to disorientation.
- *Consider the welfare of the animals* researcher are surveying at all times, and review procedures as their use them. For example, find a trapped small mammal dead and in an advanced state of rigor mortis, it may be because they are entering traps in the heat of the late afternoon not all nocturnal animals are fully nocturnal. In this case, traps may need to be checked twice per day, or closed during the day, to avoid further deaths.
- *Wherever possible, ensure that any disturbance to the environment of animals is not permanent.* Walk on existing tracks when ever possible, avoid trampling or breaking vegetation, and replace any logs, rocks or leaf-litter moved when searching for animals.

In addition there are specific issues relevant to particular methodology, below is a summary of the major issues, for humans, plants and animals.

Plants

- *Beware of thorns and prickles* when touching or handling plants.
- *Do not eat plant parts* as many area poisonous, and some poisonous plants resembles non-poisonous ones so misidentification is possible.

- Be aware of the possibility of *allergic reactions* to plant parts and take care when handling.

Birds

- *Avoid close-range observation* wherever possible during breeding and feeding. In extreme cases, a bird may permanently abandon a nest site if it feels threatened by humans observing too closely. Aggressive or disturbed bird behaviour in terms of loud repeated calls or frantic flight may indicate that researchers are too close to a bird's nesting site. Consult the field guides for details on when birds are likely to be breeding.
- Remember that some birds nest on the ground or in vegetation very close to the ground. *Be careful where researchers walk* when pushing through dense shrubs so that they don't disturb nests.
- *Avoid overuse of mimicry calls as* they can distract birds. Also avoid prolonged exposure to spotlights during nocturnal surveys.
- *Do not disturb nest sites* as birds may abandon eggs or chicks.
- *Do not remove eggs from nests* under any circumstances.

Mammals

- Avoid spotlighting the animals for prolonged periods as it may damage their eyes or cause temporary blindness. They could become completely disorientated and cause themselves harm.
- Care should be exercised when trying to attract an animal using call playback as it may leave it open to predation by another species. Avoid prolonged exposure of animals to calls played from the tape.
- When positioning *pitfall traps* are aware of the potential water run-off in case of rain and the amount of shade in the area. Place a flat stone in the bottom of pitfall traps to prevent animals from drowning when it rains.

Alternatively use pitfall traps with several holes drilled in the base to allow rain to run out.

- Place *traps* in a sheltered location away from the morning sun, so that it does not get too hot before the trap is checked each morning. Ensure that traps have enough shade and will not fill up when raining particularly pitfall traps.
- Always provide bedding inside traps such as plenty of dry leaf litter or shredded newspaper and more in cold weather. Generally, from October to December it can be cold enough to be a concern but does depend on where the survey is conducted and any possible sudden shifts in the weather.
- Cover the rear part of the cage trap with a sung fitting plastic bag to act as barrier to rain and cold conditions. Make sure the bag is positioned so that it will not collect water or flap in the wind.
- Make sure the cage trap is secure on the ground if positioned on a slope, so it is not likely to move while and animal is trapped within. Tent pegs and light rope are useful for securing traps in place not necessary for trapping in flat areas.
- Be aware that not all nocturnal mammals are fully nocturnal and may enter traps during the day, where they will be susceptible to heat stress. Traps may require checking during the day or need to be closed in the morning and reopened in the late afternoon if researchers suspect this may be occurring.
- Insectivorous bats are very susceptible to heat stress. Harp traps must be cleared before they are exposed to direct sun. Bats held during the day because they were cleared from a trap in the early morning must be held in appropriate conditions by an experienced and licenced person. They must be released that night at the point of capture.

REPTILES AND FROGS

- *Follow the protocol for handling frogs as described in wild life conservation conventions*
- Frogs should only be handled when necessary. They can become susceptible to fungal infections when skin secretions are removed. *Over handling* can also cause overheating and drying, especially if they struggle. If an animal becomes dry, pour 5 to 10 millimeters of fresh water over it, holding it in cupped hands. Wash hands without soap to reduce contaminating the frogs with chemicals present on the hands or use a new pair of disposable gloves for each hand, when handling frogs is in the body of water where the frog was found.
- Once a frog is caught, place it in a plastic bag with a small amount of fresh water in it. Tie it so that there is still air in it and so the frog cannot get out while it is being identified.
- Ensure that we cause the minimum of disturbance when using the frog call method. Avoid trampling vegetation at the waters edge and muddying the water.

Standardized Data Collection Methods

To get the maximum benefit the survey we should endeavour to keep the survey effort constant between all sites. If survey methods are not the same across the whole survey area it may result in not being able to achieve our aim or reducing the usefulness and credibility of our results. It is possible to get away with slight.

Three phases of data management need to be considered:

Stage 1. Collection of data.

Stage 2. Data sorting and storage.

Stage 3. Dissemination of data

These stages can be used as a guide to ensure that all aspects of our data management are dealt with to assist in achieving our aim. Alternations to data collection methods, but major changes to the survey plan should be documented in detail describing what changes were made, and why. Prior to and during the survey the need for everyone to have a commitment to consistency and accuracy in data collection needs to made clear to team leaders so they can ensure team members understand this and follow through.

Stage 1. Collection of Data

- Before the survey begins decide what identification reference sections will be used and ensure that everyone uses the same ones.
- Results should be recorded in full being sure to include relevant extra comments. It is best to use pencil in the field as it is waterproof. Pencil sometimes does not photocopy well so we may have to go over our data sheets in pen.
- Document all survey sites on a master map as accurately as possible, even if no animals were present. It is a good idea to make copies of the map of our survey area for each discipline. Then all mammals sites, for example, will be on one map and reptile sites on another.
- Make sure that all data sheets are completed correctly at the end of each survey day. All team leaders should be instructed to return all completed data sheets at the end of the survey. For data sheets not completed for long periods of time, and hence of no use to anyone.
- Check consistency throughout the survey, that is, that data sheets are filled out in the same manner and data collection methods used are the same throughout the survey.

Collecting Voucher Specimens

What is a Voucher Specimen?

A voucher specimen is a part of a plant or whole/part of an animal collected from our survey area and kept as a permanent record that was found in the area. Plant samples are collected and sent to the National Herbarium when they have not bee recorded in that area before. Animal vouchers are collected and sent to the Indian Museum when they have not been recorded as occurring in our survey area. To be able to collect plant and animal specimens we need to be licenced. A person or group that is licenced should know the procedures and what is required when collecting, storing and sending sample specimens to ensure that is done correctly.

Need for Collecting Voucher Specimens

Sample specimens make up a part of the data from scholars survey and so form an important part of the natural resource database of any region. The specimens become a permanent resource for future research as well as being used to verify future specimens.

Who should Collect Voucher Specimens?

All researchers investigating in any survey should be qualified to collect samples specimens for their relevant discipline. They will also know when it will be necessary to collect specimens. It should be decided if the team leader will be asked to lodge the specimens and who will take responsibility for them. During the planning for biodiversity survey it is best to contact the appropriate organizations where vouchers should be sent. Biodiversity will need to know what material to collect for plants and animals likely to be found in the survey area that are of significance, as well as quantity.

Stage 2. Data Sorting and Storage

The process of transferring data from field sheets to the natural resource data storage at an agency or institution should be as easy and error free as possible. All data that is collected should go to the relevant agency or institution, but researchers may wish to keep simple records on their own computer. This can be done using a spreadsheet or database package. If using a spreadsheet package, the field data collection sheets could be used as models. This maintains a consistency of how data is viewed making data analysis easier. The data researchers have collected is very precious and it is advisable to keep a few copies in different locations.

Stage 3. Dissemination of Data

Where researchers' data goes is crucial to ensure that they make the best use of it. It is one of the ways to ensure that they get support for the next survey, as people can see an output for all their effort.

Who gets Copies of the Data Collected?

The aim and objective scholars set for the survey should help determine where to send the data. But following are some suggestions. Every government agency, institution or expert that has given support, time or advice towards the survey, should be sent a copy of the data in report form and the raw data attached as an appendix. This may encourage them to assist in a future survey and can be a selling point when looking for help or support.

- Send the report to state and Central Government.
- Always send a copy of the data to whoever has responsibility for the land scholars have surveyed and who has approved to access the land.
- They may want to send it to the local media to create awareness.
- Send a report to any government agency that we feel requires the information for planning purposes.
- Local community groups or organizations could be sent a copy to help gain support for our cause.

Results and Reporting

Reports on surveys serve a multitude of purposes ranging from pressuring politicians and advertising one ongoing work, to presenting scientific data for use in habitat management. The aim and objective we set for our survey should be able to guide us as to what sort of report to do and what to do with it. Presenting our data in a report-style document along with complete and accurate descriptions of the methods used, means that others can use it for future research. This is a good thing to aim for when making a decision about how to present our data.

As a part of survey planning we could develop a list of potential reports that might be generated to assist in receiving feedback or getting things done. This list could include: Likely target dates for release/dispatch of the report; The target audience; What is our message (s) to the target audience; What requirements does the report have in terms of data or other accompanying material.

Mammal Voucher Specimens

The taking of voucher specimens must be under the control of licence holders only. The treatments outlined below can also be carried out on animals found dead in traps, road kills and animals required to be euthanased by vest due to injury. Voucher specimens will only be useful when they are accompanied by good field records.

Preservation of Dead Animals

Mammal specimens should be slit carefully along the abdomen, and preserved initially in 10% formalin, followed by 75% ethanol. The cut should penetrate though the abdominal muscle wall but not damaged the internal organs, and requires some experience to do correctly. The cut need only be big enough to allow penetration of the preserving fluid, and should not be the entire length of the abdomen. An anterior-posterior cut is best. For larger mammals a better alternative is to inject 10 percent formalin into the abdomen, thorax and large muscle masses. Formalin must be 10 percent formaldehyde, 90 percent water preferably distilled and buffered with calcium carbonate about 1 teaspoon per litre of formalin.

Method for All Pickled Specimens

- Make slit through to abdominal cavity, making sure not to damage internal organs.
- Preserve body in formalin for 48 hours for mammals as small as or smaller than house mice; up to 3 weeks for larger vertebrates.

- Transfer body to 70-75% ethanol after draining formalin from body.
- Place the body in a glass jar or vial topped up with the ethanol.
- Label the body jar or vial with the field number. Data label should be placed inside the jar.
- Ensure the voucher specimen data form is completed.
- Do not store the specimens in the sun as this can cause fading of colours.

Freezing is a good alternative. Wrap bodies tightly in thick plastic bags. Do not cut the animal open.

Safety Considerations and Tips

- Formalin and formaldehyde are extremely corrosive to living tissue, and will cause skin to peel; wear plastic gloves at all times when handling; do not breathe vapours; only use in a well-ventilated space; avoid all contact with eyes.
- Transport formalin and formaldehyde only in clearly labeled bottles with secure, leak-proof safety caps, and inside a second, sealed container that will protect the bottle from extremes of heat and breakage on rough roads.
- Formaline and formaldehyde should never be kept in close proximity to food or cooking equipment.
- Small volumes of ethanol up to 5 liters can be bought.
- From chemical suppliers and some pharmacists. Volume greater than 5 liters can only be bought from a registered supplier by those holding a current concessional sprits permit.
- Ensure that pickled specimens are preserved in ethanol, not formalin, and that all containers are tightly sealed and do not leak. If ethanol is not available and methylated spirits have been substituted, note this in the covering letter that accompanies the specimens.
- Ensure that every individual animal has a unique number on a tag attached firmly to the body, preferably to one of

the limbs. The date and place of collection, and name of collector, should be included on the tag.

- Tags should be written on in dark pencil, most pens will wash off in alcohol. Mammals may be tagged with either heavy paper and string tags or using a plastic tag made by punching the number into a Dynamo labeller.
- Each individual sent to the Museum must be accompanied by a completed voucher specimen field data sheet. Field data sheets should be filled out in dark pencil or ink which is indelible to alcohol. Drawing pen ink and some special felt-tipped marker and alcohol-proof, but most felt-tipped markers and ball-point ink will run badly.
- Please keep photocopies of the completed voucher specimen field data sheet in a safe place, along with a record of when we sent the material to the Museum.
- Voucher specimens also should be accompanied by a covering letter which lists the specimens. Museum collection managers should be requested to complete the sheet and return it to the person nominated, in order to verify the identification of the specimen. Do not forget to provide a return name, address and phone number at the bottom of the form.

Anabat Detection

Target Animals and Habitat

Anabat detectors are systems that can make the inaudible ultrasonic calls of insectivorous bats audible to humans. The system also allows these calls to be recorded on good quality cassette recorders and analysed on a microcomputer using special software and a hardware interface. The computer display of the recorded calls permits experts to identify most species of insectivorous bats with confidence based on the characteristic shape of the call as displayed by the special software.

An optional timer can be used to start and stop the recording system at present times. Another optional extra enables fully automated overnight recording. Fro-comprehensive surveys, only a basic section procedure will be described. These options can be applied as desired.

Note that we will need to send any cassette tapes to a specialist for species identification. This can be expensive with some specialists charging more cost.

Anabat detection can be used in all habitat types.

How to Use an Anabat Detector

- Check that Anabat detector is working correctly by turning it on and rubbing dry fingers on from of the microphone. With the volume at about half setting one should hear swishing or whoosing sounds.
- One detector should be used per survey site.
- Set the detector on the ground at the center point of the transect with the microphone propped upwards at 45° to the ground.
- A 40 kHz calibration tone should be recorded for a few seconds at the start of each recording session, and at the end of recording.
- Start the tape recorder and leave it running for half an hour. Avoid moving around the detector. The sounds of rustling cloths and crunching grass and leaf litter will swamp the recording of bats.
- Recording should start within about half an hour of sunset.
- One should attempt to run an Anabat detector for as many nights as possible at each site. It is a passive method that does not interfere with animal behaviour.

Sending Anabat tapes for specialist analysis and identification

Careful information management is required after tapes have been recorded to ensure that researchers are able to relate the identifications provided by experts to their field records. If researchers are unsure of how to prepare tapes for submission to a specialist, contact them and discuss their requirements. The following notes provide an outline of the important aspects of good information management. Always make sure that researchers have given each cassette tap that they use a unique identifying number or label. There is a place on the field data sheet to record this.

Always send a copy of the field data sheet, not the original.

Some specialist identification services may agree to write the results on the copy of the field data sheet and send it back to researchers. Often, however, they will send them a printout or list of identifications against the unique identifying information that they wrote on each cassette tape and recorded on our field data sheet, together with an invoice.

Special Considerations and Tips

Always approach running Anabat detectors quietly to minimize the amount of noise recorded. Do not use Anabat detectors in rain. Do not store Anabat detectors in damp conditions.

Harp Traps

Target Animals and Habitat

Harp traps are a proven method for humanely capturing many insectivorous bat species. They can be placed at survey sites or for targeting special interest areas. Harp traps can be used in all habitat types but are most effective in forest, woodland and tall shrubland habitats. Roads are useful flight corridors for bats and provide a good and convenient place to set up harp trap.

How to Use a Harp Trap

- Check that all traps are complete and serviceable before going to harp trapping locations to erect traps. Frequent problems are the odd broken string that must be replaced, telescopic legs jammed with sand, or other mechanical problems caused by corrosion due to storage of wet traps.

- If we are trapping at a survey site, a harp trap should be placed within 50 meters of the transect line, if possible near the center of the transect.
- Onto a topographic map, record where the harp traps are placed as a permanent record.
- Harp traps at a survey site, or those placed at specially targeted locations, should be erected in gaps in the vegetation so that the string banks are freely exposed and able to intercept bats that might habitually fly through such gaps.
- To ensure that animals are not being stressed from recaptures, put harp traps out for a maximum of three nights in any one area.
- Animals captured in harp traps may be carefully removed and placed into a calico bag when ready for identification. Do not remove animals from traps until ready to identify in order to minimize stress.
- Also note that insectivorous bats can suffer heat stress if expose in sunny conditions even if they are in the capture bag of the bat trap. Harp traps should be checked within two hours of dusk, and again at intervals of 3-4 hours through the night. They should be finally cleared and closed one hour before dawn.
- Any animals caught should only be handled by experienced people who are licensed to carry this out.
- Bats should be released at the capture location as soon as possible.
- Keep trapping equipment clean at all times.

Special Considerations and Tips

- Always make sure that the trap is dry before storage.

Animal Care Issues

- When approaching a harp keep noise to a minimum so as to reduce the stress to captured bats.

- Insectivorous bats are susceptible to the heat stress. Harp traps must be checked regularly and cleared before exposure to sun.
- Do not remove animals from traps until ready to identify. Animals should be removed from the trap, identified and released at the point of capture.
- Release an animal as son as possible after identification, but at night only.
- Harp traps should be checked within two hours of dusk, and again at intervals of 3-4 hours through the night. They should be finally cleared and closed one hour before dawn. Bats should be released on the night of their capture if possible.
- If it is necessary to hold bats during the day they must be held in appropriate conditions by a licensed person. They must be released the next night at the point of capture.

Cage Traps

Target Animals and Habitat

Baited cage traps have been the traditional method for catching medium-sized mammals. Cage traps seem to be most successful in forests, shrublands, woodlands and grassland.

How to Use a Cage Trap

- Check the trap is working correctly. Trigger mechanisms often get jammed with bait or other objects.
- Set trap on a flat surface so it does not rock when the animal enters it.
- Traps should be placed along a minimum 250 metre transect at spacings of about 50 metres where visibility is low. In open country with good visibility they should be placed at intervals of between 50 and 100 metres.
- Assign each cage trap an identify number which should be written on it.

- Into a topographic map, record the trap identity number and where the pitfall trap was laid along the transect line as a permanent record.
- Tie flagging tape into plants to aid in identifying the location of each trap and write the trap number on the tape with a permanent marking pen. Ensure, where possible, that each piece of flagging is visible from the two trap locations on either side of it. In very dense vegetation it may help to place extra flagging tape at mid-points between trap locations. The trap must be removed when the survey is finished at it is polluting and may be confusing for future surveys.
- To ensure that animals are not being stressed from recaptures, put the cage traps out for a maximum of three to four nights in any one area.
- Set each cage trap in a sheltered location for example, under vegetation and away from the morning sun so that it does not get too hot before it is checked each morning.
- Always provide dry bedding material and more in autumn and winter. We can use dry leaf litter or shredded newspaper or non-absorbent cotton wool placed in plastic for further insulation.
- Animals captured in cage traps may be carefully removed and placed into a hessian bag when ready for identification. Do not remove animals from traps until ready to identify where captured in order to minimize stress.
- Animals should be removed from the trap and identified at the point of capture. Do not carry the trap to a different location, as this will cause unnecessary stress to the animal inside.
- Any animals caught should only be handled by experienced people who are licensed to carry this out.
- Clean the traps between setting if they have caught an animal.

Cage Trap Baits

- Cage traps should be baited with either fruit or meat baits.
- Often, alternate traps are baited with fruit and meat baits to ensure that both herbivores and fungus-feeders for example, potoroos and carnivores for example, quolls will be sampled.
- Any fruit is acceptable as an attraction for herbivores and fungus-feeders. Place one or two pieces of fruit at the rear of the cage.
- For carnivores, a range of baits can be used including pieces of cooked chicken, a spoonful of cat or dog food, or sardines.

Special Considerations and Tips

- Frogs, reptiles and birds are also captured on occasions, particularly where something acts to direct animals towards the trap, for example, fallen logs or fences.
- Make sure that each cage trap is identified with a number before it is put into the transect line.
- When checking and collecting cage traps, their identity numbers should be recorded as each trap is checked. This ensures that all cage traps are checked daily and all are collected after the survey is completed.
- To clean the cage traps simply set to with hose, bucket and scrubbing brush. Before storing traps away the survey, wash them out with water and detergent and then dry them.

Animal Care Issues

- Always provide bedding such as plenty of dry leaf litter or shredded newspaper and more in cold weather. Generally, from October to December it can be cold enough to be a concern but does depend on where the survey is conducted and any possible sudden shifts in the weather.

- Cover the rear part of the cage trap with a bag to act as barrier to rain and cold conditions. Make sure the bag is positioned so that it will not collect water or flap in the wind.
- Make sure the cage trap is secure on the ground if positioned on a slope, so it is not likely to move while an animal is trapped within. Tent pegs and light rope are useful for securing traps in place and not necessary for trapping in flat areas.
- When approaching cage traps keep noise to a minimum so as to reduce the stress to the animals in the trap.
- Be aware that not all nocturnal mammals are fully nocturnal and may enter traps during the day, where they will be susceptible to heat stress. Traps may require checking during the day or need to be closed in the morning and reopened in the late afternoon if we suspect this may be occurring.
- Do not remove animals from traps until ready to identify. Animals should be removed from the trap and identified at the point of capture. Do not carry the trap to a different location, as this will cause unnecessary stress to the animal inside.
- Release an animal as soon as possible after identification.

Nocturnal Call Playback

Target Species and Habitat

Playing the natural calls of some mammals through a loudspeaker can often prompt a responding call from animals living in the area. The species making the response can there be identified from its call by an experienced listener. The technique works best for medium and large territorial tree-dwelling mammals that naturally communicate by calls, such as gliders and koalas. Some nocturnal birds such as large forest owls also can be detected this way, solitary species such as some gliders may additionally use calls to warn of dangers and advertise food resource to members of the same social group. Sometimes animals will respond to calls of a different species; for example, gliders may respond with alarm calls to the calls of owl species which prey on gliders.

How to do Nocturnal Call Playback

- Call playback is conducted only at the mid-point of each transect.
- The survey should begin when it is completely dark and in the first few hours of the night. Start with 10 minute listening period to check for any mammals calling. Everyone in the team needs to be silent during this stage.
- The call back tape used must have full range of calls known for each species expected in the survey area and must not have any males and females calling to each other as this can cause distress for mammals during the breeding season. Carry out calls only for species not already recorded at the site.
- Play each species call for 5 minutes in darkness, sweep with spotlight and then listen for 10 minutes.
- Repeat for next species
- Play calls in darkness while viewing the skyline and tree canopy in all directions for incoming nocturnal mammals.
- Repeat this procedure at each survey site on as much night as possible during a survey.

Special Considerations and Tips

- Windy and rainy conditions should be avoided as it is difficult to hear any response to calls played.
- People in the team should be some distance apart so that there is more chance of hearing any animal calls.
- Only play calls for mammals that are likely to occur in the habitat our site is in. But do not assume that because the country maybe degraded that it will not be attractive to nocturnal mammals. Possums live and breed in the middle of our large cities.

Animal Care Issues

- Avoid spotlighting the animals for prolonged periods as it may damage their eyes or case temporary blindness. They could become completely disorientated and cause themselves harm.
- Use a respondent filter and dimmer switch to reduce light intensity for observations, once the animal has been spotted.
- Care should be exercised when trying to attract an animal using call playback as it may leave it open to predation by another species.
- If a mammal has already been identified by its call then it is not necessary to play the call again at that site on the same night, and it is not necessary to continue playing the call to bring it closer for visual identification as this may cause undue distress.
- Ensure that the calls we are playing are not distress calls. A territorial or mating call is likely to provoke a similar response from a member of the same species, but a distress call may provoke a distressed responses. In some cases, a distressed mammal may feel from the safety of their home territory or perhaps retreat to their nest and not feed that night.

Elliott Traps

An Elliott trap is a small (30x10x10cm) box made of aluminium which collapses flat for easy storage and transport. When set, animals enter the open end of the trap, lured by bait placed at the far end. The animals steps on a treadle on the floor of the trap, which triggers the door of the trap to close via a spring mechanism. Over the past 20 years they have become the standard trap for studying and surveying small mammals because they are easy to use, relatively inexpensive, and suitable for a wide range of species.

Target Animals and Habitat

Baited Elliott traps have been the traditional method for catching small mammals. Elliott traps seem to be most successful for small mammal trapping in forests, woodlands, shrublands and grasslands.

How to Use an Elliott Trap

– Check that trap is working correctly. Trap treadles often get jammed with bait or other objects.
– Set trap on a flat surface so it does not rock when the animal enters it.
– Traps should be placed along a 250 metre transect at spacings of 10 –15 metres where visibility is low. In open country with good visibility they should be placed at intervals of between 30 and 50 metres.
– Assign each Elliott trap an identify number which should be written on it.
– Onto a topographic map, record the trap identify number and where the Elliott trap was laid along the transect line as a permanent record.
– Tie flagging tape into plants to aid in identifying the location of each trap and write the trap number on the tape with a permanent marking pen. Ensure, of flagging is visible from the two trap locations on either side of it. In very dense vegetation it may help to place extra flagging tape at mid-points between trap locations. The tape must be removed when the survey is finished as it is polluting and may be confusing for future surveys.
– Some species are less likely to enter traps on the first or second night. Elliot traps should be placed in any one for a minimum of 2 nights, a maximum of 4 nights, to minimize stress from recapture. If we are trapping more often, we should leave a gap of at least as long as the number of consecutive days between trapping. i.e., trap 4 days, cease 4 days trap again for 4 days.

- Set the Elliott trap in a sheltered location for example, under vegetation and away from the morning sun so that it does not get too hot before it is checked each morning.
- Always provide dry bedding material and more in autumn and winter. We can use dry leaf litter or shredded newspaper for further insulation.
- Animals captured in Elliott traps may be carefully removed and placed into a calico bag while preparing for identification where captured. Do not remove animals from traps until ready to identify in order to minimize stress.
- Any animals caught should only be handled by experienced people who are licensed to carry this out.
- Clean the traps if they have caught an animal before resetting them the next night. Clean and assemble traps with care as the trigger mechanism can get damaged.
- On slopes, Elliott traps should be placed so that they are roughly horizontal, and braced against vegetation to ensure they do not roll down the slope.

Elliott Trap Baits

- Elliott traps should be baited with a rounded teaspoon of a mixture of peanut butter, rolled oats and candied honey.
- To mix up the bait use a clean bucket and our hands and mix the ingredients to the consistency of a thick paste.

Special Considerations and Tips

- Frogs, reptiles and birds are also captured on occasions, particularly where something acts to direct animals towards the trap, for example, fallen logs or fences.
- Be very careful when opening an Elliott that has been triggered and has a very heavy occupant. It may be a snake. Experienced, licensed trappers should know how to handle this.
- To clean the Elliot traps remove the pin to open the trap. Clean out the bait mixture, bedding and scats. Use a small

brush for example, nailbrush to wash trap out in water and detergent. Before storing traps away after the survey, ensure they are dry. Store Elliott traps in a collapsed, folded position with the holding pin in place.

- Make sure that each Elliott trap is identified with a number before it is put into the transect line.
- When checking and collecting Elliott traps, their identity numbers should be recorded as each Elliott trap is checked. This ensures that all Elliott traps are checked daily and all are collected after the survey is completed.

Animal Care Issues

- Always provide bedding such as plenty of dry leaf litter or shredded newspaper and more in cold weather. Generally, from October to December it can be cold enough to be a concern but does depend on where the survey is conducted and any possible sudden shifts in the weather.
- Cover the rear part of the Elliott trap with a plastic bag to act as barrier to rain and cold conditions. Make sure the bag is positioned so that it will not collect water or flap in the wind.
- Make sure the Elliott trap is secure on the ground if positioned on a slope, so it is not likely to move while an animals is trapped within. Tent pegs and string are useful for securing Elliott traps in place not necessary for trapping in flat areas.
- When approaching Elliott traps keep noise to a minimum so as to reduce the stress to the animal in the trap.
- Be aware that not all nocturnal mammals are fully nocturnal and may enter traps during the day, where they will be susceptible to heat stress. Close traps during the day by collapsing them otherwise curious mammals will pry them open. If traps remain open during the day they must be checked regularly.

- Do not remove animals from traps until ready to identify. Animals should be removed from the trap and identified at the point of capture. Do not carry the trap to a different location, as this will cause unnecessary stress to the animal inside.
- Release an animals as soon as possible after identification.

Opportunistic Mammal Sightings

Many species will be detected while traveling to and from survey sites, or outside standard survey times or survey sites. Record all of these sightings and their locations, dates and times for the whole survey on the opportunistic animal sightings field data sheet provided. This additional information is valuable as it helps to build the list of mammals recorded for each area. Some of the mammals we see outside the survey area or in the survey are out not near a survey site may provide some insight into how mammals utilize a number of habitat types.

Field Data Sheet

Codes necessary to complete this field data sheet are given in the Field Work section.

Date of First Observation

Enter the date of the first opportunistic observation on the field data sheet.

Date of Last Observation

Enter the date of the last opportunistic observation on the field data sheet.

Animal Species Code

Enter the four-digit species code from the list of species provided in the Field Worksection.

Animal Species Name

Enter the common and scientific name.

Number Observed

Enter the number of individuals observed. A code may be added after the number to indicate that the count is an estimate.

Observation Type

Enter the single letter code from the list in the Field Worksection corresponding to the type of observation made of the species.

Habitat Type

Enter the code that best describes the microhabitat

Breeding

Enter a code from the table provided in the Field Worksection corresponding to indications we have of the breeding conditions of the individuals. If we do not know their breeding condition leaves this blank.

Written Locality

Give a brief description of the observation locality.

Notes

Space is provided at the bottom of the field data sheet for we to enter any additional notes that may help interpret any of the observations.

Special Considerations and Tips

- Take care to collect all field data sheets into a single bundle of sheets at the end of the survey. A survey that extends over several days can lead to an accumulation of opportunistic sightings entered on many copies of the opportunistic animal sightings field data sheet.
- If there is more than one opportunistic animal sightings field data sheet at the end of a survey, we should give them a sequential number when they are bundled up. This will greatly assist those involved in entering records into databases.

Animal Care Issues

- Only catch and handle animals for identification if we are experienced and properly licensed to do so.
- If capture is necessary for identification, always ensure the stress to the animal is kept to a minimum by using appropriate techniques and handling the animal only for as long as need for identification.
- Always release captured animals as soon as possible at the place that they were captured and in conditions that will not increase risks to the animal for example, predation.
- When observing an animal be sure to keep quite and all movements to be slow, so as not to startle them. Unless of course we are being chased which is not likely to occur, as they will normally run from us.

Spotlighting

Target Animals and Habitats

Arboreal tree dwelling mammals are the targets of this survey method. Owls and other birds roosting for the night may also be detected during spotlighting. Such records should be entered on the spotlighting field data sheet described below.

Opportunistic sightings or hearings include all mammals seen or heard outside the recording time period of ten minutes spent at each sample area. This situation may occur when team members are walking between sample areas or setting up the transect. All mammal sightings or hearings should be recorded as opportunistic sightings on the opportunistic animal sightings field data sheet.

This method is suitable for all habitats, except for areas without trees.

How to Spotlighting

- The survey should being when it is completely dark and in the first few hours of the night. Avoid times when the full moon is out when spotlighting is known to be less successful.

- Spotlighting should be carried out along the transect at each survey site.. The transect should be walked at one kilometer per hour or less, being careful not to disturb any traps that may be around the transect.
- Spotlighting may be conducted along roads or tracks when traveling between sites. These observations should be recorded as opportunistic sightings.
- If a mammal is moving across the transect, even during the recording time period, without stopping, they are recorded as a opportunistic sighting.
- Clearly mark the start and finish of each transect line, with flagging tape ensuring that they are easily located when dark.
- All animals seen or heard that can be clearly identified within 50 meters either side of the transect should be recorded on a spotlighting field data sheet.
- All animals seen or heard outside this 50 metre distance are recorded as opportunistic.

Special Considerations and Tips

- Team members should be some distance apart to increase the change of hearing any mammal calls.
- Using call playback tapes is more likely to cause arboreal mammals to call out and reveal themselves.
- Remain quiet at all times even on the way to the next site. This will ensure that animals are less disturbed and more likely to reveal themselves.
- Take care not to disturb any diurnal birds.

Animal Care Issues

- Avoid spotlighting the animals for prolonged periods as it may damage their eyes or cause temporary blindness. They could become completely disorientated and case themselves harm.
- Use a red filter and dimmer switch to reduce intensity for observations once the animal has been spotted.

Alternatively we can switch off the powerful spotlight and use lower powered hand torches so as not to hurt the animal's eyes.

Pitfall Traps

A pitfall trap is a hole in the ground in which a bucket or length of plastic pipe is placed so that the top of the plastic is level with or just below the surface of the ground. Animals fall into the open trap and cannot climb out because the sides of the trap are too slippery. Pitfall traps are made more effective at capturing animals by using drift fences, a length of short fence which runs either side of the pit or between several pits. Animals encountering the drift fence move along its length and are directed to the trap. Pitfall traps are suitable for capturing both frogs and reptiles during the day or night, as well as small nocturnal mammals.

Target Animals and Habitat

This method is most useful for small mammals in the semi-arid and arid regions. A number of rare animals are most easily captured in pits, for example, ningauis.

How to Use a Pitfall Trap

- Pitfall traps should be greater than 40 centimeters deep unless physical limitations prevent digging to this depth. The depth required to hold a hopping mouse is unknown, but is likely to be much deeper. Shallower pits will allow some dasyurids and rodents to escape.
- The tip of bucket must not be higher than the ground surface and there must not be any holes or gaps around the outside of the bucket.
- Each hole should have a minimum of 10 meters of drift fencing associated with it.
- Lawet four pits in a line at the edge of the sample area preferably centred along the length of the transect so that the center of the pit line is near the center of the transect.
- The pits should be spaced about 20 metres apart with a continuous drift fence running along the pit line.

– The drift fence should extend about five meters past each end of the pit line. Drift fences should be made from a material that cannot be climbed by small mammals. Thick plastic, such as builder's dampcourse, is ideal. Shadecloth, fabrics or flyscreen are not suitable. Drift fences should be at least 15 cm high when erected, so that mammals cannot jump over them. They should be held in place by metal stakes or tent pegs.

– Place a few handfuls of leaf litter in the bottom of the trap to provides shelter. Beware that this also provides a shelter for snakes and spiders or other harmful animals.

– Always mark the spot where the pitfall trap is with flagging tape on a ten pet, tree or something similar that is not likely to move.

– Record the pitfall trap identifies number on the flagging tape using a black permanent marking pen and onto the field data sheets. The tape must be removed when the survey is finished as it is polluting and may be confusing for future surveys.

– Onto a topographic map, record the trap identity number and where the pitfall trap was laid along the transect lines as a permanent record.

– Animals captured in pitfall traps may be carefully removed and placed into a calico bag while preparing for identification where captured.

– Any animals caught should only be handled by experienced people who are licensed to do this.

Special Considerations and Tips

- Frogs, reptiles and flightless birds are also captured on occasions, particularly as the drift fences act to direct animals towards the trap.
- Make sure that each pitfall trap is identified with a number before it is put onto the transect line.
- When checking and collecting pitfall traps, their identity numbers should be recorded as each trap is checked. This

ensures that all pitfall traps are checked daily and all collected and the holds are filled in after the survey is completed.

- Be aware of dangerous animals being caught in a pitfall trap, for example, bull ants, spiders, scorpions and snakes. It is best to look before we put our hand inside the pitfall trap and move the leaf litter around with a stick while looking for any harmful animals.
- If other animals are caught in the trap, they should be treated with the same care we would show any mammal. If we are able to identify them, do so.
- At the end of the survey, be sure to fill in the hold the pitfall trap was in.

Animal Care Issues

- When positioning pitfall traps be aware of the potential for water run-off in case of rain, which can cause it to flood.
- Place the pitfall trap in a sheltered location away from the morning sun, so that it does not get too hot before the trap is checked each morning.
- Place a flat stone in the bottom of the pitfall trap to prevent animals from drowning when it trains. Alternatively use pitfall traps with several holes drilled in the base to allow rain to run out.
- When approaching pitfall traps keep noise to a minimum to reduce the stress to the animal in the trap.
- Take care when removing an animal from a pitfall trap to minimize harm to the animal and to ourself.
- Be aware that ants may invade pitfall traps and can attack and kill small mammals and other animals. If this occurs, spray the upper walls of the pitfall trap with a thick layer of insect repellent.
- Mammals mostly enter pitfall traps during the night. However, pitfalls will also capture reptiles during the day. If we are not using pitfalls to survey both mammals and

reptiles, ensure that the pitfalls are closed with a tight-fitting lid during the day. Pitfalls which are left open during the day must be checked during the day, to avoid heat-stress on captured reptiles.

Animal Welfare

- All survey methods that deal with the capture of animals must be supervised by a person who is proficient in and holds a current licence for which allows them to carry out these surveys. It is illegal and potentially life-threatening to the mammals involved to carry out capture techniques without the appropriate licence.
- Some of the methods used to survey mammals can cause stress to the animal and the aim should always be to reduce this factor.
- Always ensure that there are sufficient people for the survey so that all traps are inspected at first light and not left until later in the day. Remember that the animal will already be confused and concerned about not being able to get away. If the animals are left in the trap for longer than is necessary, the morning sun may cause them to overheat. In a cold situation, hypothermia may set in and the animal may die. Some animals are more susceptible to hypothermia than others. In general dasyurids are more susceptible than native rodents.
- Animals should be released as soon as possible after they are caught. Ensure that the animals are released at the capture site. If the animals are released too far away from where they were found, they could be attacked by another of the same species because they are in the wrong territory. Release animals into shelter for example, dense ground vegetation that protects them from predators and minimizes disorientation.
- Keep the time spent handling animals to the minimum required for identification to reduce stress on the animal. If it is necessary to keep animals for longer than the minimum required to identify it, for example, because

we require a second opinion on the identification, ensure that animals are kept in quiet, dark cool and ventilated conditions.

Scat Collection

Target Animals and Habitats

Scats or animal droppings have proven to be very efficient sampling techniques for mammals. The identification of the depositing species can usually be determined from the size and shape of the scat. Several useful filed guides have been published.

In addition, the species consumed by the depositing species can often be determined from examining the contents of scats. This can be done for both herbivores and carnivores, although it is very rarely done on a routine basis for herbivores. This is because there are far more plant species to be accounted for and because very highly skilled people are required to perform the identifications of microscopic plant material.

The situation is much better for predators and very useful information can be gathered on other species of animals present in an area by examining the contents of predator scats that is, hair, feather and bone fragments. This still requires expert assistance, but the rage of possible prey species is much narrower and sufficient reference material is available to ensure that accurate identifications are usually possible. Scat collection can be attempted in all habitats but may be difficult to do in wetlands.

How to do Scat Collection

Scat collection is undertaken at a standard survey site or opportunistically at any other time and place in the survey area. A useful approach is to undertake a scat collection search as part of the site reconnaissance or during the marking out of a transect at a site. We should also collect scats found in and around hair tubes, Elliot and other traps say within a 50 metre radius of each trap or those found when spotlighting.

Sending Scats for Specialist Analysis and Identification

Careful information management is required after scats are collected to ensure that we are able to relate the identifications provided by experts to our field records. If we are unsure of how to prepare scats for submission to an expert, contact them and discuss their requirements. The following notes provide an outline of the important aspects of good information management. Always send a copy of the field data sheet, not the original. Some expert identification services may agree to write the results on the copy of the field data sheet and send it back to we. Often, however, they will send us a printout of identifications against the unique identifying information that we wrote on each envelope and recorded on we field data sheet, together with an invoice.

Special Considerations and Tips

- Do not handle scats with bare hands. Many animals carry worms and other parasites which are harmful to humans.
- Do not store scats on plastics bags. They can rapidly go mouldy.
- Do not handle equipment, clothing or unprotected skin while wearing gloves which have been used to collect scats.
- Wash hands thoroughly with soap and water after collecting or handling scats and before handling food or cooking utensils.
- For comprehensive surveys, members of other teams from different disciplines can also assist by collecting scats. Ensure that they have all the necessary materials and equipment, copies of scat collection field data sheets, and have been instructed in safe handling procedures.
- Ensures that scats do not get crushed, which may make them impossible to identify.

Target Animals and Habitat

There are many mammals that leave signs of their presence. These include scats that is, droppings, and tracks, scratches on trees, diggings, shelters and bones. The previous section looked at the topic of scat collection and identification. Methods to collect information on all other forms of mammal signs are described in this section. Sign search can be applied in all habitat types.

How to do Sign Search

Collecting information on mammal sign other than scats is undertaken at a standard survey site or opportunistically at any other time and place in the survey area. We should also collect information on signs around hair tubes, Elliott traps is kept within a 50 metre radius of each trap or those found when spotlighting.

Look for track or footprints in area of soft soil or mud. Good places to look are around dams, puddles, or along the sides of dirt roads and vehicle tracks. Tracks are best seen in slanting light, so searching in likely spots in the early morning or late afternoon is ideal. Look for scratches on trees caused by arboreal marsupials such as gliders, possums and koalas. Any skulls or other bones may be collected and the species identified from a field guide or by referring the material to experts. The steps involved in collecting information on mammal signs other than scats are detailed below/ when a sign is located that we wish to record write the following details on the sign collection field data sheet

- AMG zone, Easting and Northing or the site number if we at a standard survey site ;
- Surveyor name;
- Date;
- Preliminary identification of species if known.

If we are collecting a skull or other bone fragments for later identification by experts, then clearly label the specimens with an appropriate and unique field identifier and record that identifier on the sign collection field data sheet.

Field Data Sheet

Record animal sign data on the opportunistic animals sightings field data sheet. Record the type of sign we have found using the observation type codes. Photograph the sign, if possible.

Special Considerations and Tips

- Ensure that we take reasonable precautions when handling material from recently deceased carcasses.
- Take care when handling skulls as they can be very fragile.

Animal Care Issues

- When we find signs of a mammal dwelling the best thing we can do is leave it alone. This is an animal's home or territory and disturbing these sites in any way may cause unnecessary stress.

Sending Specimens such as skulls and other bones for expert identification

Sometimes we may not be able to identify skulls or other bones and we need to seek expert advice. Before sending any specimen away for identification, we must record a unique field identification number on the specimen and on the field data sheet. If we are going to send the specimen to the Museum, fill out a voucher specimen field data sheet and submit it with the specimen. Always send a copy of the voucher specimen field data sheet.

Target Animals and Habitats

Large diameter 10-15 centimeters hair tubes are be used to detect mammals that have greater than 500 grams body weight, such as, quolls, dingos, bandicoots, bettongs and potoroos. Hair tubes with a diameter of 4 centimeters are used for mammals that weigh less than 500 grams, such as, feather-tailed gliders, pygmy possums, rodents and dasyurids such phascogales, planigales and antechinuses. Hair funnels are theoretically able to capture hairs from mammals of both sizes. However, they are more expensive to use because they must be bought from the manufacturer. In contrast, hair tubes can be made by anyone from easily available materials.

Point Count Method

The basic method that has been chosen is based on setting up a single line at each site called a transect. Birds can be identified either visually, or by their calls. This method involves identifying all the birds we see or hear while standing at a series of points along a transect a straight line through the site.

A systematic search over a fixed area and for a fixed time such as the method specified here has the added advantage of providing an index of the abundance of individuals and species, which can be later compared with other sites where the same systematic search method was used. The reliability of the abundance index can be reduced by either overestimates or underestimates of bird numbers. To reduce overestimates, particularly when several team members are observing, try to ensure that each individual bird is recorded only once. Inexperience bird watchers may find they need to refer to sections to aid identification, which will reduce the time spend watching and may result in underestimates of abundance. Hence, ensure that a least one member of the team is watching at all times.

Birds Survey

- Be small in size with two to four members in each team;
- Have one team member nominated as the scribe and therefore be responsible for accurately and consistently recording the bird observation provide by other team members; and
- Speak and move quietly and avoid sudden movements.

A count should be conducted on all transects each day the survey is running, unless weather conditions are such that it would be difficult to see or hear the birds. All transects should be surveyed in a different order each day because the birds observed will vary depending on the time of day. Bird counts are conducted at the start of first light, which is before sunrise. This is the time when birds vocalize most, and is known as the "dawn chorus". It is also a time of maximum bird movement as birds move through the bush to begin feeding. All transects should be finished within 3 to 4 hours after first light. On cold mornings keep our binoculars inside clothing when not in use. This will reduce fogging when we are looking through them. Always check where we are standing when we stop to view a bird, as ants have little respect for busy bird watchers.

Special Considerations and Tips

- Weather such as wind, heavy rain, mist, fog, haze and electrical storms are good reasons not to conduct a bird survey.
- When a bird survey transect is in dense habitats where it is difficult to see birds it may be necessary to reduce the size of the circle at each sample point, for example, in dense forest or high dense shrub where visibility is not more than 20 meters. In such cases we can survey within the 15 meters or 30 meters radius sample area.
- It may be difficult to identify some species in dull light. If a transect is in an area that is still in darkness after sunrise, like a steeps gully, then this transect should be done later in the morning as the light in the sky increases.
- Teams that include members who have never identified birds before may benefit from extra training handouts.
- Another method for training team members that can be used in conjunction with observation is recording the calls at each point along the transect, making a note of taken back to camp when all transects are done so that all team members can become more familiar with the bird calls and assess any discrepancies. Recording will work better if we have a directional microphone.
- Any bird watcher should avoid close-range inspection during breeding and feeding as this may disturb the bird. In extreme cases, a bird may permanently abandon a nest site if it feels threatened by humans observing too closely. Aggressive or disturbed bird behaviour loud repeated calls or frantic flight may indicate that they are too close to a bird's nesting site.
- Remember that some birds nest on the ground or in vegetation very close to the ground. Be careful where

we walk when pushing through dense shrubs so that we don't disturb nests.

Nocturnal Call Playback

Playing the natural calls of some mammals through a loudspeaker can often prompt a responding call from animals living in the area. The species making the response can there be identified from its call by an experienced listener. The technique works best for medium and large territorial tree-dwelling mammals that naturally communicate by calls, such as gliders and koalas. Some nocturnal birds such as large forest owls also can be detected this way, solitary species such as some gliders may additionally use calls to warn of dangers and advertise food resource to members of the same social group. Sometimes animals will respond to calls of a different species; for example, gliders may respond with alarm calls to the calls of owl species which prey on gliders.

Call playback is conducted only at the mid-point of each transect.

The survey should begin when it is completely dark and in the first few hours of the night. Start with 10-minute listening period to check for any mammals calling. Everyone in the team needs to be silent during this stage. The call back tape used must have full range of calls known for each species expected in the survey area and must not have any males and females calling to each other as this can cause distress for mammals during the breeding season. Play each species call for 5 minutes in darkness, sweep with spotlight and then listen for 10 minutes. Record on our field data sheets the calls of any animals that are heard repeat for next species. Play calls at the same volume at each site and on each night. Calls should be loud enough to attract neighboring animals, but not much louder than the call of the animal themselves. If calls are played too loudly, animals which respond may be too far away to be heard by the survey team. Viewing the skyline in all directions for incoming nocturnal birds.

Special Considerations and Tips

- Windy and rainy conditions should be avoided as it is difficult to hear any response to calls played.
- Only play calls for mammals that are likely to occur in the habitat site is in. But do not assume that because the country maybe degraded that it will not be attractive to nocturnal mammals. Possums live and breed in the middle of our large cities.

Animal Care Issues

- Avoid prolonged exposure of birds to bird call plays on tape and to the spotlight.
- Use a light with a narrow beam.
- Use a red filter and dimmer switch to reduce light intensity for observations, once the animal has been spotted.
- Care should be exercised when trying to attract an animal using call playback as it may leave it open to predation by another species.
- If the bird has already been identified by its call then it is not necessary to play the call again at that site on the same night.
- Ensure that the calls we are playing are not distress calls. A territorial or mating call is likely to provoke a similar response from a member of the same species, but a distress call may provoke a distressed responses. In some cases, a distressed mammal may feel from the safety of their home territory or perhaps retreat to their nest and not feed that night.

Bird Survey

The taking of voucher specimens must be under the control of licence holders only. Although the bird survey methods presented in the section do not involve the capturing of birds, there may be situations where we find road kills or have access to birds euthanased by vets due to injury. The treatments described below can be used to preserved bird specimens from either source and prepare them for delivery to the Museum. Specimens will only be useful when they are accompanied by good field records.

Preservation of Dead Animals

We will require a 10% formalin solution made up in the following proportions : 10 percent formaldehyde, 90 percent water preferably distilled and buffered with calcium carbonate about 1teaspoon per litre of formalin.

Method for All Pickled Specimens

- Make slit through to abdominal cavity, just large enough to allow penetration of the preserving fluids, making sure not to damage internal organs.
- Inject formalin into the large pectoral muscles
- Preserve body in formalin for 48 hours for mammals as small or smaller than house sparrow; up to 3 weeks for larger specimens.
- Transfer body to 70-75% ethanol after draining formalin from body.
- Place the body in a glass jar or vial topped up with the ethanol.
- Label the body jar or vial with the field number.
- Ensure the voucher specimen data form is completed, including the nothing of any tissue samples taken.
- Do not store the specimens in the sun as this can cause fading of colours.

Freezing is a good alternative. Wrap bodies tightly in thick plastic bags. Do not cut the animal open.

Safety Considerations and Tips

- Formalin and formaldehyde are extremely corrosive to living tissue, and will cause skin to peel; wear plastic gloves at all times when handling; do not breathe vapours; only use in a well-ventilated space; avoid all contact with eyes.

- Transport formalin and formaldehyde only in clearly labeled bottles with secure, leak-proof safety caps, and inside a second, sealed container that will protect the bottle from extremes of heat and breakage on rough roads.
- Formaline and formaldehyde should never be kept in close proximity to food or cooking equipment.
- Small volumes of ethanol up to 5 liters can be bought from volume greater than 5 liters can only be bought from a registered supplier by those holding a current concessional sprits permit.
- Ensures that specimens are preserved in ethanol, not formalin, and that all containers are tightly sealed and do not leak. If ethanol is not available and methylated spirts have been substituted, note this in the covering letter that accompanies the specimens.
- Ensure that every individual animal has a unique number on a tag attached firmly to the body, preferably to one of the limbs. Tissues samples taken from individuals should have the same number as the remainder of the body.
- Birds may be tagged with either heavy paper and string tags or preferably using a plastic tag made by punching the number into a Dynamo labeler.
- Each individual sent to the Museum must be accompanied by a completed voucher specimen field data sheet. Forms should be filled out in dark pencil or ink which is indelible to alcohol. Drawing pen ink and some special felt-tipped marker and alcohol-proof, but most felt-tipped markers and ball-point ink will run badly. A copy should be placed inside the jar or vial.
- Please keep photocopies of the completed voucher specimen field data sheet in a safe place, along with a record of when we sent the material to the Museum.
- Voucher specimens also should be accompanied by a covering letter which lists the specimens. Museum collection managers should be requested to complete the sheet and return it to the person nominated, in order to

verify the identification of the specimen. Do not forget to provide a return name, address and phone number at the bottom of the form and a stamped, self-addressed envelope.

– Ensure specimen jars and vials are packaged with plenty of padding and in a crush-resistance box.

Opportunistic Bird Sightings

Many species will be detected while traveling to and from survey sites, or outside standard survey times or survey sites. Birds may be identified opportunistically either by their call or by their appearance. Record all of these sightings and their locations, dates and times for the whole survey on the opportunistic animal sightings field data sheet provided

This additional information is valuable as it helps to build the list of birds recorded for each area.

Special Considerations and Tips

– Take care to collect all field data sheets into a single bundle of sheets at the end of the survey. A survey that extends over several days can lead to an accumulation of opportunistic sightings entered on many copies of the opportunistic animal sightings field data sheet.

– If there is more than one opportunistic animal sightings field data sheet at the end of a survey, we should given them a sequential number when they are bundled up. This will greatly assist those involved in entering records into databases.

Animal Care Issues

– Any birdwatcher should avoid close-range inspection during breeding and feeding as this may disturb the bird. In extreme cases, a bird may permanently abandon a nest site if it feels threatened by humans observing too closely. Aggressive or disturbed bird behavior loud repeated calls or frantic flight may indicate that we are too close to a bird's nesting site.

– Remember that some birds nest on the ground or in vegetation very close to the ground. Be careful where we walk when pushing through dense shrubs so that we don't disturb nests.

Plants Survey Method

Voucher Specimens

There are two reasons to collect voucher specimens: To allow we to spend time at base camp to properly identify plant species; and for expert identification at a later time.

Both of these require plant material to be preserved. Material must be properly preserved in order to submit it to the Herbarium or other herbaria, for example, at universities, for identification. We may wish to add the species to our field herbarium for future reference or the herbarium may wish to add our material to its collection. Specimens of most plants can be preserved indefinitely by careful drying followed by storage in dry, insect-free conditions. Although lacking the freshness and colour of live material, dried specimens, with notes taken in the field, usually provide most of the features required for the identification and botanical study of the plant.

Specimens accompanied by notes taken in the field at the time of collection provide a record of the occurrence of that species at that time at the given locality. Specimens can be re-examined at a later date if there is doubt over the identification of the plant. If a record of a species is new for an area for example, an extension of the known distribution of a native species or a record of a recently naturalized species for the area, it is important to send a specimen to the State herbarium for conformation. The lodgement of labeled specimens at the State herbarium builds up the know distribution of species within the state and is a basis for future botanical and ecological research work.

A herbarium is a collection of dried plants; it may be a small private collection or a large collection. This collection contains almost a million specimens and is the largest collection of plant specimens. These specimens provide a record of past and present distribution of plant species and are used as a basis for correct identification of species and for botanical and ecological research on plants.

Learn to use the hand lens properly as it is important to use it when looking at details of hairs, glands, oil dots, leaf buds, fine veins, etc. The essential point is to hold the lens close to the eye and to bring the specimen back and forth to get it in sharp focus, rather than moving the lens. There are two advantages here : first we get a wide field of view and the greatest useful magnification that the lens can provide; and second it is much easier to get good light on the specimen, without shadows being cast on it by the lens.

How to Collect in the Field

Before we being collecting specimens it is a good idea to look at all of the plants of the species growing around the site. If there is a lot of variation in the plants, such as leaf size or shape, then we should collect a few specimens to show this variation don't lump all of the material together unless it is all from one plant, but make separate collections. We should take care that we do not remove the only plant in the area, especially if we need to collect the whole plant. Specimens should include as many parts of the plant as possible to help identification and be useful as a plant collection. We should aim to collect an average-looking branch or stem about 20 to 30 centimeters long, with the leaves and flowers attached to the stems. Take note of the following points :

- In the absence of open flowers, buds should be included as they can be cut open to show the number and arrangement of the flowers parts ;
- Developing fruits can be cut open to show the structure of the ovary;
- If variation in leaf shape and size is apparent, specimens should include different parts of the same plant to represent this variation;
- For plants with loose bark, a section of bark should be collected; and eye but water may be strongly

coloured, often red or green. Collection should be made by scooping some water with the algae into a glass jar.

Large Attached Algae including Seaweeds

Collections should include a small piece of rock or whatever they were attached to.

Preservation of Plants

Pressed and Dried Collections

Most collections held in herbaria are preserved as this is the easiest method of handling and storing plant collections. The following sections apply to pressed specimens unless indicated otherwise. Occasionally specimens are allowed to dry without pressing, so as to retain the natural orientation of the organs such as mosses, fungi leaves of flowers on the stem or because the specimens are too bulky to press satisfactorily.

The size of a specimens is usually governed by the size of the sheet of paper on which it is going to be stored. Specimens sent to herbaria for identification should be about 30 centimeters long as herbarium sheets are usually about 43 centimeters long x 28 centimeters wide. Smaller sheets are not used since they encourage the collection of inadequate specimens. However, cards with in 20 centimeters x 13 centimeters may be used for field-reference collections as they can be conveniently taken into the field or stored in a card file. They are not recommended for specimens that are to be used for a long-term, systematic study of specimens as occurs in the major herbaria. For plants with large leaves or massive fruits, do not limit the size of the specimen in the name of convenience. It is more important to have a complete and therefore useful specimen rather than conform to arbitrary rules on the accepted size of the collection.

Specimens are best pressed as soon as possible after collection, before wilting and shriveling occurs. Most plants may be kept in sealed containers such as plastic bags for up to a day if it is not convenient to press immediately. If specimens need to be kept for longer periods, or if the conditions are very warm, then plants should be placed in an esky or refrigerator. Generally we recommend pressing the day's collection of the day's field work. However, some plants show such rapid wilting, particularly of the flowers, that any delay should be avoided as far as possible.

Method

Specimens should be pressed flat and dried between sheets of semi-absorbent paper such as newspaper. Papers with a glossy surface should be avoided because they are not absorbent enough to aid drying. The plants should be carefully laid out between the sheets, as their shape at this stage largely determines their final appearance. The flowers should be spread out with the petals carefully arranged, wilted leaves should be straightened and unnecessary shoots of excessively twiggy shrubs may be cut away. Leaves should be placed so that both the upper and lower surfaces are visible on the specimen. Flowers with copious nectar may go mouldy very quickly if excess nectar is not shaken off before pressing.

When plants are very uneven in thickness, sheets of spongy plastic foam polyurethane or similar about 1 centimeter thick, placed between the newspaper sheets help to distribute pressure evenly across the specimen. If foam sheets are not available, several thicknesses of folded newspaper may be used. Care must be taken to ensure "damp spots" do not develop in the press. When using foam sheets it is advisable to circulate warm air around the press or change the drying papers more frequently.

Specimens are best pressed with moderate pressure, preferably in an arrangement that will permit as free a circulation of air as possible. This can be achieved by strapping the pile together in a press, that is, between frames made from sheets of heavy non-bending cardboard, hardboard, plywood, pegboard or, best of all, a lattice of wood or metal strips for example, weldmesh. Supplies of suitable materials can usually be obtained from packaging and cardboard manufacturers, who will cut materials to suitable sizes, or from hardware and building suppliers. The press frames should be the same size as or a little larger than the drying papers. The two sides of the press should be strapped together withhold leather belts or ropes looped at one end; the other end can be passed through the loop, pulled tight and secured with a couple of half hitches. It is better to use two ropes, one at each end of the press, to even out the pressure across the specimens. Pressing small numbers of specimens by placing sections or other weights on the pile of specimens is not recommended as specimens quickly go mouldy without air circulation.

The paper should be checked for dampness and changed when necessary. As the number of changes require will vary with the original water content of the plants and with the weather conditions, no exact guide can be given. Most plants should dry in less than ten days. For the first few days the paper should be changed daily, but after that time the frequency of changes needed depends on relative humidity levels. In tropical and wet conditions, daily changes will be necessary throughout the drying period but in drier conditions, the final one or two changes need only be given at intervals of three or four days. Used paper should be discarded, or thoroughly dried again before reuse.

When in the field for an extended time, drying can be aided by placing the pressed plants in a warm, sunny position during the day. In reasonably dry climates, drying is aided by securing the presses to the roof rack of a vehicle while driving in dry daytime conditions. If available, a hot-air fan directing air around the press will assist drying. Drying cabinets with a forced circulation of warm air are used in large herbaria to shorten drying times and to reduce the need to change drying papers, but these are not necessary for small numbers of specimens.

A few species regularly turn black on drying, but in general, brownish or blackish colours in the completed specimens, or the growth of fungi, indicate that drying was too slow, often because the papers were not changed frequently enough in the early stages of drying.

Preserving Special Groups

- Mosses, lichens and other plants which cling tightly to rocks, branches or soil Allow the plant to dry inside a paper envelope without pressing.
- *Orchids:* Orchids requires particular care when pressing due to their delicate flowers. The flowers at least one should be spread out evenly so that the flower parts face the paper surface without creases or folds never allow the parts to fold up or stick together. We may also cut off each part of a flower and spread them on the same piece of paper and the press. Alternatively, we can make a wet collection as outlined below.
- *Grasses, Ferns and Other Small Herbs:* These should be collected with roots attached and have the soil carefully shaken off before pressing.
- *Small Algae:* If the algae are to be stored for more than a couple of days, a preservative needs to be used. This has traditionally been formalin added to the water to make a 5 per cent solution. However, formalin is extremely toxic and care should be taken not to breathe its vapour. Such a solution should not be sent through the post or by courier. A safe option is to add sufficient concentrated alcohol or methylated spirits to the water containing the algae to make a final solution of 70 per cent alcohol. It is preferable to concentrate the algae first for example, by filtering so that the amount of liquid stored is much less and it is easier to find the organisms. The jar should be clearly labeled with the preservative used.
- *Large Algae and Other Water Plants:* Dried specimens can be made by floating the specimens in water, sliding a stiff sheet of paper underneath and floating the algae onto the paper. Cover with a non-adhesive layer such as cheese cloth or wax paper to stop the paper sticking to it during subsequent pressing of the specimen.
- *Succulent Plants:* Plants such as cacti, figs species of Ficus and mistletoes drop their leaves entirely upon drying or remain alive for an excessively long period in the press.

This is overcome by killing the plant before pressing, either by freezing the specimen for a few hours, dipping it in boiling water for a few hours, dipping it in boiling water for a few minutes, or by using a microwave oven depends on the type of oven and the specimen itself but it usually about two minutes. Succulent materials is 'done' when it has a flaccid, water-soaked appearance.

When the cell tissue has been killed, absorbent papers must be changed at least daily for the first few days and complete drying in the case of cacti may take more than a month. Alternatively we can make a wet collection as outlined below.

Wet or Spirit Collections

Very fleshy or delicate structure including fragile or structurally complex flowers and fruit, small annuals, some aquatic species or small algae may be preserved in a liquid preservative rather than by drying, in order to maintain the original shape and structure of the plant. Specimens preserved in solution in air-tight glass jars are often referred to as 'spirit collections' or 'wet collections'.

Method

Specimens should be put into glass jars with screw-top lids. Glass jars are preferable to plastic, as they seal more tightly and it is easier to see the contents. Most material can be satisfactorily preserved in seven parts of ethanol ethyl alcohol available from most pharamacies with three parts of water to give a 70 per cent solution, usually referred to as 70 per cent alcohol. The type of preservative used should be clearly labeled. This information, together with the collection number for the material should be written in pencil on a small piece of paper since many inks are alcohol soluble and placed inside the container. An additional label should put on the lid of the container, also written with pencil.

Colours fade quickly in these solutions, so we should make comprehensive notes and photograph the colours of flowers and fruit. The specimens tend to become flaccid so it is usually advantageous to have these collections to complement pressed and dried specimens, rather than relying on them as the only animal.

Field Data Sheet

Physical Environment

This section of the plant quadrat field data sheet asks we to describe the location and environment of the quadrat. It is important to fill out this section of the field data sheet independently for the quardrat and not to repeat what is already recorded in the site field data sheet covered in Recording Site Details sectionlet.

Vegetation Structure

Layer

Stand off to the side of the quadrat, so that we can see the whole quadrat and look at the vegetation as a whole. What we can see is not a uniform blanket of green but a number of layers that vary in density. Looking carefully, we should see that each i.e. dense more green layer is composed of the upper fringes of some plants, for example, the upper leaves of trees. The less dense layers are composed mainly of stems or trunks. Different types of vegetation will commonly have different numbers of layers. For example, a moss garden or grassland may have a single layer, but a grassland may also have a moss or lichen layers as well as a grass layer. At the other extreme a rainforest may commonly have four layers or even more, especially in the tropics. We might like to have a look at different vegetation types in our local area to get used to different numbers of layers. Of course it is rarely this clear cut and we will need to use our own judgement to decide how many layers our quadrat has.

Cover Abundance Scale

Treating each plant canopy in each layer as opaque, estimate how much of the ground is vertically undermeath the canopy for each layer. By using both the percent cover and the cover abundance scale together, it is possible to estimate the total sunlight falling on each successive layer. This is a important part of the microhabitat occupied by both plants and animals.

Dominant Plant Species

List the species that dominate the canopy in each vegetation layer, either by name or by number as listed in the next section of the field data sheet.

Vegetation Species

Subdivide the quadrat into a grid to make it easier to keep track of which plants we have already counted. If we are only looking at trees, then the grid squares can be 5 x 5 metres but if we are looking at small herbs then 1 x 1 metre grids are easier. Working within each grid square develop a search pattern to suit ourself that completely covers the gird, for example a zig-zag track. As we move across our quardrat try not to crush the plants underfoot; this becomes extremely important if we plan to revisit the site later.

Number

As we find each different plant enter the details in the first available row in the table. Each row is numbered so that we can refer to the plant by number so that we can refer to the plant by number at a later date. The second page of the field data sheet continues the numbering system. If we require more numbers then use another cop of the second page and replace the numbers to keep in sequence.

Plant Name

When we being survey work or move to a new area, we should identify each different plant that we find using a field guide. Once we have done this we will quickly become familiar with the common plants and will only need to use the field guide for the more uncommon plants. If in doubt always check.

After looking at the field guide, if we are still doubtful to collect a voucher specimen. Keep a specimen for ourself and collect a specimen for identification at a herbarium. Until and will only need to use the field guide for the more uncommon plants. If in doubt always check. This number can be entered in the second-last column on the plant quadrat field data sheet. A "notes" column has been incorporated, which we can modify to suit the purposes of our particular survey.

Percentage Crown Cover

Choose an average-looking plant from the quadrat and estimate the percentage cover from the crown of the plant. The rows show similar percentage cover for plants with varying leaf sizes large to small, left to right. We can look upwards to silhousette the leaves against the sky or downwards to look at the area of shadow cast on the ground midday will give the most accurate results.

Cover Abundance Scale

Treating each plant canopy in each layers as opaque, estimate how much of the ground is vertically underneath the canopy for each plant species. By using both the per cent cover and the cover abundance scale together, it is possible to estimate the importance of each plant species within the vegetation. This is an important part of the microhabitat occupied by both plants and animals. There are two methods of recording species frequency for plant surveys depending upon the accuracy required for our particular survey: counting the total number of plants in the quadrat; or assigning a descriptive term to the frequency, that is, rare, occasional, frequent and common.

Identification of Plants

Identification of plants is a skill which we develop with practise and there really is no substitute for it. We can speed the learning process up by doing one of the many short courses available. Alternatively, the easiest starting point we have found, is How to identify plants by G. Harden and J.Williams. This publication is essentially the course notes for a plant identification course. There are many other instructional texts available, which will expand our skills.

Line drawings are used in preference to photographs in most publications because they better illustrate obscure features of the plant. For those who develop their skills and interest in identifying plants, there are more technical reference works called "floras".

Setting Up a Field Herbarium

If our survey includes repeated identification of the same species then it becomes very useful to set up a field herbarium. This is a small collection of dried specimens of the plants that we have collected from our study area and identified. The. second and subsequent times we identify the same species, the field herbarium provides a quick and easy was to identify material. Also it is easy to pick out what collected material is new to us and needs to be identified carefully.

Setting up a field or even a home herbarium entails protecting preserved specimens against damage caused by repeated handling. Specimens should be attached to large sheets of art paper in the size of 43 centimeters x 28 centimeters using archival-quality tape or thread, for example, Scotch Magic Tape. Dental floss can be used to attach bulky or woody specimens by puncturing the sheet on either side of the specimen, threading the floss through and tying the ends together in a simple reef knot. The use of archival tape is faster than dental floss and is easy to remove if the underside of the specimen needs to be examined. Ordinary stickly tapes are unsuitable as the adhesive breaks down, becoming tacky and detached. Small pieces of material that may have become separated from the specimen for example, seeds can be placed in small plastic bags and pinned to the sheet.

One disadvantage of mounting specimens is that it can make parts of the specimen inaccessible for examination, so it is essential that this be borne in mind during specimen arrangement and mounting. For example, easily removed mounting media should be used; specimens should be strapped to the sheet, rather than glued all over; and the specimen should be carefully arranged before it is attached so that all parts are visible.

- Do not tape over flowers or small fruits;
- Remove any loose soil from the underground organs;
- Leave sufficient space on the lower right-hand corner for the specimen label;
- Arrange the specimen to display as many features as possible, for example, upper and lower leaf surfaces, inner and outer aspects of flowers
- Larger specimens should be arranged diagonally across the sheet, or placed in a "V", "N" or "M" form if they have been pressed in that particular shape;
- For small plants it is usual to mount more than one on a sheet, distributed evenly with the largest towards the base.

The plant name and information from the plant voucher collection field data sheet should be written on a permanent label stuck to one corner of the herbarium sheet the bottom right-hand corner being the standard or sometimes annotations may be written directly on the sheet or card. Cards 20 centimeters x 13 centimeters are a suitable size for personal reference sets of identified materials but are unsuitable for research collections. Consideration will have to be given to protection of our field herbarium from pests and moulds. Check the further reading section for more information on this subject.

To identify a plant properly we will need to take the relevant identification sections into the field or bring part of the plant to the identification sections. It will be very difficult in most cases to identify the plant from memory or a photograph. In most instance it is necessary to examine the parts of the flower, fruit and leaves in detail.

Plant Conservation

To conserve the biodiversity of species, collection of all plants should be undertaken with care. Initially we suggest that you we common plant names know to we; later we can begin to learn how to identify plants according to their scientific names.

Method of Collection

Collect several pieces of the plant that we wish to identify, we should try to include at least : Leaves ; Flowers ; and Fruits. It is best to collect larger pieces of plant with these different parts attached to a stem. Store them in a plastic bag with a label that list: The survey site where they were collected; What the whole plant looked like for example, a tree, shrub, vine or herb; The height of the plant; and Flower colour, as this will fade if it is stored for longer than a day.

Keep the plastic bag as cool as possible–at least in the shade–and work on identifying them as soon as possible. If we are not able to work on the plants until later, we can store them in the refrigerator for a day or two. Longer storage will require pressing and drying the plants.

Identification

Identification of plants is a skill that we will develop with practise. The first few attempts will take the longest as we learn our way around the field guide. We will find that some groups are easier than others, for example, many people find grasses very difficult. When we hit these "difficult" groups we may be satisfied with a genus name rather than a full species name. In this country, many plants do not have a common name so we generally use the scientific name as well as the common name, if possible. A scientific name as well as the common name, if possible. A scientific plant name takes the following form : Example, for the Hard-leaved Scribbly Gum : Genus name, species name Eucalyptus sclerophylla. If we have difficult deciding which gum our plant actually is, then we can use the term : Eucalyptus species don't worry too much initially, as we gain practise we will become more through in our identifications.

Invertebrate Survey

Invertebrates, animals without backbones include a diverse range of classification groups known as taxonomic groups such as protozoas single celled animals, worms, insects crustaceans, and mollusks, ranging in size from minute to over two meters in length. Organisms are classified scientifically in a hierachically series of groups with the smallest group being species. Species which are most like each other are grouped together in a genus. Similarly genera are grouped into families, families into orders, orders into classes, classes into phyla into kingdoms. Invertebrates occur in almost every terrestrial habitat except the extreme Polar Regions. Whilst this section currently deals with terrestrial species, the boundary between aquatic and terrestrial environments for invertebrates is often blurred and some terrestrial invertebrate groups require water at some time during their lifecycle.

Invertebrates are the biggest group of animals found throughout the world, with species that have been describing being in excess of 1.4 million. This does not take into account the large number of invertebrates, which have not yet been described. In India there are an estimated 329,000 species of macro-invertebrates those that can be seen without magnification, and a further 105,000 species of microorganisms, but only 5588 species of vertebrates such as frogs, reptiles, birds and mammals.

Invertebrates are slowly becoming known to science, and we have just begun to realise their enormous diversity. There are fascinating variations between mouthparts, legs, winged and wingless forms, in camouflage and in their reproduction, ecology and habitat requirements.

Invertebrates are highly adaptable and found in a huge variety of habitats. They occur on plants, leaves, in flowers, stems, in grass, beneath bark, in fungi, moss, plant and leaf litter, under rocks, fallen logs and in caves and soil. Undertaking the survey techniques described here provides the opportunity to become familiar with the world of macroinvertebrates–those invertebrates which can be seen without magnification.

The survey method we use will be influenced by which invertebrates groups we wish to collect data on. This section describes general techniques that will locate a wide range of macro-invertebrates, that don't require much specialized equipment, and that can be used easily by non-experts. There are a large number of targeted techniques for surveying specific invertebrates groups, such as butterflies, beetles, and flies, and some of them are presented in the section called insect Survey Methods.

The methodology outlined in this section relates to the live capture and release of macro-invertebrates. It is important to ensure the welfare of all plants and animals when undertaking biodiversity surveys. Without the access to appropriate experts for identification, it is pointless to use collection methods that kill many invertebrates.

The aim of undertaking the baseline survey is to obtain information on the types of macro-invertebrates found within the area. The capture and release methods used can help to build confidence and skills in finding and identifying macro-invertebrates, and demonstrate which groups of species can be found in which types of habitats, at a particular time.

The huge diversity of invertebrates means that this section cannot hope to provide instruction in the identification of invertebrates, other than one of the major groups, the insects. There are many good reference sections which can provide information for identifying macro-invertebrates.

Aims

- To learn simple techniques for surveying and identifying macro-invertebrates.
- To determine the macro-invertebrates groups present in various habitats.

Where to Survey

Everywhere, there are very few places where invertebrates can't be found. Look in the air, on plants, under logs and rocks in leaf litter, in soil, on dead animals and on other invertebrates themselves. They are most numerous in the warmer months but can also be found in the depths of winter, although we just might need to search a little harder. In using the collection techniques below, we should search as many different habitats within our survey site to find as many macro-invertebrates as possible. The two important aspects of the baseline survey are to become familiar with both surveying and identification.

Animal Welfare

- Invertebrates and other animals should be released at the point of capture.
- Do not handle invertebrates unnecessarily, as many have delicate body parts or protective skin secretions that can be easily damaged. Very little is know about whether handling causes stress to invertebrates, so it is best or err on the side of caution by restricting handling.

Need for Invertebrates Survey

Although some invertebrates are obvious and well known many others are cryptic and often overlooked. They are easy to find and don't require animal ethics permission to study. Every backyard, local park and remnant of bush is teaming with invertebrate life. The importance of invertebrates to the functioning of ecosystems includes : Plant pollination-insects play a significant role as pollinators of flowering plants. Soil formation and maintenance of soil fertility. Decomposition -Invertebrates play a crucial role in breaking down and recycling of dead plant and animal material. Food web-invertebrates are the principal source of food for many amphibians, reptiles, birds and mammals.

The main reason for surveying invertebrates is to have a list of what is present, and to create more awareness about their amazing diversity. When we survey invertebrates as well as plants, birds mammals, frogs and reptiles at the same time and at the same sites we begin to understand how invertebrates interact with habitat and other species. If we choose to only survey invertebrates using the methods in this section we can produce a list of what is present in the area at a particular time.

Aim of Invertebrate Survey

The aim of invertebrates survey will determine what survey methods are most appropriate. We may already have an interest in a particular group of invertebrates-spiders, beetles or butterflies, for example - and seek out some of the sections that will help us identify them and instruct we in specific techniques for surveying them. Alternatively, we may wish to become familiar with the wider diversity of invertebrates by using some general survey techniques described in this section. To do a survey with the aim of detecting biodiversity changes due to natural or human-induced disturbance requires professional scientific planning, and is not dealt with in this section.

Identification and Choice of Methods

Although it is relatively easy to go out and collect large numbers of invertebrates at a single go, it's not a very useful thing to do. That collecting effort only generates meaningful information when those invertebrates can be identified to some taxonomic level, and the closer to species level the better. For the non-specialist, the task of identifying invertebrates is made even more difficult by the lack of easy to follow comprehensive guides, and the specialized language which is used to distinguish between closely related species.

However, we shouldn't let this put we off surveying for invertebrates. We will be probably able to identify invertebrate specimens to some higher taxonomic level, such as class, order of family, without expert assistance. We may even be able to identify some animals to species level if they are relatively distinctive and have been described in easily accessible publications.

If we decide to collect invertebrates because we are unable to identify them or have them added to a permanent collection. It is important to note here that how much an organization can process may be constrained by their available resources so they may not be able to lodge or identify species if their costs for doing so cannot be covered. Where collection methods in this section result in the death of individual, morally and ethically we must consider whether the data we are collecting is worthy of this death. As a general rule, do not use survey techniques that kill large numbers of invertebrates unless we have professional assistance in their identification and have arranged to lodge the specimens with a museum of public collection.

If we are likely to encounter problems getting expert assistance in identification, use survey methods that involve only live capture. We can then release the animals after we have identified them to whatever taxonomic level is within the capabilities of the team. We may wish to photograph some animals for further identification at a later date. We might choose to kill and preserve a representative of each type we suspect of being a separate species, and have those specimens professionally identified.

Limitations of Methods

Different invertebrate species require different techniques to survey them, depending on the habits of the species and the habitat it prefers. For example: We would not expect to find earthworms in a light trap set to attract night-flying insects.

Pitfall Traps

Pitfall trapping is an effective means of sampling the ground-active invertebrates of an area. A pitfall trap is simply a container sunk into the ground so that the top of the container is level with the ground surface. Ground-active animals fall into the trap while moving about, and generally cannot escape but this will depend on the depth of the trap, the slipperiness of its sides, and the abilities of the species. Small barriers like drift fences are often set up around the trap to directly any passing animals towards the trap. Pitfall traps for invertebrates can be either 'dry' or 'wet'. 'Wet' pitfall traps contain a fluid which kills and preserves the animals which fall into the trap, and are not dealt with here. The method outlined below is for dry pitfall traps which capture animals alive.

How to Set up Dry Pitfall Traps

- Pitfall traps for invertebrates can be much smaller than those used for mammals or reptiles. We can use a simple can, jar or icecream container.
- Dig a hole the size of the container, ensuring the top of the container is level with the ground surface. There should be no holes or gaps around the outside of the container into which animals may fall.

- Layout four pits in a line at the edge of the sample area, preferably centred along the length of the transect so that the center of the pit line is near the center of the transect.
- Place a few handfuls of leaf litter in the bottom of the trap to provide shelter.
- Cover the pit with a roof raised approximately 15mm from the ground to ensure rain and litter do not enter the trap, and discourage small reptiles and mammals.
- Baits can be used in the trap to attract some kinds of invertebrates, but is not necessary. Rotted fruit or meat can be used. If our use baits at one site, use them at all sites for a systematic survey.
- Drift fences can be used with the traps and can consist of 50-80 mm high strips made out of any smooth material that cannot be climbed by most invertebrates metal or plastic is ideal. These barriers should be dug a short way into the soil and radiate from the trap in a cross pattern, with the trap at the center to ensure a large coverage of area for obtaining invertebrates. If we use drift fences at one site, use them at all sites for a systematic survey.
- Always mark the spot where the pitfall trap is with flagging tape on a tent peg, tree or something that is not likely to move.
- Record the pitfall trap identification number on the flagging tape using a black permanent marking pen and into the field data sheets. The tape must be removed when the survey is finished as it is polluting and may be confusing for future surveys.
- Into a topographic map, record the trap identify number and where the pitfall trap was laid along the transect line as a permanent record.

Special Considerations and Tips

- Take care when handling invertebrates, particularly spiders and centipedes. Many bite or sting, and some may be venomous and could cause injury. We should wear hand protection.

- Use a stick or forceps initially to move around the leaf litter in the trap, checking for any venomous species.
- If we can't identify an invertebrate, consider photographing or sketching it for later identification.

Animal Care Issues

- Always ensure there are no gaps around the edge of the pitfall into which animals might fall.
- When removing pitfall traps from the ground, always fill in the hole left behind.
- Release captured invertebrates as soon as possible in the cover of leaf litter or some other ground covering to reduce the likelihood of predation.
- If capture is necessary for identification, always handle the animal only for as long as needed for identification. Use small paint brushes to move small macro-invertebrates without damage.
- If we are finding many dead invertebrates in traps exposed to much sun, it may be to heatstress, and we may need to check traps early in the morning or twice daily.

Opportunistic Invertebrates Sightings

Many invertebrates will be detected while traveling to and from survey sites, or outside standard survey times or survey sites. Record all of these sightings and their locations, dates and times for the whole survey on the opportunistic animal sightings field data sheet provided in the field worksection. This additional information is valuable as it helps to build the list of invertebrates recorded for each area.

Special Considerations and Tips

- If we cannot identify an invertebrate, consider photographing it or sketching it for later identification.

Animal Care Issues

- Take care when catching and handling invertebrates for identification. Use small paint brushes to move small invertebrates around.

- Always release captured and in conditions that will not increase risks to the invertebrates for example, predation

Systematic Searching

Simply searching an area thoroughly giving attention to all potential hiding areas or types of microhabitat is an effective method for revealing the diversity and abundance of invertebrates species within an area. A systematic search is a search of a fixed area with a fixed level of search effort, and ensures that the results can be compared between sites and between surveys conducted at different times where the same area and effort levels were used. This method is suitable for all habitats.

Methodology

- Before we start the survey decide how many people will search each sample area and if it is more than one, to divide the area equally for each person to search. Then decide how long each sample area will be surveyed for. This will then set the standard effort for each site so ensure that we document it for future reference if we wish to repeat the survey.
- Search all the microhabitats available in our sample area, including leaf litter, the top layer of soil, the base of plants, under rocks and in the accessible foliage, fruits and flowers of trees and shrubs. Trace along the branches of shrubs to the leaves and back again. They are often more common on the underside of leaves than on the top surfaces.
- When we locate invertebrates, note its habitat associations and behaviour, for example the species of plant it is found on, whether it is feeding on the plant, how deeply it was buried in soil, whether it was in moist or dry leaf litter, whether it was flying or resting.
- Freshly eaten leaves or the presence of faeces indicate that an invertebrate may be near by.

- Searching at night with a torch or fluorescent lamp can reveal nocturnally active invertebrates.

Special Considerations and Tips

- Take care with handling invertebrates, particularly spiders and centipedes. Many bite or sting, and some may be venomous and could cause injury. We should wear hand protection.
- If we cannot identify an invertebrate, consider photographing it or sketching it for later identification.

Animal Care Issues

- Always replace material disturbed during the search as best as possible, including rocks, bark and leaf litter. Try to leave the site much as we found it.
- If captured is necessary for identification, always handle the animal only for as long as needed for identification. Use small paint brushes to move small macro-invertebrates without damage.
- Always release captured and in conditions that will not increase risks to the invertebrates for example, predation.

Light Traps

Light traps are a good way of live-trapping invertebrates that are attracted to lights at night. Most will be insects such as moths and beetles, but lacewings, bees, wasps, spider and flies may also be attracted. The trap consists of a vertical sheet held taut between two poles, and illuminated by an ultra violet (UV) light. Invertebrates fly or crawl towards the light, and may be picked off the vertical sheet or another sheet spread out under the trap.

Working of the Trap

- The trap consists of a vertical sheet held taut between two poles with a mercury vapor bulb suspended above the sheet as the UV light source. A ground sheet spread out under the trap will make it easier to see and collect any specimens that land on or crawl along the ground near the trap. Insects may be also found in the surroundings vegetation.

- The light can be run from dusk and operated right through till morning, as different species might be active at different times of the night. It should be manned constantly, or examined at least hourly. Always note the start and finish time when the light trap was running.
- Light traps are best used on moonless or overcast to semi-overcast nights as a full moon can lessen the effects of the UV light. Light rain may even increase the collecting but may break the lamp.
- When using a portable generator to run the UV light, an in-link choke is needed to regulate the power supply. Some light traps are designed specifically to collect for long periods.
- The generator needs time to warm up, before turning on the lamp, as this may stop the generator from cutting out when the lamp is first turned on. If for some reason the lamp is turned off and then turned on straight away it may not work; this is due to the temperature of the lamp, so just wait a while and it should come back on.
- Once we have finished collecting, make sure that the sheet have no specimens still on them, especially when changing locations; we may inadvertently carry an insect from one location to another.

Special Considerations and Tips

- Take care when handling animals, particularly spiders and centipedes. Many bite or sting, and some may be venomous and could cause injury. Consider wearing hand protection.
- If we can't identify an invertebrate, conser photographing or sketching it for later identification.

Animal Care Issues

- Release captured animals as soon as possible in the cover of vegetation. Avoid releasing animals to close to the trap, as they may return to it.

- If captured is necessary for identification, always handle the animal only for as long as needed for identification. Use small paint brushes to move small macro-invertebrates without damage.

Aerial and Sweep Nets

Target Animals and Habitats

There are a variety of differently designed nets for the collecting of terrestrial insects. Insects may be netted while flying, from their resting positions on plants or even on the ground. These techniques require an aerial net. Other insects can be obtained from vegetation by bushing flowers, foliage or grasses with a strong framed sweep net. Collecting insects using a net will often lead to injury or death of the insect. Surveyors should ensure that material collected in this manner is humanely killed and correctly preserved. Before carrying out such methodology, we should ensure that we have access to the appropriate skilled persons to both identify and retain collected material.

How to Use Nets

Aerial Nets

These nets usually have a round hoop. The best to capture a specimen is to bring the net up from behind with a quick sweep forward, then continuing forward and then sharply backward whipping the bag so that the net loops around the hoop thus trapping the specimen until we can transfer it to the required container e.g. killing jar.

Sweep Net

A sweep net is similar, but often with the outer edge made of calico to protect the net. The rim, which is usually triangular, is passed across vegetation in a sweeping action, like a broom. This will dislodge insects that may not have been seen, especially amongst container. Large specimens may be removed by hand, smaller species with the aid of an aspirator.

Pooter (aspirator)

A pooter, also called an aspirator, is a device consisting of a glass vial, a rubber stopper with two holes in it, and two glass or metal tubes extending out of the stopper. One tube's inside end is covered in gauze to prevent dirt etc. Being sucked into our mouth and has the outer end extended by rubber or plastic tubing to approximately arms length.

How to Use a Pooter

The tubing is held to the mouth while the second tube is positioned close to the insect being collected. With a quick breath in, the insect is sucked into the glass vial; it can then be transferred to a killing jar by removing the stopper.

Beating Trays

– This method can provide a high number of specimens in all stages of the insect's life cycle i.e. eggs, nymphs, larvae, pupae and adults.
– Simply hold a stretched out cloth preferably white or an umbrella up side down under the branches and beat the bush with a stick or give it a firm shake.
– The insects and other invertebrates that fall onto the cloth, are best collected with an aspirator; be sure to look for very small insects.
– Once the insects have been collected, those to be kept will need to be pined and preserved.

Batted Traps

What will we Collect ?

– Mostly flies and beetles

How does it Work ?

– The rotting bait will attack a variety of insects, which inter the slits in the sides of the bottles. The insect travels up to the top section where they are unable to escape.

- This trap can be left out for a week or more but under good conditions will fill quickly and therefore should be replaced.
- To empty this trap it should be labeled then taken back to base where it can be placed into a freezer for 24 hours, then taken apart and the specimens sorted.

Barrier Trap

What will we Collect ?

- Generally beetles and other fást flying insects. Also a range of animals that are collected form the ground. Lighter insects like butterflies would probably fly over or around this trap also acts as a pit fall trap.

How does the Trap Work ?

- Barrier trap or flight interceptors are similar to a malaise trap. This trap is simply a sheet of black plastic or netting stretched out between two poles e.g. to mato stakes, therefore forming a barrier.
- A trough placed at the base PVC pipe cut in half at ground level will work in two ways, firstly the pipe will collect those insects that have hit the barrier and will also act as a pit trap, collecting ants, beetles and other animals that come across the pipe and fall in.
- The trap is positioned where flying insects are expected, these insects hit the sheet and fall down into the trough that contains a mixture of equal parts of 70% ethanol and water. The trough may sit on the ground, but if the rim is positioned at ground level, if then doubles as a pit trap.
- Specimens are then collected as yellow pans, using a small tea strainer take out as many specimens as we can, place into zip lock bag, then with a larger strainer pour off all of the solution empting the specimens collected into the same zip lock bag and label.
- Any insects that fly into the barrier usually drop straight down into the container below, which is filled with a fluid

that will kill and preserve the insects or spiders until they are collected.

Yellow Pans

What will we Collect ?

- These will attack flies, beetles wasps, bees, ants, grasshoppers and other groups of insects and sometimes spiders.

How does the trap Work ?

- Yellow pans are basically, yellow cereal bowls 170 x 45 millimetres which attract and range of insects and other animals.
- A yellow pan collecting kit should contain tea strainers small and large size, a number of plastic zip-lock bags and water enough to fill pans.
- The yellow pans are usually placed in a series of 10-15 pans at various intervals along a creek, filled to the rim with water, and a very small amount of detergent.
- Place the pans along a creek, shoreline, or in a forest, especially near different microhabitats e.g. seeps or a shaded overhang at regular intervals. Note. They of course can be placed anywhere.
- After 24 hours we should have a collection of small files, wasps and other small insects. Collect the specimens from the pans by pouring the contents through the tea strainer, then flick the back of the tea strainer to deposit the specimens into zip shut.
- Put all zip lock bags in a plastic container and add a little ethanol to the container.

Malaise Traps

What will we Collect?

- This trap is designed to catch insects, which attempt to fly over an obstacle, for example files, wasps, and moths.

How does it Work ?

- This traps works as a barrier for flying insects. When confronting an obstacle, insects will either fly over or around the obstacle. This trap is designed in the form of a tent but with open sides and a wall running down the center, one end is higher than the other, with a hole in the central wall at the highest point.
- The colour of this trap is also important, the bottom half is black which makes the trap hard to see and the top half is white, the highest point is placed towards the setting sun. insects are generally drawn to light, therefore they will be directed towards the highest part of the trap where the collecting container is attached to the outside, the container may be either wet or dry, dry specimens can be pinned, while specimens collect in 70-80% ethanol will have to preserved in ethanol. The container should be emptied daily but may be left for up to a week.
- This trap can be suspended over creeks or set with poles across the flight path of flying insects. The traps netting should also be examined for larger specimens. E.g. moths, butterflies, and wasps.

Light Traps

What will we Collect ?

- A variety of insects moths, beetles, etc are attached to the UV light, they fly or crawl into the light.
- Other insects can also be collected from light traps for example, lacewings, bees, wasps and flies.
- A sheet spread out under the trap will make it easier to collect crawling insects and those that don't land directly onto the sheet, also insects may be found in the surrounding vegetation.

How does the trap Work ?

- By emitting ultra violet (UV) light, the trap consists of a vertical sheet held taut between two poles with a mercury vapour bulb suspended above the sheet as the UV light source.

– Insects of many orders, for example moths (Lepidoptera) beetles (Coleoptera) and bugs (Hemiptera) are attracted. These and other insects land on the sheet and surrounding area, from there they can be collected and placed into a killing jar.

– We may also find that spider and other insect predators e.g. robber flies (Asillidea) will be found on the sheet. A ground sheet spread out under the trap will make it easier to see and collect any specimens that land on or crawl along the ground near the trap.

– Once we have finished collecting, make sure that the sheet (s) have no specimens still on them, especially when changing locations; we may inadvertently carry an insect from one location to another resulting in a wrong distribution.

– Light traps are best used on moonless or overcast to semi-overcast nights as a full moon can lessen the effects of the UV light. Light rain may even increase the collecting but may break the lamp.

– The lamp set-up described above, if not manned all the time, should be examined at least hourly. The light can be run from dusk and operated right through till morning. It should be mentioned that at various times between dusk and dawn different species might be active. Always note the start and finish time when the light trap was running.

– When using a portable generator to run the UV light, an in-link choke is needed to regulate the power supply, use a R650 generator that will run for approximately 3.5 to 4 hours. Some light traps are designed specifically to collect for long periods.

– The generator needs time to warm up, before turning on the lamp, as this may stop the generator from cutting out when the lamp is first turned on. If for some reason the lamp is turned off and then turned on straight away it may not work; this is due to the temperature of the lamp, so just wait a while and it should come back on.

Pitfall Traps

Pitfall trapping is an effective means of sampling the ground-active invertebrates of an area. A pitfall trap is simply a container sunk into the ground so that the top of the container is level with the ground surface. Ground-active animals fall into the trap while moving about, and generally cannot escape but this will depend on the depth of the trap, the slipperiness of its sides, and the abilities of the species. Small barriers drift fences are often set up around the trap to directly any passing animals towards the trap. Pitfall traps for invertebrates can be either 'dry' or 'wet'. 'Wet' pitfall traps contain a fluid which kills and preserves the animals which fall into the trap, and are not dealt with here. The method outlined below is for dry pitfall traps which capture animals alive.

Dry Pitfall Traps : Procedure of Settings

- Pitfall traps for invertebrates can be much smaller than those used for mammals or reptiles. We can use a simple can, jar or icecream container.
- Dig a hole the size of the container, ensuring the top of the container is level with the ground surface. There should be no holes or gaps around the outside of the container into which animals may fall.
- Layout four pits in a line at the edge of the sample area, preferably centred along the length of the transect so that the center of the pit line is near the center of the transect.
- Place a few handfuls of leaf litter in the bottom of the trap to provide shelter.
- Cover the pit with a roof raised approximately 15mm from the ground to ensure rain and litter do not enter the trap, and discourage small reptiles and mammals.

- Baits can be used in the trap to attract some kinds of invertebrates, but is not necessary. Rotted fruit or meat can be used. If we use baits at one site, use them at all sites for a systematic survey.
- Drift fences can be used with the traps and can consist of 50-80 mm high strips made out of any smooth material that cannot be climbed by most invertebrates metal or plastic is ideal. These barriers should be dug a short way into the soil and radiate from the trap in a cross pattern, with the trap at the center to ensure a large coverage of area for obtaining invertebrates. If we use drift fences at one site, use them at all sites for a sytematics survey.
- Always mark the spot where the pitfall trap is with flagging tape on a tent peg, tree or something that is not likely to move.
- Record the pitfall trap identification number on the flagging tape using a black permanent marking pen and onto the field data sheets. The tape must be removed when the survey is finished as it is polluting and may be confusing for future surveys.
- Onto a topographic map, record the trap identify number and where the pitfall trap was laid along the transect line as a permanent record.

Methods of Checking Dry Pitfall Traps

- Animals capture in dry pitfall traps may be carefully removed and placed into a dry container eg jars in shade while preparing for identification.
- Pitfall traps may get disturbed by small mammals such as bandicoots and antechnius, birds, reptiles or frogs looking for a feed of invertebrates. There is not much we can do about this, other than check traps as often as possible at lest twice in a 24 hour period and repair any disturbed traps.
- Pitfall traps can be left out for a few days, but should be checked regularly daily or every second day to ensure

that live invertebrates are released soon after capture, do not die from heat stress, and so that the captures can be recorded before the traps are visited by vertebrates looking for a free feed.

- Many ground-active macro invertebrates are nocturnal. In summer and in locations without shade, traps should be checked early in the morning to avoid heat stress killing captured animals.

Special Considerations and Tips

- Take care when handling invertebrates, particularly spiders and centipedes. Many bite or sting, and some may be venomous and could cause injury. So we should wear hand protection.
- Use a stick or forceps initially to move around the leaf litter in the trap, checking for any venomous species.
- If we can't identify an invertebrate, consider photographing or sketching it for later identification.

Animal Care Issues

- Always ensure there are no gaps around the edge of the pitfall into which animals might fall.
- When removing pitfall traps from the ground, always fill in the hole left behind.
- Release captured invertebrates as soon as possible in the cover of leaf litter or some other ground covering to reduce the likelihood of predation.
- If capture is necessary for identification, always handle the animal only for as long as needed for identification. Use small paint brushes to move small macro-invertebrates without damage.
- If we are finding many dead invertebrates in traps exposed to much sun, it may be to heatstress, and we may need to check traps early in the morning or twice daily.

Opportunistic Invertebrates Sightings

Many insects will be detected while traveling to and from survey sites, or outside standard survey times or survey sites. Record all of these sightings and their locations, dates and times for the whole survey on the opportunistic animal sightings field data sheet. This additional information is valuable as it helps to build the list of invertebrates recorded for each area.

Special Considerations and Tips

- Take care with handling insects, particularly spiders and centipedes. Many bite or sting, and some may be venomous and could cause injury or even death..
- If we can't identify an insect, which we consider photographing it or sketching it for later identification.

Animal Care Issues

- Take care when catching and handling invertebrates for identification.
- If capture is necessary for identification, always ensure the stress to the insect is kept to a minimum and handle the insect only for as long as need for identification.
- Always release captured insects as soon as possible at the place that they were capture and in conditions that will not increase risks to the invertebrates for example, predation

Systematic Searching

Searchers of a fixed area with a fixed level of search effort is an effective method for revealing the diversity of insect species within an area. The method describes below is suitable for documenting the presence of insects in a certain area. Or we can systematically search particular habitat types. Systematic searching provides data about the habitat preferences and activity patterns of insects as we". as the variety. This method is suitable for all habitats.

Methodology

- Before we start the survey decide how many people will search each sample area and if it is more than one, to divide the area equally for each person to search. Then decide how long each sample area will be surveyed for. This will then set the standard effort for each site so ensure that we document it for future reference if we wish to repeat the survey.
- Search all the microhabitats available in our sample area, including leaf litter, the top layer of soil, the base of plants, under rocks and in the accessible foliage, fruits and flowers of trees and shrubs. Trace along the branches of shrubs to the leaves and back again. They are often more common on the underside of leaves than on the top surfaces.
- When wc locate and invertebrates, note its habitat associations and behaviour, for example the species of plant it is found on, whether it is feeding on the plant, how deeply it was buried in soil, whether it was in moist or dry leaf litter, whether it was flying or resting.
- Freshly eaten leaves or the presence of faeces indicate that an invertebrate may be near by.
- Searching at night with a torch or fluorescent lamp can reveal nocturnally active invertebrates.

Special Considerations and Tips

- When searching an area we may come into contact with various animals such as spiders, scorpions ore snakes. Take extreme care and wear appropriate hand protection.

Animal Care Issues

- Always replace material disturbed during the search as best as possible, including rocks, bark and leaf litter.
- If captured is necessary for identification, always ensure the stress to the animal is kept to a minimum by using appropriate techniques and handling the animal only for as long as needed for identification

– Always release captured animals as soon as possible at the place that they were capture and in conditions that will not increase risks to the invertebrates.

Frogs and Reptile Survey

Opportunistic Reptile and Frog Sightings

Any species of frog and reptile that we encounter during travels to or from survey sites and that we can positively identify should be recorded on the opportunistic animal sightings field data sheet. Species seen just outside the sample area or outline of systematic sampling time should be recorded as opportunistic records. Frogs can often be identified by the characteristic calls made by males during the breeding season. Opportunistic identification of frogs their calls can be add significant information to the study. Opportunistic sightings for frogs might also be done following rain or during peak summer activity, particularly around water bodies such as tanks or dams. Frogs and reptiles can also be seen during spotlighting for nocturnal mammals. Record all of these sightings and their locations, dates and times for the whole survey on the opportunistic animals sightings field data sheet provided.

Special Considerations and Tips

– Take care to collect all field data sheets into a single bundle of sheets at the end of the survey. A survey that extends over several days can lead to an accumulation of opportunistic sightings entered on many copies of the opportunistic animal sightings field data sheet.

– If there is more than one opportunistic animal sightings field data sheet at the end of a survey, given them a sequential number when they are bundled up. This will greatly assist those involved in entering records into databases.

Animal Care Issues

– Only catch and handle animals for identification if we are experienced and properly licensed to do so.

Systematic Frog Call Recognition

The males of most frog species call to advertise their location to potential rivals and mates during the breeding season. The species of the caller can be identified from the call. By recording those calls, we make a permanent record which can be checked by others or compared at leisure to reference tapes of frog calls made by experts. Bear in mind that many species may congregate at a single water body, and many individuals of one species may drown out the calls of a few individuals of another species. The method can be applied at points along the transect, for example, the center points of each quadrat or in significant patches of habitat for frogs, away the sample area.

The method described below gives we the option of basing our work on the spot call recognition during a fixed 30 minute listening time, or making a recording for 30 minutes which will be analysed later to identify calling frog species. We could use both forms of information gathering in one session at a site. If we do that, then we should do this in parallel. That is, in the same 30 minute sampling time, simultaneously do not the spot identifications and run a tape recorder. That way, one set of records can back-up the other, and the same physical conditions can be guaranteed to apply to all the species information for that session at that site.

The species we detect may depend on the time we record. Some species are more diurnal, calling most at dawn and dusk or during the day after rain, while others call more at night. Hence, to detect a good range of species present, it's best to make more than one recording at any one location, one at dusk one a few hours later. Remember also that different species will breed in response to different environmental conditions, so that several visits to a single location over spring, summer and autumn may enable we to detect a wider range of species.

Methodology

- Establish a fixed reference point at our chosen survey site.
- Listen for a 30 minute period.

- For systematic frog call surveys always commence recording at the same time relative to sunset if we are surveying over more than one day.
- Note the identity of frogs from their calls if we are skilled at this.
- Alternatively, record frogs calls for this period on a portable cassette recorder using an external microphone and seek expert assistance to identify the recorded frog calls, or use one of the call identification tapes to identify recoded call later. Always note the name of the person identifying the calls. The quality of the recoding will be critical to good, accurate and complete species identification. For that reason, use good quality cassette tapes and an external microphone – the internal microphones on cassette recorders are usually not of high quality.

Sending Frog Call Tapes for Specialist Analysis and Identification

Careful information management is required after frog calls are recorded collected to ensure that we are able to relate the identifications provided by experts to our field records. If we are unsure of how to prepare call recordings for submission to an expert, contact them and discuss their requirements. The following notes provide an outline of the important aspects of good information management. Never send the original field data sheet. It is irreplaceable. We can send a copy if the specialist is agreeable. Some expert identification services may agree to write the results on the copy of the field data sheet and send it back to us. Often, however, they will send us a printout of identifications against the unique identifying information that we wrote on each envelop and recorded on our field data sheet, together with an invoice.

Special Considerations and Tips

- The quality of recordings is critical to the success of this method. Use the best possible quality cassette recorder, microphone and cassette tape.

- Doing some test recordings using the equipment we will take into the field is also a good idea.
- If we are using a number of cassette recorders and microphones, each should be given an identifying number. This will enable us to trackdown faulty equipment and take account of differences in the sensitivity and frequency response of combinations of components.

Animal Care Issues

- Ensure that we cause the minimum of disturbance when applying this method. Avoid trampling vegetation at the waters edge and muddying the water.

Day Time Search

An intensive daytime search of habitat will reveal reptiles and frogs sheltering under logs, leaf litter, rocks, bark, trees and bushes, around the base of in-stream vegetation, and in rocky cervices. Human rubbish such as corrugated iron sheeting and log piles often provides shelter for larger species. If the area is approached quietly, animals can also be observed sunning themselves on tree trunks and other exposed locations. Animals can be captured by hand for closer identification, a skill that requires quick wits and quick reflexes. Some species are distinctive enough to be identified without handling the animal, and it is best to avoid handling large goannas and most snake species for safety reasons.

It is important that our search is conducted quietly, so that animals are not forewarned of our presence, and is not destructive of animal habitat. Ensure that all objects we move are returned to their original place immediately. All opportunistic reptile and frog sightings or hearings should be recorded on an opportunistic animal sightings field data sheet. Opportunistic sightings or hearings include all animals seen or heard outside the recording time period of 30 minutes spent at each sample area. This situation may occur when team members are walking between sample areas or setting up the transect.

Methodology

- In the sample areas, we can do a daytime search for frogs and reptiles.
- For a period of at least half an hour, search through the sample area looking under movable rocks, logs, loose bark and raking over the leaf litter. It is best to conduct the search relatively early in the day when reptiles are not too active. Otherwise, it will be very difficult to identify them before they disappear.
- Record the identity of the species of reptile and frog that we encounter during the search period.

Special Considerations and Tips

- It is useful to appoint someone as data recorder so that they can concentrate on accurate record-keeping and searchers can concentrate on searching for and identifying reptiles and frogs.
- When searching an area we may come into contact with venomous animals such as spiders, scorpions and snakes. Take extreme care and wear appropriate hand protection.

Animal Care Issues

- Avoid damaging the habitat that we are searching. Replace any disturbed rocks, logs, bark or leaf litter as best we can.
- Do not handle or touch any animals unless there is someone present.

Nocturnal Streamside Search

In habitats with well-defined watercourses, we should include a nocturnal search for frogs. Some reptiles may also be encountered during these searches, particularly nocturnal reptiles such as geckos. Some species of frogs and reptiles are more active during the day, while others are more active during the night. While night time searches may be more difficult simply because of the greater difficulty of searching with limited light, they may reveal species which were too well hidden to be detected during a daytime search. If the location is approached quietly with a spotlight, animals in exposed locations can sometimes be observed prior to searching. An intensive search of streamside habitat will revel reptiles and frogs sheltering under logs, leaf litter, rocks, bark, trees and bushes, around the base of in-stream vegetation in water and rocky cervices. Animals can be captured by hand for closer identification, a skill that requires quick wits and quick reflexes. Some species are distinctive enough to be identified without handling the animal, and it is best to avoid handling large goannas and most snake species for safety reasons. It is important that our search is conducted quietly, so that animals are not forewarned of our presence, and is not destructive of animal habitat. Ensure that all objects we move are returned to their original place immediately.

The sensitivity of frogs to microhabitat conditions determines their seasonal and daily activity patterns. Many of the more common and well-known species enter breeding phases in late spring and early summer and their activity peaks accordingly. The activity patterns of other frog species are poorly known. Some species seem to breed only after very specific rainfall events.

All opportunistic reptile and frog sightings or hearings should be recorded on an opportunistic animal sightings field data sheet. Opportunistic sightings or hearing include all animals seen or heard outside the recording time period of 30 minute spent at each sample area. This situation may occur when team members are walking between sample areas or setting up the transect.

Methodology

- Nocturnal streamside searches should be conducted at each watercourse to target frogs.
- A standard survey will involve approximately one person-hour of searching for active, basking or calling frogs for 200 metres along the drainage line, stream, or river.

- When identification is not certain voucher specimens should be taken by a licensed person using humane methods approved by an Animal Ethics Committee. Advice on voucher specimen preparation is provided in the Field Worksection.
- Incidental observations of fauna other than reptiles and frogs should be recorded on opportunistic animal sightings field data sheets.
- The identify and numbers of reptiles and frogs detected should be recorded as soon as identification is made.

Special Considerations and Tips

- It is useful to appoint someone as data recorder so that they can concentrate on accurate record keeping and searches can concentrate on searching for and identifying reptiles and frogs.
- When searching an area we may come into contact with venomous animals such as spiders, scorpions and snakes. Take extreme care and wear appropriate hand protection.

Animal Care Issues

- Always replaced material disturbed during the search as best as possible.
- Only catch and handle animals for identification if we are experienced and properly licensed to do so.
- If capture is necessary for identification, always ensure the stress to the animal is kept to a minimum by using appropriate techniques and handling the animal only for as long as need for identification.
- Always release captured animals as soon as possible at the place that they were captured and in conditions that will not increase risks to the animal.

Reptile and Frog Voucher Specimens

Only licence holders may take voucher specimens. The treatments outlined below can also be carried out on animals found dead in traps, road kills and animals required to be euthanased by vets due to injury. Voucher specimens will only be useful when they are accompanied by good field records.

Preservation of Dead Animals

Herpetology specimens should be slit carefully along the abdomen, and preserved initially in 10% formalin, followed by 75% ethanol. The cut should penetrate through the abdominal muscle wall but not damage the internal organs, and requires some experience to do correctly. The cut need only be big enough to allow penetration of the preserving fluid, and should not be the entire length of the abdomen. An anterior-posterior cut is best. For turtles, cut at the base of each leg, not through the carapace. A better alternative is to inject 10 per cent formalin into the abdomen, thorax and large muscle masses. Formalin must be 10 percent formaldehyde, 90 percent water preferably distilled and buffered with calcium carbonate about 1 teaspoon per litre of formalin. Freezing is a good alternative. Wrap bodies tightly in thick plastic bags. Do not cut animal open.

Method for All Pickled Specimens

- Make slit through to abdominal cavity, making sure not to damage internal organs.
- Preserve body in formalin for 48 hours for herpetofauna as small or smaller than house mice; up to 3 weeks for larger vertebrates.
- Transfer body to 70-75% ethanol after draining formalin from body.
- Place the body in a glass jar or vial topped up with the ethanol.
- Place data label inside jar and include field number on label.
- Ensure the voucher specimen data form is completed, including the noting of any tissue samples taken.
- Do not store the specimens in the sun as this can cause fading of colours.

Safety Considerations

- Formalin and formaldehyde are extremely corrosive to living tissue, and will cause skin to peel; wear plastic gloves at all times when handling; do not breathe vapours; only use in a well-ventilated space; avoid all contact with eyes.
- Transport formalin and formaldehyde only in clearly labeled bottles with secure, leak-proof safety caps, and inside a second, sealed container that will protect the bottle from extremes of heat and breakage on rough roads.
- Fromalin and formaldehyde should never be kept in close proximity to food or cooking equipment.
- Small volumes of ethanol up to 5 liters can be bought from chemical suppliers and some pharmacists. Volumes greater than 5 liters can only be bought from a registered supplier by those holding a current concessional spirits permit.
- Ensure that specimens are preserved in ethanol, not formalin, and that all containers are tightly sealed and do not leak. If ethanol is not available and methylated spirits have been substituted, note this is the covering letter that accompanies the specimens.
- Ensure that every individual animal has a unique number on a tag attached firmly to the body, preferably to one of the limbs. Tags for snakes and legless lizards should pierce a fleshy part of the tail. Tissue samples taken from individuals should have the same number as the remainder of the body.
- Tags for reptiles and frogs should be those supplied by the Museum.
- Each individual sent to the Museum must be accompanied by a completed voucher specimen form. Form should be filled out in dark pencil or ink which is indelible to alcohol. Drawing pen ink and some special felt-tipped marker and alcohol-proof, but most felt-tipped markers and ball-point ink will run badly.
- Please keep photocopies of the completed voucher specimen field data sheet in a safe place, along with a record of when we sent the material to the Museum.

- Voucher specimens also should be accompanied by a covering letter which lists the specimens. Museum collection managers should be requested to complete the sheet and return it to the person nominated, in order to verify the identification of the specimen. Do not forget to provide a return name, address and phone number at the bottom of the form.
- Ensure specimen jars and vials are packaged with plenty of padding and in a crush-resistance box.
- Send voucher specimens by road courier, not by Post or air freight.

Systematic Area Search Method

Day searchers of a fixed area with a fixed level of search effort is a widely accepted and effective method for revealing the diversity of species of reptile and frog within an area. The method described below is suitable for documenting the presence of species of reptile and frog at a fixed position along a transect. It allows the maximum degree of confidence that the species recorded can be related to the site conditions and thus add to accumulation of knowledge about habitat preferences and activity patterns of species in relation to season and prevailing weather.

A systematic search over a fixed area and for a fixed time has the added advantage of providing and index of the abundance of individuals and species, which can be compared with sites where the same systematic search method was used.

This method is suitable for all habitats.

All opportunistic frog and reptile sightings or hearings should be recorded on the opportunistic animal sightings field data sheet. Opportunistic sightings or hearings include all animals seen or heard outside the recording time period of 30 minutes spend at each survey site. This situation may occur when team members are walking between sample areas or setting up the transect.

Methodology

- A standard 0.5 hectare area i.e. 50 x 100 meters forms the basis of the systematic area search method for reptile and frog survey. Search areas should be set out approximately in the center of the survey site. If there is a already a transect set up for birds or mammals for example, then we can use part of that transect to measure out 100 meters.
- The search should be standardized to one person-hour at each site.
- Surveys should be conducted by mid-morning when temperatures are sufficient to ensure that reptiles are not so active that they cannot be identified or captured by a licensed person before escaping.
- Surveying should not be conducted on very overcast or rainy days.
- This survey techniques entails active searching of potential reptile and frog microhabitats. Active or basking reptiles will be identified by sight or by keying out, using a field guide. When possible animals should be captured by a licensed person to ensure positive identifications are made.
- Sheltering species can be detected by moving fallen logs, litter, loose, shedding and fallen bark, rock outcrops and other likely hiding places.
- When identification is not certain voucher specimens should be taken by a licensed person using humane methods approved by an Animals Ethics Committee.
- Incidental observations of fauna other than frogs and reptiles should be recorded on opportunistic animal sightings field data sheets.
- The identify and numbers of reptiles and frogs detected should be recorded as soon as identification is made.

Special Considerations and Tips

- It is useful to appoint someone as data recorder so that they can concentrate on accurate record-keeping and searchers can concentrate on searching for and identifying reptiles and frogs.

- When searching an area we may come into contact with venomous animals such as spiders, scorpions and snakes. Take extreme care and wear appropriate hand protection.

Animals Care Issues

- Always replace material distributed during the search as best as possible.
- Only catch and handle animals for identification if we are experienced and properly licensed to do so.
- If capture is necessary for identification, always ensure the stress to the animal is kept to a minimum by using appropriate techniques and handling the animal only for as long as needed for identification.
- Always release captured animals as soon as possible at the place that they were captured and in conditions that will not increase risks to the animal for example, predation.

12

Bioregional Planning

Today, governments and communities rely mainly on their national parks, and other special protected areas, seed banks, tissue collections, zoos, and botanic gardens to maintain the diverse life-forms found in the world's many habitats. These mechanisms and tools comprise vital components of any national biodiversity strategy. However, new policies are required that secure these advances and provide concepts and means that account for shrinking wildlands, increasing pressures for food and fibre production, urbanisation and infrastructure development, and declining government budgets for reserve management. The challenge now and in the future is to protect biodiversity wherever it is found, from parks to farms, commercial forests, coastal zones, fishing areas, and people's backyards. Genetic, taxonomic and ecological diversity manifests itself throughout whole ecosystems. Where impoverished, restoration is needed. Where used as biological resources, concern for sustainability and equity are more and more demanded.

The debate over management and governance of biodiversity and biological resources is further stimulated by several additional driving concerns: First, sustainability—public agencies and private land owners are well aware that they need to shift their management practices to attain greater productivity and sustainability in ecological, social and economic terms. Second, equity—they are already attempting to incorporate national goals, like endangered species, air and water quality, into their own management practice locally, but call for new and creative mechanisms that will make this integration of goals equitable, efficient and practical. And third, participation – few opportunities for meaningful citizen, corporate and indigenous community involvement in decision making and planning for people's home territories exist. They want to be involved in the debate, negotiate their common vision and help formulate and implement plans for action. For a start, stakeholders – those that live in, use, depend upon, have impact upon, or otherwise care about the place and its multiple values – need access to the planning and implementation process and a flow of science-based, socially sensitive, and economically solid information that is helpful to all constituents. For bioregional management to become a reality, we need to face the scales of justice: on the left we have pressure upward to shift management to ecosystem scales; on the right, there is downward pressure to negotiate authority and responsibility for natural resources to those that are stakeholders in these bioregions. As we shall see, there are other scales to be balanced.

Frentz et al. (1995) noted that on this latter approach, there are presently over 600 experiences under way presently in the United States, without counting all the many earlier and on-going programs that share many of the same principles and objectives. Some are led by Federal and State government agencies, others by communities and non-governmental organisations. Their use of the terms "ecosystem" or "bioregion" embrace units of land and aquatic space at different geographic levels, according to the problem being addressed e.g., water supply and the associated watershed; range of migratory wildlife; etc., the biogeographic system in use such as ecoregions, biogeographic provinces; etc., or community sense of place in terms of native communal lands and nations; northern Rockies; southern Appalachians; etc.. Bioregional management and bioregions We'll refer to these four approaches collectively as *bioregional management* to simplify our task and emphasise how they contribute to an integrated framework for planning and action. The unit of management will be called here a *bioregion.* While there is no internationally agreed definition of a bioregion, the following is offered for our purposes: *a geographic space that contains one whole or several nested ecosystems, characterised by its landforms, vegetative cover, human culture, and history, as identified by local communities, government agencies, and scientists.* Bioregional planning also lacks an agreed definition, but we can consider it to be: *an organisational process that enables people to work together, acquire information, think carefully about the potential and problems of their region, set goals and objectives, define activities, implement projects, take actions agreed upon by the community, evaluate progress and refine their approach.*

Characteristics of Bioregional Planning

By synthesising and integrating the four approaches, a general framework for bioregional management can be deduced, the key characteristics of which are: large, biotically viable regions ranging from 1,000 to 1,000,000 sqare kilometres. Strong personal or institutional leadership and management a structure that features one or more core zones or protected areas and interconnecting corridors, all lying within a matrix of mixed land use and ownership that contains resources and the potential for investments and management that foreshadow economic sustainability full involvement of stakeholders the plans are socially acceptable information is widely available, is-scientifically solid and comprehensive research and monitoring activities are geared to supporting decision making functions, shifting technologies and practice towards sustainability and supporting adaptive management. The process and debate is informed by science and technical knowledge and information the processes is committed to adaptive management impoverished habitats are restored with local plant and animal materials public agencies and private interests cooperate to utilise available skills and to build capacity as needed by the group efforts are made to horizontally integrate public agency programs, capacities and budgets and to vertically integrate the various layers of government interest in cooperating with other similar programs internationally.

Challenges to Bioregional Planners

Local government, resource planners, community leaders and other stakeholders can anticipate three imposing challenges when they decide to establish a bioregional management program. Few existing organisations possess the human skills, facilities, institutional and financial capacity to manage entire regions of countryside. For example, the capacity to inventory and map resources, communicate effectively, negotiate and plan, monitor and evaluate progress will be found scattered among existing mandates, programs and budgets. Coalitions of stakeholder groups can integrate their capabilities to build that new capacity. Where vital skills or facilities are absent from the bioregion the options include cooperation with distant institutions, or the development of missing capacity in an existing or new organisation locally.

Typically missing are mechanisms to convene the various parties that have an interest in the future of their region. The latitude by which government agencies may negotiate with other stakeholders is often limited by central office country-wide administrative procedures. Corporations may not feel free to participate in discussions that will reveal their strategies and plans. Local farmers and foresters may feel inadequate to deal with these players because of a perceived lack of access to the process, of information, and of confidence that anything useful will actually happen. Responses include training opportunities for all stakeholders on how to participate, negotiate their case, build scenarios for the future and evaluate options. Information about the region, its ecosystems, economy and culture, presented in a way useful and accessible to all those involved is a must. Leadership typically starts with an individual, a person with a vision, the personality and inter-personal skills to work with people throughout the community, and thick skin to persevere through endless debates until acceptable levels of consensus are reached and people get on with implementation of an agreed program.

And, in all cases, a whole array of organisations and institutions already litter the landscape, be they in health, agriculture, transportation, sewage and waste disposal, water management, schools and education, civil society service groups and political governance. The configuration of these multiple and distinct mandates and their corresponding programs and budgets, create a maze of institutional obstacles, but also a panoply of opportunities. From our experience thus far, stakeholders can capture the skills, use of equipment and facilities, financial grants and other opportunities offered by these programs. Develop the capacity to manage develop leadership for the program; give careful consideration to social and governance issues as part of overall program management; use authority judiciously to foster cooperation; redistribute power over land and resources to develop authority and responsibility in the bioregion; identify and assess availability of human, technical, facility and financial capacities in bioregion and fill in the gaps; use and build upon existing capacities wherever found; develop the capacity to work with global change, long and short term natural cycles, and catastrophic events; Foster stakeholders as co-managers leaders, planners and policy-makers should get to know stakeholders, their concerns, interests and perspectives; focus initial action on a few issues of interest to the widest possible set of stakeholders in the region; link conservation and restoration activities with local goals and objectives for socio-economic development; give local residents and communities access to decision-making processes and the skills needed to participate fully in the development and implementation of democratically managed bioregional programs; to keep negotiations fair, give all stakeholders information of equivalent value; give stakeholders incentives to encourage their full involvement in and commitment to the process and program; ensure that stakeholders receive a fair share of benefits; in areas of multiple jurisdictions, try to develop coordination mechanisms that do not immediately challenge an institutions' or a nations' mandates or sovereignty; honor all commitments made during negotiations; implement projects promptly that respond to community needs; instances, however, some level of central authority will be required to represent society as a whole and future generations and oversee compliance with appropriate standards and norms. Neither extreme of totally central or entirely community management appears to be ideal or practical.

Focus. While analyzing global and regional ecological, economic and social issues to define the bioregion and its program of work, keep the focus on issues of common concern to all interested local parties. *Conservation and Development.* Balancing conservation and development goals keeps bioregional programs relevant to local residents and resource dependent people and industry. A preoccupation with conservation alone can alienate potential partners and isolate the area from the regional economy. *Funding.* Welcome international funds to support the launch of bioregional programs, but shift the balance of funding as soon as possible to a mix with internal sustainable funding sources. *Sovereignty.* Establish cooperative arrangements among central, regional and local government entities and private interests that respect national and institutional sovereignty and existing mandates. *Centralized command*-and-control approaches to natural resources management and ownership will have to gradually give way to co-management and co-finance arrangements. *Time and Adaptability.* The time needed to develop new ideas and techniques needs to be balanced with that required by managers, communities and other interested parties to adapt their respective social norms and resource use techniques. Some of the cycles of nature, like migration pattens, are predictable, so managers can easily prepare for them. But others, such as natural disasters, stride unannounced, while still others occur over intervals too long to fit into even long-sighted plans. Bioregional programs need to anticipate and prepare for natural disasters and humanitarian assistance demands while dealing with normal daily projects and activities. *Old versus New Institutions.* Where pre-existing institutions can be adapted to lead, or contribute specific capabilities to the bioregional program, then on balance, their reform may by more efficient than the formation of new organizations. Alternatively, where established bodies cannot or choose not to evolve, or where needed skills and capabilities are not available in any existing institution, then a new body may be justified to fulfil that role. Can bioregional management enhance biodiversity's chances? What can be observed where such programs are being implemented is that: A larger and more complex pool of skills and tools are serving to help protect, restore, inventory, collect, study, appreciate, utilise and educate about the region's biotic wealth. Greater numbers of individuals and group could be involved in charting the region's future and in reconsidering how their own livelihoods need to be harmonised with the region's resources.

Public and private organisations and landowners in their regions are negotiating new agreements to promote cooperative planning and implementation, including co-management and co-finance options. Participants demand, and learn to employ data, information and local, traditional and scientific knowledge in their process of decision-making. Stakeholders develop the skills and access needed to participate in planning and management activities of the region. The need for careful protection and management of selected sites for their ecological functions is understood and accepted, while at the same time greater emphasis is given to improvements in the sustainable utilisation of biotic resources and the equitable sharing of costs and benefits. Over time, bioregional programs give greater emphasis to policy, legislative and administrative considerations and place increasing weight upon social, cultural and economic factors. Easy technical and expert fixes are gradually held suspect. Participants favour incentives and other positive mechanisms to foster their cooperation in regional programs. They also recognise the role of government agencies and regional authorities to provide long-term continuity, support and commitment to the aims of the program and national goals. The 'bioregional' contribution to planning emphasises the supremacy of natural units over other jurisdictional areas, including political divisions. It puts a premium on ecological integrities and on expansions of spatial vision, and vigorously promotes informed public participation. On the latter count, the very accessibility of competently focused historical reconstructions can directly assist the planning process. Tracing through the most specific antecedents of current approaches effectively grounds what may be perceived as a wholly novel and even threatening perspective in more familiar territory, thereby empowering a wider cross-section of the community. We identify six inter-related themes and characteristics: *consensus* - albeit seldom as complete or as passionately articulated as the principle demands on the pivotal significance of natural resources, especially water resources, in the driest of the inhabited continents; *anxieties* - about the causes and effects of environmental degradation, particularly soil erosion and water erosion; *innovations* - in management frameworks, designed to accommodate an augmented range of parameters, including environmental factors; *interchange* - between rural and urban communities on the one hand and their elected representatives and deputed experts on the other; *engagement* - by committed individuals and groups; *exemplars* - of new planning approaches here and overseas.

The United Nations Educational, Cultural and Scientific Organisation founded in 1946 has made a large contribution to regional planning initiatives, including a promotion of integrated river basin management in the 1950s. In the early 1970s, UNESCO's World Heritage Convention emphasized the universal natural and cultural values of 'sites' meriting a kind of supra-national status; the move recognised collective responsibilities which transcended the geographical accidents of politically-defined sovereignty.This seemingly nebulous idealism was to become a potent weapon in the armoury of new environmentalists. It demanded that the purview of regional planning be stretched to cater for these special areas. How to do it was no less problematical than selecting what should be protected, but UNESCO's 'Man and the Biosphere Program' (1979) devised a *biosphere reserve* strategy which encouraged the building of a network of sacrosanct places around the world. Well-received elaborations recommended that in each case the core should be protected by a surrounding buffer zone and a well scrutinised transitional area further out; only limited protection-related activities would be permitted within the buffer zone. Thus the Program advertised the worth of bioregional planning.

Integrated Strategic Planning

At a somewhat smaller geographic scale but of possibly greater significance to bioregional planning, has been the evolution of strategic planning at the metropolitan, sub-metropolitan and local scale, and sometimes in regional circumstances as well. The language of economics must be introduced to the bioregional discourse. There needs to develop broad community and inter-governmental agreement about the relative importance of biodiversity and bioregional planning to ensure the evolution of a shared understanding of desirable outcomes and of priorities - the national strategy is a first step. Protect and enhance the biological diversity and ecological integrity of terrestrial, freshwater, coastal, estuarine, and marine environments. The strategy goes on to outline a total of 40 goals addressing: the natural environment; economic development; natural resources; social development; urban development; and infrastructure.

It goes on to address initial implementation issues. There is no judgment at this point as to whether or not the attention to biodiversity can be regarded as sufficient; or whether or not it satisfies the expectations about the impact of bioregional planning that many might hold. But the issues are there, on the table, as part of the planning process, to be addressed and resolved along with the many other issues affecting and arising from the region's development. What has already become apparent through the planning and consultative process is that further work needs doing on a number of aspects of the regional strategy.

The draft is out for public comment. This consultation provides the opportunity for debate on the relative weight given in the draft strategy to biodiversity questions, to economic development objectives, to use of land and resources, and to other factors. Resolution of the debate over which values are the most important will be critical. The power of the argument for a bioregional approach and for maintenance of biodiversity in such situations cannot simply be the power of a scientific or even moral stance. It needs to be a power based on wide understanding, through information and through knowledge, of the implications for the future of the region of giving, or not giving, this issue a standing above others. This is often a debate between short-term and long term values - jobs versus the sustainability of the environment.

If economics has become and is to remain the basic language of assessing the merits of how resources are to be utilised, then bioregional planning must be capable of engaging in economic assessment of the value to society of maintaining biodiversity. That's not easy, because the techniques in economics for measuring long-run rates of return, and social rates of return where non-monetary costs and benefits must play a major part, are not well developed, not widely understood, and where they do exist are not widely used. "What is a 'bioregion'? Almost by definition, bioregional planning must involve local communities. As the Global Biodiversity Strategy such as WRI, IUCN, UNEP; 1992 points out: "A bioregion is a land and water territory whose limits are defined not by political boundaries, but by the geographic limits of human communities and ecological systems. Such a region must be *large* enough to maintain the integrity of the region's biological communities, habitats, and ecosystems... It must be *small* enough for local residents to consider it home... A bioregion is also defined by its people. It must have a unique cultural identity and be a place in which local residents have the primary right to determine their own development. This right does not, however, imply an absolute right... Within a bioregion lies a mosaic of land or aquatic uses... Within the ecological and social framework, governmental, community, corporate, and other private interests share responsibility for coordinating land-use planning for both public and private land and for defining and implementing development options that will ensure that human needs are met in a sustainable way".

Critical to the concept of a bioregion is the sense of community "ownership" - of being part of a "community of common interest". A bioregion must take account of biophysical features and the social context in which those features occur. The concept of bioregional planning is seen to be quite new, but in fact as Powell (1993) identifies, precursors of bioregional planning shaped the ways in which we develop the country. Although in some cases a bioregion might fall within a single, relatively small catchment coinciding with a local government area, it will more likely spread across existing political and land tenure boundaries-boundaries which local, state and national administrators find hard to relinquish. Since the UN Conference on Environment and Development the Earth Summit, held in Rio de Janeiro in June 1993, governments of different countries have repeatedly espoused their commitment to community consultation and involvement. Nowhere is that involvement more important than in the development and implementation of bioregional planning. But community consultation means different things to different people. It could be noted that when governments undertake community consultation they may simply aim to give information to those in the community who are involved or they may seek to gather information or to gain feedback on ideas developed within government. In some cases community consultation may be designed as a dynamic information sharing process. While each of these has its place, the expectations of both local communities and community interest groups is that they will be given opportunities for meaningful participation in decision-making from the planning stages of a project onwards.

There has been discussion about the need to ensure that governments at all levels adopt some "standards" for community consultation. More than once the idea of developing a Charter of Consultation has been proposed. Such a Charter, itself developed through widespread consultation, would serve to define for governments and community participants just what the principles of proper consultation should be. The development of a Charter of Consultation is indeed a concept whose time may have come and one worthy of further consideration. In order to undertake consultation effectively, those who are to be involved need When applied to bioregional planning, this can perhaps best be achieved through a stepwise process in which community groups and interested individuals play an ongoing role: Step Ecologists, mapping experts and those concerned with water catchment and other scientific research would work with State and local governments and local communities to define the most appropriate natural boundaries upon which to begin bioregional planning and implementation of programs for the identification and maintenance of biodiversity.

A public involvement campaign would clearly define and interpret the concept of biodiversity, at the same time as it invited participation on identifying key elements of biodiversity in the region. Community groups, industry representatives, relevant State agencies and local government officials would be approached to provide input, probably to a government funded detailed survey of biodiversity values within the bioregion. Working with relevant scientists, individuals and interest groups within the community would then use agreed methods to collect information about the biodiversity of the region, thus completing a 'rapid biodiversity assessment'. It is likely that partnership agreements may be struck between one or more levels of government. Once biodiversity values have been identified, focus groups would be held to discuss threats to that biodiversity. The focus would be upon the actual threatening processes, rather than any particular mechanism which might be adopted to halt the threat. Focus groups would work with experts provided by scientific institutions, and all three levels of government, to ascertain what needs to be done to stop the threats degrading biodiversity values. Focus groups would be helped through experienced facilitation and dispute resolution where necessary, to reach agreement on how to use the expert advice, and the information they have collected themselves, to protect the biodiversity in their region. Achievable targets on the way to achieving their goals would be agreed.

Local and State government officers, who have been involved in the planning process through focus groups would then incorporate the strategies into their work plans and strategic plans. Local landholders and industries would be encouraged to incorporate the strategies in their day-to-day activities. Local and State government officers would incorporate consideration of the agreed targets in their annual reporting, and support community biodiversity "watchdog" groups who would monitor the effectiveness of the strategies, and initiate their amendment as necessary. The "watchdog" group would also take account of new information about biodiversity values, and maintain a watching brief on biodiversity values. The results of ongoing biodiversity monitoring would be reported regularly through State of Environment or other reports by local, State and National governments.

Existing community networks designed to disseminate information about various aspects of biodiversity, have a role to play in maintaining and supporting local "watchdog" groups. Land management and planning has traditionally occurred using a cadastral base. The division of land into ownership parcels, largely without any ecological context, is a planning legacy that strongly impacts upon our ability to plan biodiversity conservation and management in an effective way. Equally, until recently the landscape scale has been the poor relation of biodiversity as it has been much overshadowed by the fixation of attention at the species and genetic levels. Effective, long-term land management that will conserve ecological processes, crucial to the capacity of the land to sustain human communities, needs what Barker (1992) terms a "paradigm shift". That paradigm shift is a bioregional basis for biodiversity conservation and management.

Courier (1992) defines a bioregion as "a land and water territory whose limits are defined not by political boundaries, but by geographical limits of human communities and ecological systems". A major advantage of the bioregion is its comparatively large size. Management of small units of area tends to be labour intensive and largely tactical in orientation. Biological integrity and ecological health are difficult to maintain. Large units of area, however, extend the available management options to maintain biological integrity and ecological health. A bioregional approach to biodiversity management, where each bioregion has a mix of land or sea use, including a range of protected area types, must become a keystone for part of any biodiversity conservation strategy. Policy for the conservation of landscape diversity is the most essential, yet most neglected, of all biodiversity conservation concerns. This situation is due, in part, to the complexity at the landscape scale. This complexity is not only environmental but also a reflection of the diversity of stakeholders which are involved. National legislation and international treaties and conventions have been largely directed at protection at the species and genetic levels. Some efforts to include communities have been made but, with rare exceptions, landscapes have been ignored by mainstream conservation efforts. Changing to a landscape scale Land management decisions are primarily made at the level of the ownership parcel. Most often, the land manager makes the decision to thin trees, clear-fell, remove shrubs, or drain a wetland. What is needed is the integration of government strategies and policies with informed local knowledge about the health of our ecological systems. The key role of indigenous knowledge is especially emphasized here.

Decisions made in one part of the landscape matrix often spread effects across the landscape. To provide a context for regional and local planning, and to anticipate these impacts, it is necessary to understand broader landscape patterns and processes. Managers must develop responsive and accountable relationships with the land. Freudenberger and Freudenberger (1994) propose that our relationships with the land must be informed by sound ecological understanding, and that the concept of ecological sustainability must be incorporated into social justice concepts, otherwise justice is short lived. Without a caring and understanding relationship, for and between humans and nature, whole communities and the landscapes on which they depend will become degraded. A much broader approach to the conservation of biodiversity must prevail. A focus on the landscape scale is a key factor in an overall bioregional approach to land management. Most of our conservation problems need to be addressed in that context, rather than at the species level. Processes which endanger species largely arise from landscape degradation, especially vegetation removal or fragmentation. Certainly, the approaches taken by landscape ecologists in emphasizing the importance of metapopulations, *i.e.* populations in a landscape framework, are proving valuable.

Walton et al. (1992) hold the view that conservation of biodiversity MUST focus on management at the landscape level, especially in an increasingly anthropogenic environment and in the face of global change. Conservation of biodiversity must focus on management at the landscape level, especially in an increasingly anthropogenic environment and in the face of global change. What is needed is a bioregional planning framework where conservation is predicated upon access to a scientifically credible classification of the environment into landscape units or bioregions. Brunckhorst and Bridgewater (1994) reported that a bioregional approach across all jurisdictions which also reflects a cultural identity with the environment, provides an integrative framework for cooperation on the preservation of biodiversity. Existing narrow views on the implementation of regional planning need to be overcome. Integration of current ecosystem-based management, watershed management, wildlife management, community based programs, land/sea management and resource use, and ecologically sustainable development will be required for true integrated regional planning. Ultimately a synthesis of desirable and culturally meaningful criteria are required in a bioregional planning framework to integrate environmental and sustainable development objectives. A bioregional framework for planning and management should both reflect the natural environment and human society living within it. Partnerships need to be fostered to address ecological, cultural and economic issues at the scale of the regional landscape.

Protected Areas in a Bioregional Framework

The protected area estate is perhaps the most valuable asset for the maintenance and management of genetic, species, community and landscape diversity, as well as key ecological processes. As the Global Biodiversity Strategy, Courier (1992) notes: "..., protected areas linked by corridors become means of maintaining functioning natural or nearnatural ecosystems over large regions".

Protected areas are obviously most valuable in conserving genes and species habitats, and for maintaining key ecological processes. The roles of protected areas go far beyond that of simply reserves. They provide valuable functions often beyond their boundaries, for example as a catchment for the provision of clean water, as a propagule bank for rehabilitation, as a refuge for recovery of species populations and communities, as a site for replenishment of sites which have lost species through local extinction, and as a site for future provision of genetic material for medicinal and horticultural interests. Brunckhorst and Bridgewater (1994) outline some of the potential roles for existing protected areas in habitat restoration and recombinant ecology for enlarging, redesigning and linking protected areas. Protected areas and their management Establishing a national protected area system with the aim of: maximising the conservation of biodiversity; viewing protected areas as part of the landscape matrix, rather than as islands putting into place an integrated management framework for those landscapes will be a major step in moving from a cadastral to a bioregional approach.

Bridgewater and Walton noted that recent moves by the World Conservation Union (IUCN) to define protected areas by management objectives is strengthening the role protected areas could play in the bioregional approach. IUCN has established six categories: 1. Protected area managed mainly for science or wilderness protection 2. Protected area managed mainly for ecosystem protection and recreation 3. Protected area managed mainly for conservation of specific natural features 4. Protected area managed mainly for conservation through management intervention 5. Protected area managed mainly for landscape and seascape conservation and recreation 6. Protected area managed mainly for sustainable use of natural ecosystems The basis of categorisation is by primary management objective. There are, in fact, a wide variety of potential primary management objectives for protected areas.

Biosphere reserves, a model of the bioregional approach The concept of protected areas within an integrated landscape management framework is not new. The Biosphere Reserve concept, formulated to wed conservation and sustainable development, has been developed and championed under the UNESCO Man and the Biosphere Program (MAB).

Biosphere Reserves are recognised areas of representative environments which have been internationally designated within the framework of UNESCO's MAB Program for their value to conservation through providing the scientific knowledge, skills and values to support sustainable development.

Biosphere Reserves are nominated by national governments but must meet agreed criteria and adhere to a minimum set of agreements before being admitted to the worldwide network. In particular, each Biosphere Reserve should perform three complementary functions: a biodiversity conservation function with a focus on conserving a representative sample of major ecosystems; a development function with a focus on humans in the biosphere, emphasising an integrative role for local communities; and a logistical function such as combining conservation research, education, training and monitoring.

Biosphere Reserves are a special kind of conservation area-traditionally a nested series of zones with differing management intensities viz., core area, buffer zone and transition area), designed to include people within an overall conservation framework. They may comprise any mix of terrestrial and marine elements. About 90 per cent of all Biosphere Reserves are covered by some form of designated Protected Area *e.g.*

national park and nature reserve, although eight per cent have no nationally designated areas associated with them.

Bioregional Planning Frameworks

Lambert et al. (1996) propose a bottom-up approach to bioregional land classification, planning and management. The reality is, however, that government plays a vital role in the collection, compilation, storage and analysis of complex national and regional data sets and in deriving consistent boundaries and descriptions of regions. We question the capacity of the local community to undertake these functions. Local communities can play a role in providing a context for this information, and for driving the onground implementation.

Some Guiding Principles for Policy Formation

Protected areas, and elements of remnant vegetation connected by corridors or "greenways", can provide the reservoirs of biodiversity, and avenues for adaptive responses to environmental change. If protected areas are isolated, their purpose is ultimately defeated. Corridors reduce landscape fragmentation and assist in the rehabilitation of degraded areas. An understanding and appreciation of ecological infrastructure must thus be at the heart of the bioregional approach. Boundaries exist in time and space. Many have been created and maintained by human intervention. Despite the fact that we have imposed boundaries with reckless abandon, we know very little about them. Their impact on the flora and fauna is poorly understood and largely unappreciated. The risk of losing biodiversity, and disrupting ecological functions, is greatly increased when natural patterns and boundaries are altered. So let us use naturally defined boundaries as our major bioregional management paradigm.

Dynamic systems require dynamic management. The challenge is to incorporate natural levels of spatial and temporal heterogeneity into management. Conservation managers need to be able to identify landscape infrastructure, and pay due attention to its key elements in management plans. This is an area where considerable research appears necessary to characterise the structure and, especially, function of such landscape elements, in particular how they relate to ecological system processes at various scales. The challenge for the bioregional approach lies in developing policies which generate realistic assessments, establish priorities, support the necessary institutional reform, discourage further degradation, provide incentives for improvement, and create the wealth to pay for the conservation of biodiversity.

The bioregional approach must link ecology with economics, sociology and politics and ensure that good policy intentions do not yield inappropriate results. Agency and institutional missions must be clearly defined by tasks, not by organisational or professional subcultures. Performance and achievement must be evaluated on a regular basis. In the development of policies for the bioregional approach, the strategic intent of those policies must be clear and unambiguous. That intent should: capture the essence of success; be stable over time; and set a target that is deserving of personal effort and commitment. Clearly, conservation by legislation needs more thought and less emotion. We must be cautious that the heady atmosphere of a political victory is not a prelude to a nightmare of legal or administrative entanglements, which squander badly needed resources and distort issues. We must maintain a perspective on medium to long term goals, which discourage quick, easy and temporary solutions - not be demeaned by Odum's (1982) "tyranny of small decisions". If we are truly resolved to conserving biodiversity, then we must make the leap to a landscape scale, and to plan our protected areas within a matrix of sustainably Ecological Forestry, the growing and use of trees, must be based on bioregional preceptors.

Sustainable forestry must be based on the ability and willingness of our nation to pay for the delivery of outputs from the forest-wood,water, landscape, soil protection, biodiversity, recreation, solitude experiences... Hence, ecological sustainable forestry can only be undertaken within strategic planning concepts. Forestry is about the growing, management and sometime harvesting of trees. For most users, healthy trees are preferred, even by non human users, and as tree species prefer specific combinations of soil, water, elevation, aspect and climatic regimes, the best growth is limited to suitable sites. Forests cannot be grown on all lands in India. A healthy robust forest is best able to withstand shocks of the unexpected, and to survive, and continue to protect those dependent others, as well as to recover from the adversity. Forests are structural entities, as, by their presence, trees provide a diverse range of habitats for other organisms-flora, fauna and micro-organisms. But trees also provide one of the best solar energy sourced materials available for use for buildings, furniture, paper products and for use as fuel. For that reason we need to continue to develop systems to favour the selective management of specific forests for a preferred outcome, including wood production for as long as wood is required by society.

Each forest could be managed for a single purpose, but there is no need to be that parsimonious, and forestry has become a leader in bio-diverse regional planning in India, that is, in providing multipurpose forests, usually with a dominant managerial outcome. This talk will concentrate on those multiple purpose forests, of all types, used by US, for the dominant outcome of timber production. In that context, bioregional planning must take account of that dominant end use as well as the tree species/natural resource matrix that leads to good growth and healthy forests, to determine suitable species compatible with user objectives. A future vision for the forest industry interlinked in farming, coupled with the commercial and other benefits that additional trees can provide for soil, water and habitat in rural area. How the process is managed to achieve that vision in space and in time, without disruption to land holders, industry, communities or infrastructure. Recognition by city dwellers and government of the benefits that increased forest cover will bring to the rural environment, and how they may contribute to the vision. The vision of "Forestry in Bioregional Planning" that is emerging in India. It is based on: defined geographic boundaries of long term forested lands with regional community and industry input and management, leading to local value adding processing for domestic and export markets, and establishment of additional forests and plantations, mostly on private lands, involving Landcare. The question of sustainable forest management must be tackled in a pragmatic manner. Criteria and indicators are being developed by which progress can be assessed. These criteria and indicators will evolve, but initially will focus on basic biophysical resource systems-soil, water, carbon balance. Planned product sustainability. maintenance and enhancement of biodiversity, especially habitats. Equity, community involvement and managerial systems. Voluntary environment management systems may be used to ensure progress, and to communicate results assessed against stated objectives. These will require independent verification by audit for larger forest managers.

Tourism

The natural resource needs of the tourism industry can be observed as follows: a full range of managed settings within a region, from the highly developed to the basic information at various levels and styles on the natural resources of areas, especially on their wildlife and their interactions with the environment information on management practices education and training in knowledge and interpretation of the natural environment.

Currently, the industry pushes any notion of intrinsic environmental responsibility onto the narrowly defined ecotourism segment of the industry. Indeed the National Ecotourism Strategy is written exclusively for this very small segment. Capitalising on the interdependence of tourism and bioregional planning The way of incorporating the requirements of tourism into bioregional planning is to seek tourism industry involvement and support for the development of representative systems based on bioregional biodiversity. Information on the distribution of animals and plants remains an on-going goal for ecological science a healthy national taxonomic enterprise is vital management decisions based on state-of-the-art science will become outmoded as new information and ideas accrue:

accordingly we must take an adaptive approach to management scientific input in the form of standard protocols for data-collection is required The scientist as analyst The jump from data base to the biological definition of regions requires analysis to detect repeated patterns. Essentially this involves the development of classifications - of 'plant associations', 'formation types', 'biogeocoenoses', 'synusiae', 'ecological communities' and so forth. The crucial thing is that the outcome of such analyses must identify units which have meaning on the ground, and which are useful in answering whatever set of questions motivated the analysis in the first place. The scale at which such associations are defined will reflect the initial set of questions and, although there may well be a most appropriate scale for that set of questions there is no universally 'correct' scale - our classifications impose a discreteness, determinism and degree of simplification on an essentially continuous, stochastic and complex real world.

In the present context the usefulness of regions defined by such analytical processes is to be judged in terms of the manageability of the resulting units, a general impression of commonsense, and a comparability with units elsewhere. There are many different approaches to this analysis, some more transparent and obvious than others.

Special Situations

The easiest and least equivocal ecological classifications are those based on some underlying environmental feature which is biologically overwhelming.

Hydrological Regions

A second clearcut and deservedly popular approach in the analysis of spatial data is that which bases categories on hydrological patterns. The most successful and best known bioregional approach to conservation and management.

First however assiduously 'good' scientific information on ecological classification, biotic distributions or the mechanisms underlying patterns of biodiversity, is sought the results will never be the last word. Accordingly laws, regulations, conservation decisions or protocols made on the basis of these data need continuous oversight and the potential for modification in the light of new information, ideas or theories must be reserved. It is frequently observed, as the protagonists in some conservation debate square up to each other in courtroom or media conference, that each is able to trundle out its selected scientific 'experts'. These advocates may well give contrary interpretations of the same situations. This is absolutely in line with what we would regard as a healthy scientific enterprise but it may lead to great frustration on the part of the body politic, legal or industrial.

The temptation, given this basic uncertainty and open-endedness inherent in the scientific method, could be to discard science and its products as a basis for decision making. This would be a grave error. First science remains the most objective and powerful tool we have for dealing with the material world and, second, contemplate the alternatives: political expediency, economic rationalism, graft and corruption, separately or severally come to mind.

The organised scepticism of the scientific community can act in another positive way within conservation decision making. Scientists make very good critics of the spoken and written word cutting through rodomontade and persiflage to challenge underlying principles, data and interpretations. In the bureaucracy of a democracy such scepticism, properly organised, can lead to better informed decisions, if not faster ones. But democracy is not anarchy and scientific input, however welcome and constructive, needs to be organised if decisions are to be made within any reasonable time span.

So, for the non-scientists present, bear in mind that scepticism and wrangling over interpretations are endemic to the scientific method and should be viewed as one of the great strengths of the approach. Accordingly beware of absolute statements presented as science. For the scientists present, bear in mind that the continuous bet-hedging and carping criticisms that are part of science may project such an image of the profession to some of its users that we run the risk of being sidelined. Scientists necessarily approach issues and statements in a constructively sceptical manner this scepticism should be viewed as a healthy and essential part of a democratic decision-making process proper scientific scepticism is inappropriately used as a tool in political debate scientific absolutism should accordingly be looked on with suspicion The scientist as consultant Non-scientist and scientist most commonly come into contact professionally when the latter acts as consultant to the former.

Compile Data on the Planning Region

- *Define regional boundaries* according to watersheds, biophysical discontinuities, boundaries of socio-cultural areas, administrative and political boundaries, or a mixture of these
- *Subdivide region into management units to which will be allocated types of management* these could include patches of remnant habitat, units of tenure, subcatchments, agricultural fields with different cropping regimes, wetlands, sections of floodplain, or a mixture of these. *Compile data on the biodiversity of each of the management units* (this should ideally be at more than one level, for example both vegetation communities and localities of plant and animal species; the occurrence of these features in a context broader than the region should be recognized, for example the extent to which species or communities are restricted to, or peripheral to, the region)

Assess Threatening Processes and their Effects

Assess the distribution of past, current and likely threatening processes in the region major processes will be clearing, grazing by domestic livestock, and logging; others will be sedimentation, chemical pollution, altered fire regimes, invasion by weeds and feral animals, water diversion, settlement patterns, taking of plants and animals, and gathering of fuel wood. *Identify the features that have been affected by threatening processes or are vulnerable to them* and recognize effects and vulnerability in a context broader than the region so, for example, the decline of vegetation types or species outside the region is taken into account.

Identify types of management required and mechanisms available

Identify types of management required to control extractive or destructive *land uses and other threatening*. But can involve a much wider range of approaches, including management for sustainable grazing and logging of native habitats, changes to agricultural practices, planning of waste disposal, water allocations for wildlife, fencing of remnants, and reconstruction of habitats *Review the availability, security and effectiveness of existing mechanisms to implement types of management* particular types of management can often be implemented in more than one way; a review of the alternatives produces a 'toolkit' for use in the region; this might highlight gaps in the range of mechanisms that are likely to be necessary; knowledge of the security and effectiveness of each of the mechanisms could guide choices on when they are appropriate

Set Targets for Types of Management

Decide on the areas or number of occurrences of each feature that need each type of management in most regions it will not be possible to contain all the remaining natural and semi-natural habitats in protected areas, much less strict reserves; decisions are therefore needed on how much, or how many occurrences, of each feature should be under each type of management; however, some types of management might need to be applied to all of the remaining examples of particular communities, as in regions where natural habitats are reduced to a few small remnants or where a specific management unit such as a unique wetland is the focus of conservation efforts

Decide whether and how conservation management should be applied to discrete replicates spatial replication of protective management for some features might be desirable as insurance against chance losses of one or more examples, to increase the range of biological variation protected within landscape subdivisions such as vegetation types, or to protect genetically distinct populations within a species

Decide whether effective application of a management type requires a particular spatial configuration of that type the persistence of some attributes under certain types of management is likely to depend on the size, shape and connectedness of the management units involved *Decide whether effective application of a management type requires a particular spatial configuration relative to other types* the effectiveness of some types of management will depend partly on their juxtaposition with other management approaches; strict reserves, for example, could benefit from buffer zones of semi-natural land under multiple use; local government zoning to protect estuarine habitat will be less effective if, at the regional scale, land uses in catchments promote sedimentation and chemical pollution

Select thc Specific Management

Assess to what extent the targets for each type of management have already been achieved. In most regions, some or all of the proposed management types will already be applied to some extent; the challenge for implementation is then to close the gap between the existing and targeted applications. *Identify optional management units to achieve targets* unless all the occurrences of an attribute are to be managed in a particular way, choices will be available as to the specific management units allocated to that type; the extent to which options are available can be shown by the irreplaceability of the units. *Spatial relationships between management units with the same types of management* one of the considerations for resolving the choices available for a given management type will be design criteria such as size, shape and connectedness; these will partly determine the need for ongoing maintenance activities as well as the likely persistence of the features being protected *Spatial relationships between management units with complementary types of management.* There will be many situations in which types of management require, or benefit from, a particular spatial configuration relative to one another; *Other viability and manageability criteria: condition, nature of boundaries, local and regional support, impinging land uses and natural processes* as well as spatial relationships, a variety of other criteria for resolving choices, where options are available, will determine viability and manageability.

Implement Management Types

Decide on the feasibility of applying to specific management units each of the possible mechanisms for achieving each type of management when the feasibility of applying particular mechanisms for implementing management and the extent to which surrounding activities impinge *Manage units to retain biodiversity values within them and monitor results* In many management units, the primary purpose of management will need to be complemented by on-ground activities to allow features such as communities or species to persist; for example, in protected areas, the persistence of attributes will depend to some extent on the maintenance of natural disturbance regimes and actions taken to minimize the impacts of recreation; in native rangelands or harvested forests, on-ground maintenance will include the control of weeds and erosion from roadsides; monitoring is essential to gauge the success of maintenance activities.

Manage Units to Adjust

Their influence on other units due to natural or anthropogenic processes the management of many units will influence the condition of adjacent or nearby ones, through processes such as water flow, movement of chemicals, and the spread of fire and exotic species; the movements of some animals will mean that some management units will be linked ecologically, even if they are in different parts of the region; it follows that some maintenance activities must be directed at managing these interactions.

The targets are the initial goals of the exercise. They should take into account the types of management that can be applied and the needs of specific areas for those types of management. Targets will define the options that exist in the landscape for applying particular types of management. A target requiring protection of 10 per cent of the vegetation remnants in a subregion. Targets are essential, not only to drive the processes of selection and implementation in an explicit, defensible way, but as a basis for later reviewing the success of bioregional planning. Initial targets are by no means written in stone. The need to alter them might become apparent during the selection of sites, perhaps because ground inspection reveals that it is better to dedicate resources to a smaller number of remnants with the greatest potential for rehabilitation. Similarly, the implementation stage might show that, because of unforeseen constraints, it is simply not feasible to apply a particular type of management to all the selected remnants. In these situations, the targets would be adjusted accordingly and the reasons for the changes recorded.

Considerable effort might go into selecting which of the remnants should be the focus of conservation efforts in the time frame of the goal. The feasibility of applying the type of management intended for the remnants. If this proves to be unfeasible for some remnants, alternative remnants might need to be found by returning to the selection stage or, failing this, alternative types of management might need to be applied. The second consideration is the timing of conservation action. Are some of the selected remnants likely to be lost or degraded before others? If so, these should have first call on protective management.

Bioregional Planning in India

Since 1951 the planned development is accepted as a strategy for the socio-economic development of India. This effort has attracted economists and politicians and their attention has been drawn to national programmes in sanitation, public health, nutrition, water supply and housing, but very little effort has been made to develop the environment and ecology till the end of the third five year plan. During the fourth plan period (1968-73) the concept of environment issues has been well recognized. The fourth plan document highlights the inter-dependence of living things and other relationship with land, air and water and the need for harmonious development. It could be possible on the basis of systematic identification and analysis of environmental issues with reference to economic and ecological spheres. The inclusion of environmental aspects of planning has got the attention among our planners in the fourth plan period itself.

The fifth plan (1974-79) document has not focused anything specifically on environmental planning. However, the sixth plan (1985-90) document has identified the importance of preservation and conservation of renewable resources like soil, water, plant and animal life to sustain our economic system. By that time there has arisen concern on the depletion of these natural resources and their deterioration that has caused a decline in the quality of life of the people. This is due to lack of proper consideration of costs of environmental degradation at the policy planning level, lack of long term perspectives on environmental protection coupled with commercial motive of destruction of resources base, deforestation and depletion of natural resources. Hence, this situation has forced the planners to take a new approach to development focusing on techno-environmental and socio-economic evaluation of all developmental activities in the country.

Further the sixth plan document highlights the administrative and legislative arrangements for environmental protection and indicates some of the environmental development programmes to be implemented during the sixth plan period.[2] The proposed important activities to environ-mental protection include the making of an environmental impact assessment an integral part of the entire planning process, setting up of an environmental information system, programmes for increasing public awareness about environmental protection, creation and management of biosphere reserves, monitoring and conservation of marine ecosystem.

The seventh plan (1985-90) envisages the formulation of a national conservation strategy to ensure that the goals of sustainable development could be met in future with the involvement of NGOs and the private sector. [3] The major programmes that have emerged during the seventh plan have been development of instrumentation, equipment and institutional facilities for environmental monitoring, pollution control and waste management, regulation of production and consumption of hazardous substances, environmental guidelines for starting industries and so on.

In the seventh plan period, the progress of implementation of various environmental development programme has been reviewed and identified a list of environmental issues, such as stabilization of both human and livestock population, urgency of conservation of environmental resources and the need to adopt a coordinated, highly decentralized approach in planning and management of environment with the active participation of people.[4] The eighth plan (1992-97) document emphasizes on destruction and degradation of forests and its consequence in terms of heavy toll of our soil and water resources. The increasing trend in air and water pollution is attributed to increasing contamination by heavy chemical metals and other toxic substances, due to careless industrial and agricultural practices, unplanned urban growth and industrialization. The important objectives of environmental development during the eighth plan include the protection of natural environment, regeneration and restoration of degraded ecosystems and increasing their productivity and generation of employment through these activities, de-centralization of control over nature and natural resources, understanding natural process, formulation of a national policy for the environment and an appropriate institutional and legal frame work to support the policy and ensuring coordinated and integrated government action aimed at conserving nature and sustainable use of natural resources. Further the eighth plan has envisaged the individuals and institutions more accountable to the people for their actions impinging on environment and ecosystem along with monitoring the state of environment.[5] The ninth five year plan (1997-2002) has followed the objectives of environmental management along with sustainable development concept.

It could be observed that all the major initiatives in environmental planning in India have been undertaken by the central government. Though environmental programmes are being implemented by the state agencies, local bodies,

NGOs, effective decentralization of environmental planning in the sense of formulation, implementation and monitoring of environmental plans and programmes in the state, district, local and regional level has not yet been done. Although the district and the block are accepted as a proper unit of administration of our planning programmes, planning authorities and administrative agencies are inadequate both in number and quality. It could be seen clearly that the present system of plan formulation and implementation do not have any systematic identification of locational issues. Hence, we need a concrete environment regional planning at the local levels.

The concept of environmental regional planning emerges because the planning done at the national level does not ensure its applicability at local levels and because primarily a centralized system of planning has not adequately allowed for the implementation of various aspects of environmental development progammes throughout the nation. The spatial consideration could be given more weightage rather than overall objectives of environmental protection. In India, all regions are not same with respect to climate, rainfall, temperature, wind velocity, topography, soil type, soil structure, mineral composition, vegetative cover and so on. Further, the nature and causes of environmental degradation are not the same in all parts of the country. In India; there are eleven environmental regions, viz, hill areas, islands, agroclimatic zones, biosphere reserves, wetlands, mangroves and coral reefs, protected areas for wild life conservation, urban regions, urban slums, coastal areas and ecologically fragile areas. The environment management plan suitable for one region may not suitable for another region. Hence, there is a need for specific concrete regional planning for different regions in the country.

Regional Planning for Hill Areas

In India hill areas constitute about 21 percent of the total area and 9 per cent of the total population of the country, and they are mainly concentrated in the Himalayan and Westernghats regions. Hill areas abound in forests, plants and mineral wealth and also hydro-energy resources. The conservation and proper utilization of the natural resources of the hill areas are necessary not only for their development, but also for sustaining the development of plains.

All hill areas are not same in their geographical and geological features and also their deterioration conditions. Hence, there is a need for specific regional planning at the district and block levels, particularly in hill states like Jammu and Kashmir, Himachal Pradesh, Sikkim, Manipur, Meghalaya, Nagaland, Tripura, Arunachal Pradesh and Mizoram. These states have already been covered under hill area development programmes with the objectives of socio-economic development of the hill people in harmony with ecological development of the areas. These hill areas are also found in certain regions in some states. They are the following (i) Two hill districts of Assam-North Cachar and Karbi Anglong (ii) Eight districts of Uttar Pradesh - Dehradun, Pauri Garhwal, Tehri Garhwal, Chamoli, Uttar Kashi, Nainital Almora and Pithoragarh (iii) West Bengal Kurseong and Kalimpong (iv) Tamilnadu the Nilgiris district and (v) 163 taluks of western ghats, including 62 taluks of Maharashtra, 40 taluks of Karnataka, 29 taluks of Tamilnadu, 29 taluks of Kerala and 3 taluks of Goa. Thus the hill areas are concentrated in different parts of the country. In the hill areas, regional planning should discourage the practice of mono-cultivation and shifting cultivation with a view to prevent the loss of biodiversity. There is a need to prevent deforestation and soil erosion in hill areas.

The regional planning for hill area development should allow the human and livestock population upto their level of carrying capacity. A proper control and monitoring of tourism activities since it degrades the environment by causing accumulation of non-degradable wastes. Regional approach could be considered in hill area development with objectives like better water and land use, control of soil erosion through watershed management, silvi-culture development, replacement of annual crops with perennial shrubs and tree plantation in steep slopes, development of high value and low value crops and marketing and processing facilities for horticultural and species crops.

Regional Planning for Islands

Islands are found in India notably the Andaman and Nicobar Islands and the Lakshadweep Islands. They constitute a fragile ecosystem. The Andaman and Nicobar Islands have tropical ever green forests with wide genetic diversities; the Lakshadweep Islands are characterized by their coral reef aquatic ecosystem. These islands have distinct features in terms of natural resources and the needs and problems of these islands have various ecological aspects. Hence, there is a need to formulate specific regional plans for these islands. These Islands are also found around the adjoining areas of Cochin. The interaction between sea, river and land has developed a peculiar topography to Cochin area. Watershed, formed by canals and the lagoons of the Cochin Estuary of the Vembanad lake constitute about 30 per cent of the surface area of Cochin. The river Periyar empties into the Cochin channel and has created to a number of islands lying in the backwaters. There is a need to formulate specific regional planning for Cochin and its Islands ecosystems. Due to environmental quality, this region is affected by urban solid and liquid waste of substantial quantum through automatic emission. A successful implementation of island regional planning requires the following features:

- Proper periodical evaluation of sea erosion level
- Participation of Islanders in preservation and conservation of island ecosystem
- Regulating the activities of tourism
- Planning could be done with the appointment of experts in the field of occenography, geology, fishery, aquaculture, forestry and environmental sciences
- Carrying capacity of the earth should be considered in location of activities in the space
- Caution over the activities of aileners and terrorists
- Prevention of coastal pollution

- Preservation and conservation of life of aquatic organisms
- Identification of specific needs of each area in an island and formulation of appropriate action oriented development plan
- Possible ways of preventing destruction of coral reefs and breeding habit of the aquatic organisms

Agro-climatic Regional Planning

In India all agricultural lands are not the same as far as environmental factors are concerned. Hence, there is a need for specification of agroclimatic regions on homogenetic factors such as climatic factors, soil types, rainfall, temperature, topography, water resources and so on. In India, 15 agroclimatic regions are identified.[6] They are (i) Western Himalayan region (ii) Eastern Himalayan region (iii) Lower Gangetic Plains (iv) Middle Gangetic Plains (v) Upper Gangetic Plains (vi) Trans Gangetic Plains (vii) Eastern Plateau and hill region (viii) Central plateau and hill region (ix) Western plateau and hill region (x) Southern plateau and hill region (xi) East coast plains and hill region (xii) West coast plains and ghats region (xiii) Gujarat plains and hill region (xiv) Western dry region and (xv) The island region. Thus, there is a wide spread variation in distribution of different types of agroclimatic regions in different parts of the country. Hence, we need specific regional agricultural planning rather than sectoral planning. There is a wide variation in cropping pattern, irrigational infrastructure, extension services, marketing services, and productivity of crops in different agroclimate regions. Agro-climatic regional planning emphasises on crop specification and diversification of particular crops in a particular region. Hence, agro-climatic regional planning could be formulated, taking into account the following factors:

(a) Assessment of the potential for growth and diversification of crops in each region

(b) Identification of regional specific cultivation practices

(c) Analysis of the role of local efforts in achieving sustainable agricultural practices in different agroclimatic regions

(d) Identification of regional specific crop diversification in different zones in an agroclimatic region

(e) Assessment of developing acquaculture, horticulture, sericulture, fishculture, and sylviculture in different agroclimatic regions in India

(f) Assessment of developing social forestry, community forestry, village forestry and farm forestry in different agro-climatic regions in India

(g) Preparation of a development profile of each region with a view to increase production and productivity of agriculture and its allied activities

(h) Efforts to discourage the mono-cultivation and shifting cultivation practices

(i) Scope and potential for developing water resources particularly in drought prone regions through watershed development approach and rain water harvesting techniques

(j) Different type of agricultural support system according to cropping pattern and productivity in different agroclimatic regions

(k) Revegetating waste lands through indigenous species of plants and cultivation of commercial crops in waste lands of different regions

(l) Regional agricultural planning machinery should function on the basis of regional administration rather than revenue administrative set up

Regional Planning for Biosphere Reserves

Biosphere reserves are areas of terrestrial and coastal ecosystems which are internationally recognized within the framework of UNESCO's main and biosphere programmes. Twelve biosphere reserves have been set up in the country so far not only aim to protect representative ecosystems, but also serve as laboratories for evolving alternative models of development. The biosphere reserves are the Nilgiris with 5520 sq.km, Nanda Devi with 5860.69 sq.km, Nokrek with 820 sq. km, Manas with 2837 sq kms, Sunderbans with 9630 sq.km, Gulf of Mannar with 10500 sq.km, Great Nicobar with 885 sq. km, Similipar with 4374 sq.km, Dibru-Daikhowa with 765 sq.km, Dehang Debng with 5112 sq.km, Pachmarhi with 4926.28 sq.km and Khanchanjunga with 2619.92 sq.km.

Biosphere reserve aims at recognizing the conservation of natural areas and the genetic material they contain. This helps to assess ecological changes and performance of ecosystem in supporting life support system.[7] In other words, 7 biosphere reserves are internationally recognized terrestrial and coastal ecosystems. They provide scientific knowledge, skills and human values to support sustainable development.[8] There is a need for specific regional planning for terrestrial biosphere reserve ecosystem ad coastal ecosystem. The regional planning for development of biosphere reserves should focus on the following objectives.

- possible ways of preservation and conservation of species diversity, genetic diversity and ecosystem diversity
- Maintenance of essential ecological processes and life support system
- Possible ways of sustaining life of mangrove vegetation
- Protection of flora and fauna in aquatic system particularly in coastal ecosystem
- Ensuring the utilization of living resources and ecosystems in which they are found sustainable
- Intensive conservation and management of coral reef in the Andaman and Nicobar Islands, the Gulf of Mannar, the Gulf of Kuttanch and the Lakshdweep Islands
- Controlling forest fires with a view to protect and conserve both natural and man made forests
- Maintaining a viable wild life population for scientific, economic, aesthesic, cultural and ecological values

– Preserving areas of biological importance as a national heritage for the benefit, education and enjoyment of the people
– Ecological restoration of existing natural habitats and migratory route of wild animals
– Assessment of wet land for national importance
– Preparation of thematic mapping to explain forest types, species composition, crown density of forest cover and other land use in different biosphere reserves.
– Collection, storing and retrieving biosphere data for regional planning in different biosphere reserve areas
– Providing areas for ecological and environmental research.

Regional Planning for Wetlands

Wet lands are among the highly productive ecosystems on the earth and therefore they need to be conserved as they provide multiple services like water purification, regulation of flows, fisheries, habitats for plants, animals and micro organisms, opportunities for recreation and tourism, absorbing water and therefore reducing flood risks. Usually wetlands are transitional areas between aquatic and terrestrial ecosystems.[9] Wetlands consist of areas of marsh, fen peatland of any type where water is static or flowing. It also includes fresh, brakish or saltish area, including areas of back water.[20]

India has vast stretches of wetlands. Many of wetlands in India have been converted from agricultural to industrial purposes and they are affected by industrial effluents, sewage, household wastes and sedimentation due to ecological degradation in catchments area. Wetland ecosystems play a vital role in flood control, recharging of aquifers, regulating water quality, reducing sediment load and pollution abatement. They have elements for agriculture and breeding for water fowls. India has 16 wetlands of national importance, located in 16 states. The regional planning approach to wetland development includes the following objectives:

- Identification of possible ways of flood control mechanisms through wetlands
- Assessment of wetlands' resources and their loss in time series
- Formulation and implementation of management action plans of identified wetlands
- Scientific exploitation and use of wetland resources and their conservation and development
- Monitoring implementation of the programme of conservation management and research on wetlands
- Preparation of an inventory of wetlands in different regions in the country
- Action plans to control weed, siltation, pollution abatement, development of fisheries
- Promotion of environmental awareness and rational use of wetlands
- Biological methods of conservation of wetlands
- Periodic survey of socio-economic activities in wetlands and their consequence in sustainability

Regional Planning for Mangroves and Coral Reefs

Mangroves are salt tolerant plant communities occurring in sheltered coastline areas such as bays, estuaries, lagoons and creeks. They consist of swamp forest land within and its water spread areas. They stabilize the share line and act as a barrier against the encroachment of land area by sea. It is estimated that the total area of Mangroves in India is 6740 sq.km, having about 59 species and 41 genera belonging to different families.[11] Mangroves are distributed in nine states. Of which Sunderbans of West Bengal (4200 sq.km) and the Andaman and Nicobar Islands (1190 sq.km) cover the major portion of the Mangroves in India. It could be noted that forest land in the Mangroves is used for agriculture. Large tracts of mangroves are being cleared for agriculture, aquatulcutre, industrial use, construction of harbours leading to degradation of mangroves. Further they are deteriorated in terms of silt in the form of terrestrial and alluvial sediment, crude oil and petroleum derivatives sewage pesticide industrial wastes. Hence, there is a need for specific regional planning to preserve and conserve mangroves. It should be based on the following considerations:

- Identification of biodiversity status of different mangroves
- Preventing misuse of resources in mangroves by establishing appropriate control mechanisms
- Possible ways of protecting halophytic vegetation
- Protection and conservation of endangered species and scope to multiply their numbers
- Control of tourism activities on mangrove vegetation
- Preservation of breeding characteristics of shallow water living organisms
- Educating indigenous people about the possible ways of preserving and conserving the mangrove environment
- Prevention of mushroom growth of shrimp culture in and around mangroves since they pollute mangrove environment
- Advice on research and training on mangroves and related ecosystems
- Review of progress in implementation of mangrove conservation programme and updating status of the mangrove environment

Regional Planning for Coral Reefs

Coral reefs are diverse and vulnerable ecosystem characterized by a complex inter-dependence of plants and animals. They are massive limestone structures built up through the constructional cementing process and depositional activities of the animals of the class Anthozoa as well as other calcium carbonate secreting animals. Coral reefs are the centres of high biological productivity, sites of carbon dioxide sink and sources of huge deposits of calcium carbonate. They provide many natural raw materials of pharmacological importance including life saving drugs.

In the Indian sub continent, the reefs are distributed along the East and West coasts at restricted places. Fringing reefs are found in the Gulf of manner and Palk Bay as well as the Andaman and Nicobar islands. Platform reefs are seen along the Gulf of Kutch and Atoll reefs are found in the Lakshadweep Archipelago. [12]

Increasing human population and anthropogenic pressure have severely affected coral distribution and bio-diversity. Natural calamities cause considerable damage to the coral reef structure through direct and indirect means. Global warming leads to adverse impact on the survival of the coral reefs. The bleaching phenomenon of 1998 in the Indian Ocean is reported to have caused considerable damage to the coral reef in the Indian coast. Further, some factors causing damage to coral reefs include industrial settlements, over exploitation of fish resources, pollution, dredging of coral sands for cement industry, causing considerable loss of flora and fauna, deforestation and reclamation of wetlands and mangrove areas. Hence, we need proper planning to protect and conserve our coral reefs. The regional planning to conserve coral reefs in India requires the following strategies:

(i) A proper environmental planning unit should monitor the status and exploitation level of coral reefs in different coastal areas of India.

(ii) Mangroves committee should monitor the activities of over commercial exploitation of coral reefs.

(iii) People should be prohibited to enter coral reef areas in the coastal ecosystem.

(iv) Periodic survey on bio-diversity status of coral reefs in different coastal ecosystem must be undertaken.

(v) Efforts should be made to forestall or avert the accumulation of pollutants around the areas of coral reefs.

Regional Planning for Protected Area Network for Wild Life Conservation

There is a growing need to protect wild life in India, due to rapid growth of human and livestock population. The government of India has announced some of the areas as protected ones. There are 88 national parks and 490 wild life sanctuaries in the country covering an area of 1.53 lakh hectare sq.kms. Protected areas in mountain, deserts and coastal regions supporting large population of endangered species like snow leopard, red monkey, musk deer, hungul, great Indian bustard, chinkara and black buck require special protection through devices of regional planning. Planning for protection of endangered species should be brought within the framework of regional planning. It should be based on focusing the following objectives:

(i) Ecological restoration of existing natural habitats and migratory roots of the animals

(ii) Developing scientific and planned management for conservation of endangered species in their own areas

(iii) Efforts to multiply the number of endangered species through captive breeding

(iv) Ensuring maintenance of a viable population of rare and threatened species of wild life for scientific, economic, aesthetic, cultural and ecological values

(v) Allocation of special financial assistance to the protected areas

(vi) Ensuring maintenance of studbook endangered species of animals in captivity

(vii) Consideration of important biological and practical parameters necessary in planning protected areas

Urban Regional Planning

There is an increasing trend in air, water and noise pollution in the urban areas of the country. The need for urban regional planning emerges in consequence of urbanization, dispersal of economic activity, changes in technology, pattern of demand, supply and consumption that constitute the dynamic elements. Hence, there is a need to delineate planning regions in urban areas. The regional planning for urban areas require the following features:

- Solid and liquid waste management
- Traffic and transportation management for control of vehicular pollution
- Prevention and control of industrial pollution
- Substitution of inputs
- Public participation in prevention and control of pollution
- Green belt for abatement of pollution
- Environmental improvement of urban slums
- Urban basic services for the poor
- Proper spatial location of activities
- Measures to prevent concentration and accumulation of industries in urban areas
- Periodical conduct of waste minimization studies, workshops, seminars, etc.
- Adoption of clean and safe technologies in the process of production
- Identification of areas that need restoration of air quality and their prioritization
- Identification of nature and extent of pollution control need
- Assessment of effectiveness of pollution control programmes
- Water quality monitoring

Regional Planning for Coastal Areas

India has a long coastal area. Hence, there is a need to protect the cultural, aesthetic and ecological values of coastal areas. At the same time, the coastal activities should not mar the principles of environmental protection and development.

Hence, we need specific regional planning to protect our coastal ecosystem. Planning to protect coastal ecosystem requires the following features:

- Measures are to be taken to reduce coastal pollution.
- Practicing of shrimp-culture should be discouraged as it pollutes the coastal environment in the form of discharge of polluted water from the shrimp ponds.
- The life of offshore living organisms should be protected against the impact of discharge of toxic substance in the coastal ecosystem.
- Over-crowding of industries in coastal areas should be avoided.
- Life and habitats of fishermen population in coastal areas are to be regulated.
- Construction activities in coastal areas are to be restricted.
- Sea food industry is to be arranged properly.
- Fishing activities are to controlled during breeding seasons.
- Cyclone protection barriers should be developed to protect the coastal population during the times of cyclone.
- Proper planning, coordination and monitoring are to be undertaken for aquaculture, fish culture and shrimp culture activities in the coastal regions.
- Locomotion and breeding habits of sea shore living organisms are to carefully protect.
- Coastal area development could be monitored with the appointment of experts in the fields of oceanography, marine biology, biology, sociology, economics and so on.
- A specific grass root level regional planning machinery should be created with devices of inter-sectoral coordination.

- Possible ways of developing tidal power with the help of safe and eco-friendly technologies are to be explored and adopted.
- Climatic changes, tidal changes and their consequence on life and habitat of coastal population are to be monitored periodically.
- Sea transport particularly in ports, harbours and beaches are to be properly regulated.
- Scenic beauty of the coastal environment should be protected by regular surveillance of accumulation of pollutants in and around the coastal environment through the activities of human and livestock population.

13

Gender and Biodiversity

Women's work saves Biodiversity, as they play primary roles as land and resource managers. This and other major contributions of women to maintaining life on Earth was recognized globally with the celebration of the International Women's Day on March 8. The International Women's Day has been recognized since the early 1900s as women in an emerging industrial world clamored for better working conditions and pay, and voting rights. This year's event coincides with the celebration of the International Year for Biodiversity and provides an opportunity to highlight women's contributions to nature and biodiversity.

Women's significant contributions to nature and biodiversity Conservation have long recognized in traditional culture. Women and men play distinct roles in conserving, celebrating and sharing biodiversity. Women currently account for 60 to 80 percent of all food production in developing countries. The nurturing nature of women encompasses a wide circle, covering family, community and the environment, as they care for the very resources that sustain their families. In many countries, women secure seeds for planting, thus preserving a wide variety of crops best suited for local conditions. They are closely attuned to the environment and are aware of the best conditions for planting and harvest. Their knowledge of plants and trees allows them to choose leaves and roots with medicinal properties to treat wounds and ease discomfort.

In developing countries, women take care of the environment like they care for their own children. Women are connected to nature. That is why we call nature as mother," says Rodrigo Fuentes, executive Director of the ASEAN Centre for Biodiversity. The International Year of Biodiversity calls 2010 for equal rights and opportunities for women to provide progress for all. Gender equality and rights also translates to strides in poverty alleviation and Sustainable development. This coincides with the theme of the International Year for Biodiversity, which is "Biodiversity is life. Biodiversity is our life." Providing equal rights for marginalized sectors, such as indigenous groups, youth, and women in particular, should provide everyone with the opportunity to participate in the conservation of biodiversity, and thus the preservation of life itself.

Women in rural communities all over the world contribute to sustaining food production. While the diverse tasks of women are extremely difficult to quantify, their role in securing household food cannot be disputed. In addition to being responsible for the farm and the household, women are the custodians of the family's food basket. Studies have shown that women in India are the major food producers in terms of value, volume and quality. An interesting study of the gender division of labour has shown that women do 37% of sowing, 59% of interculturing, 60% of the harvesting, 59% of threshing and 69% of all the work involved in tending farm animals. Generally men and women work together in a complementary way although difficulties may arise when decisions have to be made about whether to produce for market or home consumption. Rituals and seed selection in the finger millet growing areas of southern India, mustard is the first plant to flower. This is the moment when Gowri Pooja, the ritual for the Goddess Gowri, is celebrated. Gowri is the goddess of water and a symbol of fertility. Flowers of the mustard plant are taken home and worshipped. In some parts of Karnataka young unmarried girls sing songs of praise to the goddess and collect contributions from the public. They also bring fresh soil and make an image that symbolises fertility.

After the ritual, the image is immersed in water to reaffirm the connection between plants, soil, water and fertility. Women play a major role in conserving seeds: they decide the quantity and variety of seed to be saved and the method of preservation. Seed selection by women is a continuous activity, starting the moment the crop begins to flower. Working in the fields, they observe the plants and decide which seeds to select. They identify plants of good quality on the basis of size, grain formation and their resistance to pests and insects. To cover the risk of

drought, women select enough seed to see them through two seasons. Seed storage when the selected heads of grain are brought into the threshing yard, women welcome the first cartload with a pooja, or ritual. On the last day of threshing women worship the mounds of grain and a portion of each mound is given to the poor. A gift of grain is also presented to the families who have helped with the harvest.

Before the seeds are carried away for storage, women ritually invoke the forces essential for a good crop in the next growing season. This ritual is an important part of seed preservation. Water is symbolised by a winnowing pan, protection from pests by certain leaves, soil fertility by cow dung and weeds by grass. Some of the leaves used in the ceremony have insecticidal properties. Lakkli leaves Vitex negundo, for example, are used when paddy seeds are stored; neem leaves are used as an alternative when lakkli is unavailable. In some cases paddy seeds are mixed with the seeds of field bean dolichus lab and mustard to help preservation. Tur is mixed with sand for the same reason. Sometimes seeds are stored above the kitchen where the smoke helps to keep the pests away. Ceremonial germination tests Each region has its own way of testing the quality of the seeds. In the sorghum region farmer families celebrate a festival when they finish threshing. In this festival people express their gratitude to the planets for the crops they have received and, for three days, the women worship symbols that represent the planets. At the end of the ceremony they make small mounds from the different seeds they have harvested and sanctified water is poured on each heap. Women believe that the water will carry away only the good quality seeds. If nothing is carried away they will look for other seeds.

This quality test is followed by a germination test or Negilu Pooje. In the finger millet growing areas this test takes place at Ugadi or the Hindu New Year. Nine seed varieties of cereals, pulses and oil seeds are put into the shell with good manure. These are worshipped and inspected after seven days. If there are only few sprouts or if these are too small, the seeds of that particular variety is not considered suitable for the coming agricultural season. The farmer will then exchange or borrow seeds until a suitable replacement has been found. It is considered unfavourable to purchase seeds for money.

Sowing sacred seeds A few days before sowing, the seeds are taken out of storage. They are dried and those damaged by pests are discarded.

Directly before they are sown the women take the seeds to the house deity and worship them. On their way to the field the woman carrying the seeds will visit and make offerings to the seven village goddesses, known as the 'Seven Sisters'. These seed offerings are later on collected by the poor. Women also worship the draft animals and the farming implements that will be used for sowing. None of the above procedures are followed for the seeds of high-yielding varieties bought at the market. While the local varieties are considered sacred, the highyielding varieties are regarded as impure. They are sent directly to the field where only the men are responsible for sowing them. Rains and food availability during a participatory activity with women farmers an agricultural calendar was drawn according to the rains. Each two weeks the rains are indicated with a different name. This calendar was compared with seasonal food availability and the jobs done by men and women. The women indicated the seasonal availability or lack of food.

According to this calendar the agricultural season begins with the ashwini rain, which is the signal to farmers to start preparing the land. Crops are sown at the time of the krithika and rohini rains. Reserves of lentils and cereals for consumption are usually low for both humans and animals during the growing season. At this time the entire farming community depends on local biodiversity for a supply of greens, fruits, vegetables, tubers, potatoes, sweet potatoes and bamboo shoots. During the following period of magge malai or big rain and the ubbe and utharai male rains, the fields and forests are full of green plants. Most of these are considered unproductive weeds by modern agriculture. The role of rural women as food producers is thus directly related to the diversity of local food crops and to the surrounding bio-diversity. The local market only has cereals and pulses to offer in the months after the harvest. For farmers biodiversity is manifest in both cultivated foods and wild plants. Women's specialized knowledge of wild plants includes leaves, fruits, berries, nuts, seeds, edible roots and pulses. Diversity in cultivated foods is under threat, however, due to modern agriculture. Part of the traditional genetic material is replaced and natural resources that serve as biodiversity reservoirs are affected. These threats are particularly dangerous to the people who rely on natural resources for their survival during several months a year.

The Gender Perspective

Men and women play different roles in dealing with biodiversity resources. They perceive, use, and manage these resources differently, and have different knowledge and skills in relation to resource management. Across the globe, women's role as wild plant gatherers, home gardeners, plant domesticators, herbalists, and seed custodians is well recognised. For example, the women farmers of the Garo Hills in India manage, on average, 35 species of seeds and have a thorough knowledge of their preservation and diversification. However, the enormous contribution made by women in conserving and managing biodiversity, and their capacity to do so, is rarely recognised. The lack of gender-differentiated perspectives on biodiversity research and documentation is a hindrance to this recognition.

Considering both women and men's perspectives in biodiversity research gives a complete account of biodiversity characteristics, people-plant relationships, and the causes of biodiversity loss. This, in turn, may help formulate gender sensitive biodiversity policies, strategies, and guidelines for the effective conservation and management of biodiversity.

Climate change brings another dimension to biodiversity management. The consequences of the impact of climate change on biodiversity could be different for men and women. Particularly in developing countries, women may face greater challenges in adapting to environmental change, because they do the majority of agricultural work and have limited access to other income generating opportunities. In addition, they are usually responsible for providing daily household commodities such as food, fodder, fuelwood, medicine, clothing, shelter, and much more. With men engaging more and more in non-farm economic opportunities, women are often left with an increased workload. Therefore, paying serious attention to both women's and men's knowledge, use, rights, and needs with respect to biodiversity is important and can help meet the goals of the Convention on Biological Diversity with regard to resource conservation, sustainable use, and the fair and equitable sharing of the benefits arising from the management of biodiversity resources.

Gender and Biodiversity

The importance of biodiversity to individuals varies according to gender. Based upon the social roles of and power relations between men and women, gender is shaped by culture, social relations, and natural environments. For this reason, we need to incorporate gender dimensions into our understanding of biodiversity and its conservation, sustainable use and the sharing of benefits. Gender roles affect economic, political, social and ecological opportunities and constraints faced by both men and women. Recognizing women's roles as primary land and resource managers is central to the success of biodiversity policy. For example, women farmers currently account for 60-80% of all food production in developing countries, but gender often remains overlooked in decision-making on access to and the use of biodiversity resources. Just as the impact of biodiversity loss is disproportionately felt by poorer communities, there are also disparities along gender lines. Biodiversity loss affects access to education and gender equality by increasing the time spent by women and children in performing certain tasks, such as collecting valuable resources and services such as fuel, food and water. To conserve biodiversity, we need to understand and expose gender-differentiated biodiversity practices, gendered knowledge acquisition and usage. Various studies demonstrate that projects integrating gender dimensions generate superior results. Gender considerations are not solely a women's issue; instead, this outlook could yield advantages for whole communities and benefit both sexes. CBD developed a Gender Plan of Action that defines the Secretariat's role in stimulating and facilitating efforts on national, regional, and global levels to promote gender equality and mainstream a gender perspective. The Millennium Development Goals emphasize clear linkages between gender equality, poverty alleviation, biodiversity conservation and sustainable development. Such insights should be included into our outlook and approach to reversing biodiversity loss, reducing poverty and improving human wellbeing.

Fast Facts

Exposing and understanding gender differentiated biodiversity practices and knowledge of women and men enhances biodiversity conservation, Biodiversity conservation efforts become more effective and efficient when women and vulnerable groups are empowered to participate as equals in:, information sharing and generation, education and training, technology transfer, organizational development, financial assistance and policy development.

Gender Dimensions of Biodiversity Management

It is important that in all these aspects of a community-centred integrated gene management strategy, gender roles in all approaches to genetic resources conservation, sustainable use, and equitable sharing of benefits are given attention. Women in many developing countries are the primary seed selectors and savers. Their contributions to the evolution of a biodiversity conservation ethic should be fully recognised in any system which is designed to operationalise the equity in benefit sharing provisions of CBD. If we promote a worldwide community-centred integrated gene management strategy, we will soon stop hearing about disappearing species and vanishing landscapes and habitats. None of the policies and procedures. By recognising that conservation efforts represent a continuum, with rural and tribal families performing a vital function in preserving precious genetic variability in important plants and farm animals, we will be able to attend to all the links in the conservation chain. Community conservation is a value-added link in the conservation system, since local families not only conserve but also add value to the conserved material through selection and information.

If the twentieth century was a period of understanding and chronicling threats to biodiversity and bioresources both in land and water, let us hope that the twenty-first century will be one where the threats are terminated and benefits harnessed for a better common present and future for humankind. The gender division of rights, responsibilities, works, and knowledge is taken as a point of departure to examine and explain the multiple roles of women and men as resource users managers. Employing research with gender as a key variable helps to clarify the indefinite boundaries of the household and the family, and the complex ways in which family, household, and community and ecosystem are linked.

Although researchers and policy makers are increasingly aware of the importance of gender issues, research is often 'gender neutral'. That is, researchers employ units of 'household' or 'community' uncritically and do not deconstruct the heterogeneous power and gender relations among those units. Similarly, technologies and macro-policies are viewed as gender-neutral. It is critical to move to a deeper analysis, one which facilitates proactive action which informs and supports mechanisms for positive social change. For example, investigations are needed as to how processes of increasing commercialisation and economic growth impact women and men's access to use and control of plant genetic resources differently. How are women and men involved differently in the collection and production and distribution and consumption of these resources and derived products? Research which directly investigates these issues and increases understanding of the power/knowledge relations of men and women can help to develop gender sensitive interventions and policies and create a space for women to improve their position and bargaining power.

The Indian government has taken up formal types of conservation practices such as the Reserved Forests and the National Parks and Sanctuaries, and the National Gene Banks. It has also tried to involve local stakeholders in the management of the forest resources in a limited way through the establishment of Joint Forest Management (JFM) Committees. The JFM experiment started as a trial in West Bengal and has been adopted by nearly all the state governments. Initially the programme did not envisage the participation of women, but after a decade the JFM rules say that both men and women from the same family have equal ownership of the resources from the intervention, or the non-timber forest produce.

India has diverse ecological regions each with a unique system of agriculture. Examples from different regions of India on the diverse cropping patterns and multiple and mixed cropping practices as means of ensuring food security highlighted the local agricultural diversity and the role of the community in in situ on-farm conservation. There are regions where traditional crops are disappearing due to the intrusion of commercial agriculture and due to changes in the food habits of the people. This loss of diversity has implications for gender roles and gender relations and subsequently it impacts on food security.

FAO, Gender and Biodiversity

The Gender , Equity and Rural Employment Division of FAO is carrying out activities to enhance rural people's food security and promote sustainable management of agro-biodiversity by strengthening the capacity of institutions in the agricultural sector to apply approaches which recognise men and women farmer's knowledge in their programmes and policies. The main activities include: Training: how to record and document local knowledge and how to use gender analysis and participatory methods for both research and action processes, Technical assistance for research on gender-based differences in farmer's knowledge related to agro-biodiversity conservation e.g. use of medicinal plants, local seed management etc Technical assistance to enhance communication and the exchange of information about the value of local knowledge in agriculture within and between communities, as well as with institutions that interact with farmers and with policy-makers.

Community Conservation

Yet, as the example above illustrates, successful conservation efforts are as much about people as they are about well-intentioned rules and regulations. Environmental problems often demand local solutions derived from community initiatives. Depending on the way natural resources are used, local people can have a positive or negative impact on programs. To maximize the chances that conservation initiatives and projects will succeed, those at the community level need to be invested in both the concept and the approach. This means their participation in decision-making processes and in the evaluation, monitoring, and management of natural resources and the environment.

The conservation of biodiversity relies on the involvement of the full community—both women and men—whose interests and perspectives regarding natural resources may differ. Often, an ostensibly gender-neutral initiative may in fact be biased against women. For example, in some cultures, women are discouraged from participating, or are dominated by men, in meetings to determine plans that will affect their day-to-day activities surrounding natural resource use. Yet if women cannot participate in decision making, they may not engage in executing plans that will have a positive, long-term benefit in preserving natural resources. Because men's and women's roles and impact on the environment may differ, devising a gender-neutral initiative is not enough. Rather, there must be an explicit effort to understand these roles and then design a program that will accommodate the activities of both women and men.

Another reason to avoid genderneutral initiatives and programs is that they often overlook men's and women's unique knowledge of natural resources that can be valuable to conservation programs. In Rwanda, for example, agricultural researchers used the knowledge of women farmers to develop new varieties of beans for an extension project. The yields produced by women were consistently top performers, in part because of their adaptation to the local agro-ecosystem and sustainable fit within the natural resource-based livelihood system. Men and women will have different points of view and ideas for solutions about the same problems. To understand how gender shapes activities affecting the environment, it is useful to look at who uses resources and how, who is affected by resource use, and who has the authority to make decisions about resource use.

Different Perspectives

A gender analysis is an important step in understanding men's and women's different perspectives on natural resource use. This will help development planners design inclusive projects that break down assumptions about gender roles and barriers to women's participation. An immediate benefit of conducting a gender analysis is the more accurate delineation of men's and women's roles. Women's work is often invisible or not recognized, cloaked under the catch-all phrase of "housewife."

Men, in contrast, have more descriptive titles, such as "fisherman" or "farmer," that clearly delineate their relationship to natural resources. A simple way to obtain an equally clear understanding of what women do is by asking them to define specific tasks of their "jobs." This will reveal the extent to which they work with the environment. A "housewife" might tend to the garden.

Knowing Different Priorities Makes all the Difference

In Kenya, local men involved in planning a fuel wood tree planting project assumed that women would fulfill their traditional role of providing water for seedlings. After the seedlings were distributed, the men discovered that the women were unwilling to do the extra hours of water-collecting required by the project. Furthermore, the women were not particularly interested in the trees designated to be planted. The failure to consult women in the planning phase of the project meant that their concerns were ignored. Not surprisingly, they were indifferent to its success, and the seedlings died for lack of water. However, the second phase of the project incorporated women's interests by providing the trees they preferred. They then agreed to help, and this time the project was successful.

Women often play leadership roles in promoting an environmental ethic. In addition, women's contributions to environmental management usually take place at the local level, where decentralized action on environmental issues is most needed and decisive. Yet it is typically men who dominate in formal leadership roles and positions. Women's formal involvement is scarce at all levels, from local positions to the ranks where official environmental policies are determined. This institutional bias is exacerbated by the fact that organizers often overlook women. In addition, men may be jealous or dismissive of women leaders. These biases against women can compel them to develop their own areas of leadership. This is the case in Brazil, where men dominate in institutional politics. Women, more active in non institutional politics where they can skirt the cultural norms that keep them out of formal leadership roles, are the heart of social movements that often address environmental concerns, such as pollution and water scarcity. Project implementers can foster greater women's participation by working with community members to establish a women's association of leaders and professionals or to tap into women's groups. For example, project organizers working with protected areas in Guatemala recognized that women leaders were not working collaboratively. To address their isolation and strengthen the potential of women's contributions for this project, it was recommended that a women's association of leaders and professionals be formed. Project implementers can offer professional development training for women. In situations where women are moving into jobs traditionally held only by men, training can help prepare men to work professionally with women. Also, new areas that have yet to be characterized as "women's work" or "men's work" – such as growing organic shade-grown coffee – provide opportunities to employ equal numbers of women and men. Institutional norms can result in the involuntary exclusion of women, even with the best-planned project. Media and educational materials on conservation and resource management tend to project a male bias, which sends an indirect message that undermines women's potential. This bias can be countered by ensuring that language in publications and training materials is gender sensitive and that anecdotes highlight the work and contributions of women as well as men.

Women's reticence in the presence of men is a common cultural norm that, if not acknowledged, can have the effect of excluding women from conservation projects. For example, when conducting a survey to understand gender roles and natural resources in Brazil, interviewers found that their questions, although heard by the entire family, generally were answered only by the head of the family – typically men. Women had little or no participation in the responses. However, separate conversations with the women disclosed that they felt more comfortable answering questions that were formulated by the women on the team. This simple strategy of hiring women workers can be applied to a range of activities, from community workers to extension agents. A Gender Focus For a Green Future Understanding gender roles, rights, and responsibilities is a critical part of the policies and programs that support community conservation efforts. Women and men are key stakeholders who sometimes might have conflicting interests in natural resource use and management. Understanding their respective priorities and developing initiatives accordingly are crucial to the success of long-term conservation efforts. Across the globe, women predominate as wild plant gatherers, home gardeners and plant domesticators, herbalists and seed custodians. Research on 60 home gardens in Thailand revealed 230 different species, many of which had been rescued by women from neighboring forests before being cleared. Women in different regions of Latin America, Asia and Africa manage the interface between wild and domesticated species of edible plants. This role dates back to 15,000-19,000 B.C. Women and men often have different knowledge about, and preferences for, plants and animals. For example, women's criteria for choosing certain food crop seeds may include cooking time, meal quality, taste, resistance to bird damage and ease of collection, processing, preservation and storage. Men are more likely to consider yield, suitability for a range of soil types and ease of storage. Both are essential for human welfare. In a study in Sierra Leona, women could name 31 uses of trees on fallow land and in the forest, while men named eight different uses. This shows how men and women have distinct realms of knowledge and application for natural resource management, both of which are necessary for sustainable use and conservation.

Women provide close to 80% of the total wild vegetable food collected in 135 different subsistence-based societies. Women often have specialized knowledge about "neglected" species. The majority of plant biodiversity research is not gender sensitive. This has led to incomplete or erroneous scientific results with respect to the diversity, characteristics and uses of plants, and the causes and potential responses to genetic erosion. Integrating women's traditional knowledge into botanical and ethnobotanical research, and protecting all informants' rights, are critical for improved knowledge and management.

The language used by the Convention on Biological Diversity and the Bonn Guidelines to address subjects related to indigenous and local communities is not gender-sensitive. In spite of the fact that an increasing number of experiences are highlighting the sustainable manner in which women use biological diversity, it is often true that women do so without equitable participation in the access and control of such resources. There is a tendency to ignore the natural spaces predominantly used by women in favor of those used by men, and to undervalue non-commercial mostly female production spaces in favor of commercial mostly male production spaces. Therefore, it is necessary to make visible the gender-differentiated practices and knowledge of women and men in their relations with biodiversity resources. Despite considerable efforts over the past fifteen years at national and international fora, such as the Convention on Biological Diversity, very little progress has been made in understanding the fundamental roles that women play in managing and conserving biodiversity. It is essential to recognize that women and men have particular needs, interests and aspirations, and that they make different contributions to the conservation and sustainable management of biodiversity.

Making visible the various roles women play in biodiversity conservation, sustainable use of resources and survival of the human species is only the beginning. These roles should be valued in their broader scope. Special attention should be given to the significance of women's ancestral knowledge of biodiversity and the innovations generated by both men and women in adapting and sharing traditional practices for improved management within their own and other communities. Recognizing and valuing women's knowledge and practices related to biodiversity helps us to understand the crucial step of effective participation for women in decision-making for biodiversity conservation.

Making women's decision-making role meaningful requires us to actively overcome gender inequalities at all levels, from the local to the global. Biodiversity conservation is enhanced by empowering women to participate as equals in information sharing and generation, education and training, technology transfer, organizational development, financial assistance and policy development. The exclusion of women as agents of development means ignoring half of the planet's population, which, in turn, affects the efficiency and effectiveness of the conservation of biodiversity.

Difference in Biodiversity

More diverse information on the management and ecology of plants is garnered when gendered, traditional knowledge is incorporated in basic research, ethnobotany and other studies. In Uttar Pradesh, men primarily use forest plants for fodder and mulch for agriculture. Women's uses are more related to household applications for example, medicines, tonics, cleansers, fiber, food and tools. Elsewhere, women's criteria reflect the diverse ways in which they use plant materials for example food, nutrition and culinary preferences, thatch, matmaking, fodder, fuel and leaves. In addition, women's knowledge about "neglected" species has been tapped. Improved land tenure for women can support biodiversity. Women's secure access, to land for agriculture and home-sites leads to greater on-farm habitat protection of existing biodiversity and exploration of improved varieties. Equitable access to agricultural resources and inputs can support biodiversity conservation by intensifying production on already cleared land and reducing encroachment into fragile areas. Both women and men working with plants and animals need credit, technical support and extension services in order to mitigate or reduce potentially harmful practices such as slash-and-burn agriculture and short-term fallow rotation.

A gender approach brings innovation and participatory approaches to biodiversity research. For example, by studying women's recipes, one can see the changes in a available resources over time and by season. Biodiversity research and program decisions have included broader and more diverse perspectives at local, national and regional levels, and have better reflected women's constraints, needs and preferences such as access to land and other natural resources, land use, conflict resolution, household food security during difficult economic and climatic conditions. Recognizing the importance of the knowledge possessed by women, including use, rights and needs regarding local plant biodiversity, would help to reach several objectives of the Convention on Biodiversity, which seeks the sustainable use of biodiversity components and fair and equitable distribution of the benefits derived from its utilization. With a gender perspective, intellectual property regimes would consider women's roles and could promote a more equitable distribution of the benefits derived from biodiversity and its genetic resources. Some countries are taking advantage of their compliance with the Convention on Biological Diversity's provisions to create policies and legislation that safeguard the human rights of men and women, as well as indigenous and local communities. This is a great opportunity to achieve equality and equity between men and women in their access to resources, control of their traditional knowledge, and benefits from sound management and participation in governance and decision-making.

The recommendations of women conduction enhance awareness by women of environmental issues, and the capacity of women and men to manage their environment and sustain productive resources. Disseminate information on environmental sustainability, and recognize women as 'active and equal' participants in ecosystem management and the control of environmental degradation. Assess the 'environmental impact of policies, programmes and projects on women's health and activities and eliminate negative effects. Agrarian reforms to 'guarantee women's constitutional and legal rights in terms of access to land and other means of production' World Survey on the Role of Women in Development, UN 1989. Need for 'comprehensive, reliable and unbiased statistics on the nature and role of women's contribution to food and agricultural production'. The earlier Survey in 1986 emphasised on the need to provide women the access to support services to prevent negative consequences in food production. Agenda 21, UN 1992 Take full account of women's role in data on alternative livelihoods sources tree crops, livestock, etc. Promote environmentally sound technology designed and developed consulting women. Develop research and policy analysis on impact of environmental degradation on women.

Preamble of the Convention on Biological Diversity, 1992 recognizing also the vital role that women play in the conservation and sustainable use of biological diversity and affirming the need for the full participation of women at all levels of policy-making and implementation for biological diversity conservation. Draft Platform for Action, Beijing 1995 The Fourth World Conference on Women, 1995 of the United Nations held in Beijing, called for a recognition that Women's experiences and contributions to an ecologically sound environment must be central to the agenda for the twenty-first century. Sustainable development will be an elusive goal unless women's contribution to environmental management is recognized and supported' is per Report of the Fourth World Conference on Women, 1995. Some of the relevant strategic objectives of the Beijing Declaration are: 'Encourage, subject to national legislation and consistent with the Convention on Biological Diversity, the effective protection and use of local knowledge, innovations and practices of women of indigenous and local communities, including practices relating to traditional medicines, biodiversity and indigenous technologies, and endeavour to ensure that these are respected, maintained, promoted and preserved in an ecologically sustainable manner. 'Encourage the design of projects in the areas of concern to the Global Environmental Facility that would benefit women and projects managed by women and establish strategies and mechanisms to increase the proportion of women, particularly at grass roots levels, involved as decision makers, planners, managers and scientists. Relevant articles in the draft include: Article 29: Improve women's access to and control over land and other means of production. Article 51: Recognise the daily reality of women's management of natural resources, obtaining fuel and water, managing household consumption. Article 53: Include women in decision-making regarding environment. Article 54: Include women and their perspectives in environmental policies, recognise effects of environmental degradation on women. • Article 55: Involve women in environmental management, protection, and conservation programmes.

Ecofeminism and Biodiversity

Ecofeminists argue that a strong parallel exists between the male oppression and subordination of women in families and society and the degradation of nature by similarly masculine attitudes and methods. They also explore the intersectional between sexism, the domination of nature, racism, speciesism, and other characteristics of social inequality. In some of their current work, ecofeminists argue that the capitalist and patriarchal systems that predominate throughout the world reveal a triple domination of the Global South people who live in the Third World, women, and nature. This domination and exploitation of women, of poorly resourced peoples and of nature sits at the core of the ecofeminist analysis. "While feminism is a primary entry point, women and men also come to ecofeminism through environmentalism, alternative spirituality, animal rights, and other progressive affiliations. Feminist environmentalists study gender interests in natural resources and processes based on their different roles in daily work and responsibilities. Social feminists focus on the role of gender in political economy by analyzing the impact of production and reproduction of men and women's relation to economic systems.

Gender in the conservation of protected areas two decades ago, conservation·organi-zations began integrating the human dimension in conservation practice and ecosystem management, calling for the involvement of local communities in conservation. This has been essential where people live either inside or in the buffer zones of protected areas in Latin

America and the Caribbean, and gender has been one of the critical social variables considered in the process. Gender refers to the socially constructed differences and relations between men and women that vary according to situ-ation, place, time and context, and which influence structure and decision making within communities, institutions and families. Gender analysis requires "understanding how historical, demo-graphic, institutional, cultural, socio-economic and ecological factors affect relations between women and men of different groups, which partly determine forms of natural resource management".

In relation to biodiversity and natural resources, taking a gender perspective involves understanding and integrating the relations and differences between men and women into projects. This includes the different roles, rights and opportunities of men and women con-cerning access, use, management and conservation of natural resources. It also involves considering the different ways in which environmental degradation affects men and women. Integrating a gender perspective into the conservation of protected areas con-tributes toward the following activity in the Program of Work on Protected Areas, formulated during the Seventh Meeting of the Conference of the Parties of the Convention on Biologi-cal Diversity (CBD) in 2004: "Carrying out participatory national reviews of the status, needs and context-specific mechanisms for involving stakehold-ers, ensuring gender and social equity, in protected areas policy and manage-ment, at the level of national policy, pro-tected area systems and individual sites."

Gender differences in conservation and natural resource use There are differences between women and men living in protected areas and buffer zones, and it is important that these dif-ferences be understood and considered for effective participation in the conservation and sustainable use of natural resources. These differences can be grouped into four interrelated categories. Roles and responsibilities: Generally, men play a greater role than women in the exploitation of natural resources for commercial purposes, since women also dedicate time to domestic activi-ties such as cooking, water and fuelwood collection, child care, etc. These roles can vary. For example, when men migrate in search of job opportunities, women may assume more roles and responsibilities ticipate. In the Central Selva region of Peru, the entire family takes care of the trees as well as harvesting and process-ing agroforestry products, although women are mainly in charge of the tree nurseries. Often, women have responsibilities related to the management of natural resources, but not rights over them.

Gender-responsive conservation policies and programs are those that seek to achieve biodiversity conservation success, while explic-itly taking into account both men's and women's opinions, needs, and interests. Access and control over resources: Due to economic, social, institutional and legal factors, the right to access and control natural resources, such as land and forests, is different for men and women. Women have generally expe-rienced greater restrictions, particularly with respect to independent ownership and access to land. This may partially explain why the majority of the world's poor are women who depend on natural resources for subsistence. Insecurity regarding land tenure has an effect on how much time women—and men—are willing to spend on sustain-able development practices. Lack of land and other constraints mean that women generally have fewer oppor-tunities to obtain credit and support services, unless these are provided for the explicit purpose of overcoming women's disadvantages.

Knowledge base: As a result of the dif-ferences in activities and access, women's and men's knowledge about the use of natural resources can also be different. For example, some women have become guardians of biodiversity by using their knowledge to utilize a wide variety of seeds, maintaining important in situ reserves of genetic resources or germ-plasm. This knowl-edge has been transmitted from mothers to daughters for generations. Differences in knowledge between women and men also depend on their particular social class, age and ethnic group. Public participation in decision making:

Generally women have had fewer opportunities to participate in making environmental decisions. As a result, their perceptions and interests are some-times ignored or excluded when policies are designed (Sass, 2002). The absence of opportunities is often due to cultural restrictions, women's lack of school-ing and low self-esteem, while logistical reasons may be the factor in other circum-stances. For example, in the Guatemalan Motagua-Polochic System, some institu-tions are planning to facilitate women's participation by taking care of children while their mothers attend the events.

Gender as a tool for protected area conservation In general, from a conservation perspec-tive, projects that have applied gender equity and promoted women's par-ticipation have been more effective and balanced. While each case is different, incorporat-ing gender in natural resource conserva-tion and sustainable use: Ensures that women's and men's traditional rights over resource use in protected areas are not diminished with the develop-ment of projects and programs (Martin, 2004). Helps in changing stereotypes harmful to conservation, such as the belief that women are incapable of participating together with men. It strengthens a vision in which men and women can develop the same or complementary activities in relation to nature. In the Motagua-Polochic System in Guatemala, for example, both men and women participate in fire prevention and control in three water-sheds of the Motagua River, according to their capacities.

Contributes to resolving conflicts of interest in the definition and management of protected areas, by recognizing the different interests and priorities of men and women. Helps ensure the representa-tion of diversity in environmental management and to identify local organizations responsible for it facilitates contact with external sources of financing interested in promoting gender initiatives. Increases opportunities for sustainable activities, such as those women have traditionally carried out or in which they have a particular interest. Uncovers roles and interests which are usually ignored, such as those of women, who represent slightly more than half the population in Latin America and the Caribbean.

Women have a number of characteristics that benefit conservation efforts. While it may not be easily visible to outsiders, innately women are community leaders, often the ones who organize local envi-ronmental events, and have the greatest influence on their children, to whom they can pass on environmental messages. Women are also more predisposed to use income for household expenses and they manage natural resources such as plants and fuel wood on a daily basis (Hill Rojas, 1999).

Elements to consider in incorporating gender perspective. The incorporation of a gender perspec-tive in protected area conservation is ensured through women's participation in specific activities or programs and in defining the objectives and design of conservation activities; this makes the activities or programs consistent with the interests and priorities of both men and women. In order to promote social participation in conservation and sus-tainable use processes, it is necessary to recognize, develop and revisit the concept of "the community" as a homo-geneous group. In addition to discerning the differences between men and women, it is necessary to realize that men and women are not homogenous categories either, since here too there are distinguishing social factors. The following are some basic factors to consider when incorporating gender into the design and development of conserva-tion projects.

Design of Conservation Activities

Conduct an assessment which includes a social analysis of the area's inhabitants—men and women. The assessment should identify the needs, interests, knowledge and behaviors by gender which will shape the conservation initiatives. Making gender distinctions helps to clarify the relations between humans and the environment. Consider the need to involve men in the gender discussions from the beginning, and keep it from becoming an issue exclusively about and for women. Gender workshops are recommended to raise men's awareness. Facilitate men's and women's participation in the design of conservation activities through the use of easily applicable didactic methods, such as making maps where each person draws his or her use, access and control over the natural resources. Mapping can illustrate the fact that one activity may affect women and men differently.

Development of Conservation Activities

Integrating women into existing initiatives is generally more effective than a separate effort directed at women. However, this can depend, especially in cases where there is a tradition of women working together in groups or when they are the head of household.

Suggested Themes for a Gender-Sensitive Assessment

Identify the ways women and men learn what they learn and how they use what they learn in relation to natural resources. Identify the sexual division of labor. Who does what in productive, reproductive and communal activi-ties? When and how is it done? Identify access and control of resources, goods and services, as well as benefit and cost sharing, differentiating by gender, with an emphasis on the resources that pro-tected area activities will have an effect on. Identify demographic trends: birth rates, average age that women have children and significant migratory movements that can influence con-servation efforts. Identify traditional or contempo-rary values that influence either positively or negatively the way in which men and women use natural resources.

Include quantitative and qualitative data that illustrate power and sub-ordination relations between the men and women participating in the groups. Identify gender-differentiated forms of participation in terms of quantity and quality in decision-making processes in both community and domestic spheres. Identify rules and laws that influ-ence the way in which men and women use, manage and control natural resources. Illustrate the needs, demands and priorities expressed by men and women, as well as those expressed by young and older people.

In Bolivia's Amboro-Carrasco National Parks, an education strategy is being carried out to increase awareness about the environmental benefits generated by the parks. Organizations of women and youth are among the audiences being targeted since they are considered more receptive to environmental issues. Involve women in methods to monitor project advances with respect to conservation targets and protected area management. Consider including indicators in the monitoring system to record progress toward gender equity, such as changes in power relations between men and women, number of woman participating in activi-ties, or women's access to resources. Have at least one person knowledgeable about the gender theme to provide technical assistance to the initiatives or projects. This gives the theme vis-ibility, permanence and the possibility of address-ing gender in a professional manner, although ideally, all members of the technical team should have an understanding and aware-ness of gender. Having mixed and multidisciplinary teams makes it easier to deal with gender themes. Also, it is suggested that the technical team include professionals capable of assessing human values, self-esteem and respect between family members.

Some conservation and sustainable use initiatives with gender perspective Sustainable development initiatives in protected areas and buffer zones can offer unique opportunities to promote gender equality, as long as they promote equitable participation of men and women in activities, decision making and the distribution of benefits. Watershed management initiatives that include women have had more chances of success since they are the ones primarily in charge of collecting water and thus have considerable knowledge about water resources, including water quality and reliability. In Cuatro Ciénegas, Mexico, one of the main motivations for improving the water system was so that women would not have to bring water from far distances.

Gender-sensitive environmental edu-cation initiatives have reached a larger population, since they do not exclude women and young people. Fair trade initiatives make women's input more visible in terms of knowledge, production, and protection of biodiversity, as well as guaranteeing them employment, improving income levels and securing access to technol-ogy, credit and participation in decision taking. These initiatives contemplate, for example, handcrafts made of non-timber forest products, contributing to the sustain-able use of the resources. Payment for environmental services initiatives can promote equity in terms of benefit sharing. For example, a percentage of the benefits can be set aside for activities with women. The Costa Rican Ministry of Environment established that part of the funds obtained in return for environmental services such as carbon fixing would be ear-marked for activities promoting equity in beneficiary communities.

At the request of the Zacapulco and Topón communities, in Mexico's La Encrucijada Biosphere Reserve, a project was developed to produce paper mache handcrafts to sell to the reserve's visitors. This activity has made it possible to increase environmental awareness, espe-cially in women and children, while helping supplement family income. Ecotourism initiatives have also provided a vehicle for fostering equitable gender relations, with demonstrated effectiveness in empowering women through community-based tourism enterprises. Initiatives aimed at promoting effi-cient use of energy sources have been more effective when they include women, who are the main decision makers about domestic energy use. It is estimated that women spend three times more time transporting fuel than men.

Research initiatives that incor-porate sources of non-traditional information have been more effective if they recognize women's knowledge in areas such as growing crops, crop resistance to pests, harvesting? Collection and food processing women's- and men's- knowledge about biodiversity needs to be made visible and valued. The Nature Conservancy (TNC) is a leading Conservation organization working around the world to protect ecologically important lands and waters for nature and people. Since 1951, TNC has been working with communi-ties, businesses and people like you to protect more than 117 million acres of land, 5,000 miles of river, and 100 marine sites around the world. TNC's mission is to preserve the plants, animals and natural communities that represent the diversity of life on Earth by protecting the lands and waters they need to survive.

The political ecology approach, at least as it has been applied to the study of land use decisions, has rarely given priority to the role that gender relations play in resource use decisions. The user makes choices about resources. Considerations such as seasonal fruiting of trees, game abundance, household consumption needs, market prices, labor migration, as well as the configuration of state policy and the strength of local organizations and alliances figure into the analysis, at least to the extent that they are found to affect the decision in question. The map of conditioning factors produced by this methodology can then be used to identify those particular domains within the decision environment that are subject to modification, thereby leading to more desirable outcomes such as community-based rules for game management. The merit of the approach is that it is eminently site-specific, yet also highly sensitive to the manner in which forces beyond the particular site influence local outcomes. Moreover, the findings produced by the political ecology framework not only provide a systematic understanding of the interplay of socio-environmental factors that lead to the observed patterns of resources use, but also serve to specify concrete policy interventions.

The political ecology approach, at least as it has been applied to the study of land use decisions, has rarely given priority to the role that gender relations play in resource use decisions. Yet such considerations can be easily introduced into the framework in as much as gender relations are a prominent feature of the context within which resource decisions are made. (Espinosa 1998) notes that in the Peruvian Amazon, for example, proximity to market is associated with differences in the gender division of labor, in access and control over resources, and in patterns of decision-making. The task, therefore, is to introduce explicit gender awareness into the approach, thereby producing a "gendered political ecology" framework. This point will be explored further below. Who are the multiple stakeholder groups involved in direct or indirect negotiation for resources? In what ways are their interests complementary and in conflict? How do their different levels of power and resources affect the outcomes of negotiations?. In the definition of conservation adopted here, there are always multiple users of ecosystems and resources. Resource management for conservation therefore involves direct or indirect negotiation among multiple, often conflicting, groups of stakeholders, some who reside locally and some who do not, each of whom is endowed with different levels of economic and political power.

Since natural resource use is only part of the social complex that defines a community and its gender-differentiated groups, understanding their dynamics requires an analysis of the broader historical and social context (Leach 1994: The gendered political ecology approach focuses on the material and ideological roots of Stakeholder analysis involves the identification of different groups and institutions, both formal and informal, who may affect or be affected by a resource management initiative. Agarwal 1994 notes that gender relations, including gendered sciences; gendered rights and responsibilities; and gendered participation in organizations and political activity. According to the multiplicity of women's roles viz producer, reproducer, and "consumer" leads them to integrate complex systems instead of specializing. For this reason, women may be more attentive to the ecosystem as a whole.

In many situations, women's responsibility for family subsistence and health causes them to focus more on livelihood systems and on the environment, as opposed to the more commercial orientation of men who are primarily involved in market-oriented endeavors. If so, then women could constitute key potential allies in conservation strategies based on sustainable livelihoods for local communities. This approach, which is more holistic and normative, has been advocated as an alternative to market-oriented concepts and strategies, because it focuses on the quality of life and ecosystem over the long term, and both market and non-market values.

Research is needed to assess how realistic such an approach is, under what conditions, and to what extent, gender differentiates goals, values, power and resource use practices among user groups. Analysis of gender relations and gendered resource use and management is an explicit part of any strong social analysis. This involves collecting and analyzing gender-disaggregated information on livelihood systems, rights and responsibilities, resource use, and values and attitudes regarding key resources. Where appropriate, much of this information may be gathered using participatory methods such as focus groups, resource mapping, activities calendars, and oral history interviews. Does stakeholder participation in gender-focused learning processes improve the ability of local actors to negotiate their interests in conservation? If we accept that a gender focus is useful for conservation, how can gender analysis be useful in learning processes to empower local groups? How do we expect these learning processes to translate into changes in conservation practice? How are they related to empowerment of different groups in relation to community-based resource management? In keeping with the need to "level the playing field", genderfocused learning strategies may increase awareness of the importance that women and other groups be able to sit at the negotiating table.

Because the negotiation process involves power differences, empowerment entails sensitivity to conflicts and to different perspectives within communities, while respecting community traditions and self-determination. Project analysis builds on political ecology, stakeholder and gender analysis to focus on how project activities, costs and benefits, and other outcomes differentially affect different local groups at different stages, and how the behavior of these different actors affects the outcome of the project and achievement of goals.

How are changes in resource use and management by local communities linked to biodiversity conservation? Sustainability analysis Because the negotiation process involves power differences, empowerment entails sensitivity to conflicts and to different perspectives within communities, while respecting community traditions and self determination. Community-based conservation requires responding to local felt needs, while not over-exploiting natural resources. (Bodmer et al. 1997) results that within the definition of conservation adopted here, fully protected areas are included to replenish harvested populations, for example along with managed areas. Comparative research on management of common property resources by communities around the world has demonstrated that attention to both institutional and biological parameters are essential to management success. In addition to the socioeconomic information required, community-based conservation projects need ways to monitor the biological changes associated with changing resource use strategies associated with socioeconomic changes. Analysis of biological sustainability requires basic biological inventories of key resources and habitats, and information on reproductive biology and ecology of key species, to identify sustainable and non-sustainable patterns of use and how they can realistically be adapted through management. Information is needed on harvest impacts, economic returns, and changes in institutional arrangements, as they affect different social groups, in order to project likely biological and economic outcomes under different scenarios.

Analysis of the institutional process includes attention to the somewhat unpredictable nature of politics, both formal and informal aspects of resource management institutions, and the divergent interests both within and outside communities. The framework suggests the need to combine research at different levels of analysis, qualitative and quantitative tools of data collection and analysis, and methods from the social, economic and biological sciences in order to address the broad set of questions. Moreover, the longterm nature of conservation issues requires systematic monitoring of impacts on different social groups and natural habitats over time. These challenges underscore the need to strengthen collaboration between researchers, project implementers, and local peoples to address the evolution of these complex relationships over time. The experience of building and modifying the conceptual framework has reinforced the importance of the principles of learning and adaptation as applied to conservation work. Only by engaging the commitment and creativity of a broad coalition of partners can the challenge of community-based conservation be addressed.

Innovative Approaches and Future Direction

Approaches to improve the strategies in each of the three areas namely research, policy, and programme have to be discussed. Proposals for future directions could put forth. These lead to the formulation of the final recommendations of the Consultation.

Research Strategies

There is a growing realisation that projects and programmes that involve communities and biodiversity conservation need to have special training orientations for the staff, to empower them to deal with communities. In most cases, the personnel working in such projects are science-based persons and have no training in social and gender issues. Scientists need to gain communication skills as well as a basic understanding of community dynamics. Sometimes mentoring is used as a device in community-based programmes. Multi disciplinary projects are getting more popular and are seen as being more effective in achieving a holistic goal. There are many scientist leaders in such projects, and there appears to be resistance to lateral linkages among themselves. There is a need to develop a forum for such people to exchange ideas and concerns within the project. In many biodiversity conservation projects that are placed in remote areas, the local men come forward with information. Consequently, the observation from women is not evident. The staff of such projects needs to understand that women-based knowledge is different and has also to be chronicled.

The research should not always be top down. Practices based on local knowledge should be given due consideration. Projects should be based on a ecosystem and not on a single crop or plant. Holistic projects may be more successful if not limited to a single crop or topic. There is a need to identify crops of high nutritional value and of local importance. Production of such crops needs encouragement. This will ensure food security in small yet vital areas. In many tribal societies, collection of wild food and medicinal plants make a major contribution to daily food consumption. Projects to revive such wild plants and food habits should be encouraged. Studies on gender relations and labour divisions in agriculture or in knowledge systems do not find their application in policies. There should be a forum to educate policy makers on the research findings of such projects to ensure a continuum in the research-policy environment.

Policy Strategies

The preamble to the Convention on Biodiversity (CBD) recognises women's role in biodiversity management. The Conference of Parties (COP) to the CBD has addressed the issues of sustainable use, conservation, and equitable sharing of biological diversity. Other international bodies such as TRIPS and WIPO are silent on the question of gender rights.

At national levels, governments, scientific communities, and NGOs are responsible for ensuring that the National Biodiversity Acts, Plant Rights Acts, and Intellectual Property Acts are gender sensitive. FAO has an interactive process with national governments and can prepare projects only on their request. FAO has the skills and information to prepare a guideline for gender dimensions in the management of plant genetic resources as well as a gender code for the countries. In many cases, the words 'women' and 'gender' are synonymous. Research has shown that women are primary producers, and yet at policy levels, especially for benefits and incentives, governments target men. Thus women become unpaid contributors to the family's food security. There is an urgency to address women's needs in the spheres of community rights, rewards, and recognition mechanisms. In-depth research has also blown the myth that a household is made of individuals sharing a common interest. There are differences in the power relations within a family, in terms of access to resources and decision making. These differences are also based on the caste, ethnic group, and class of the family. Experts in law and policy research are not sensitised to field situations, and women's organisations lack the special skills to deal with national and international agreements and laws. National bodies dealing with women's affairs should set up a small group of experts to integrate gender issues in Biodiversity Acts.

Delegations that represent a country's interest rarely have a gender expert and so gender issues are not discussed in international agreements. Similarly, national planning bodies and service commissions also need to be sensitised to gender concerns.

Programme Strategies

In most countries the role of women in agriculture remains invisible. Programmes should attempt to highlight the role of women in the management of natural resources. Programmes should start by documenting gender roles and relationships and then move on to action research. At times, projects deal with interventions which are not accepted by the communities readily. Strategies to make the interventions successful should also be discussed in the project plans. Partnerships between government departments, scientists, and NGOs in the field are essential for the success of programmes. It is necessary to have gender-segregated data both for the purpose of research and for policy decisions. There is resistance to the addition of data points in national Census Reports. Enumerators also need to be sensitized to gender issues in their data sheets.

When programmes are drafted, gender concerns are not given any thought. Many countries and scientists use the concept of gender as an add-on in their programmes. This is not productive to the results of the programme. FAO follows guidelines on inclusion and monitoring of gender in the programmes. Countries should incorporate similar guidelines in their national programmes. In some cases as in polio eradication, women have been the targets of the intervention. But, in agricultural interventions they are sidelined. It would be worthwhile to use women farmers as the agents of change in agricultural interventions also. Agriculture has no prestige value. If rural economies are to be revitalized, agriculture has to be seen as an important sector and it should be central in biodiversity conservation projects too.

Apart from ensuring the production of sufficient food crops, it is also essential to review whether the produce is easily available to all the local communities. Distribution of food to vulnerable regions of the country has to be a priority. Biodiversity: gender-based knowledge and livelihoods. The Convention on Biological Diversity (CBD) aims at conservation and sustainable use of biological diversity, which is defined as "the combination of life forms and their interactions with each other and the rest of the environment that has made Earth a uniquely habitable place for humans". Biological resources, both habitats and species, are being critically threatened. Key habitats are being lost at a rapid rate, and species are disappearing at a rate 50 - 100 times the natural rate. 34,000 plant and 5,200 animal species face extinction, as well as 30% of the main breeds of farm animals.

The CBD also advocates the fair and equitable sharing of genetic resource benefits. Furthermore, the CBD establishes a connection between sustainable conservation and development, and the rights of indigenous peoples and local communities. For example, plant genetic resources (PGR) are of significant importance and economic value as they represent an irreplaceable, locally adapted source of characteristics such as resistance to pests and disease and suitability to specific micro-climatic, edaphic and environmental conditions. However, Plant Generate resources are being lost at a rapid rate, in part due to the replacement of farmers' heterogeneous traditional varieties with a relatively small number of homogenous, modern commercial varieties. This loss of crop genetic diversity poses a grave threat to long-term food security. In effect, biodiversity is important in rural livelihood for spreading risks, as the reliance on a variety of genetic sources allows their agricultural systems to adapt to varying conditions, and generation of income from a range of natural resources.

Local-level biodiversity is maintained through knowledge and know-how of both women and men. In effect, because of gender-based roles in rural livelihoods, women and men acquire and transmit different and complementary knowledge: they have knowledge about different things, and different knowledge about the same things; they organize knowledge in different ways, and transmit it by different means.

Koziell, McNeil 2002 notes that this knowledge pertains to domestic plant and animal genetic resources, which are the basis of both the productivity and adaptability of agricultural systems, and also to wild and semi domesticated sources, which offer safety nets in case of food scarcity. Both sources also provide important goods for traditional medicine, as well as material for the pharmaceutical industry. Failure to target both sexes in biodiversity conservation results inevitably in a loss of knowledge of local and international relevance, as well as producing a gender bias in policies and programmes which may be detrimental to the functions that women perform. Depending on women and men's roles, their knowledge will be applied to different functions in livelihoods. It has been noted in many regions that whereas men concentrate on cash crop cultivation, women concentrate on subsistence agriculture as well as household-related tasks such as care to family members and food preparation. Thus, failure to integrate women's biodiversity knowledge in each of these functions has impact on household-level food security and nutrition.

Swiderska 2002 observed that the Biodiversity Agenda has been criticized for focusing on conserving rare species of global value, whereas less attention has been paid to biodiversity of local value that helps to sustain the livelihoods of the poor. The rural poor are the first to suffer if these resources are degraded or lost, hence their high stake in ensuring their conservation. Climate change: gender issues in vulnerability and adaptation. The objectives of the United Nations Framework Convention on Climate Change (UNFCCC) are to stabilize the concentrations of greenhouse gases viz carbon dioxide, methane, nitrous oxide, hydro fluorocarbons, perflurocarbons and sulphur hexafluoride in the atmosphere, as they are believed to exacerbate climate change and alter agricultural/eco zones. Some predict that by 2100, climatic zones in mid-latitude regions will shift between 150-500 Km towards the poles, and that due to glacial melt and thermal expansion of the sea, sea levels will rise by as much as 65 cm, threatening both coastal and low-lying areas. The frequency and intensity of extreme natural events are also expected to increase.

Adger et al 2003 notes that rising of sea levels, increasing aridity, rising frequency of disasters, among other factors, erode the quality of the natural resource base. Decrease in agricultural productivity can be expected. Climate change will exacerbate existing vulnerabilities and create new ones, thus making the fight against poverty still more difficult. Other effects of climate change will affect women and men, sometimes in the same way, sometimes differently. An increase in temperature-related illnesses and deaths can be expected; changes in the geographic range of vector-borne diseases such as malaria and dengue fever because of changes in temperature and rainfall patterns are also likely. Anaemia–resulting among other factors from malaria–is responsible for a quarter of maternal mortality.

Degradation of water sources is a usual consequence of drought and natural disasters. Children and pregnant women are particularly susceptible to diseases which thrive in such conditions such as diarrhoea and cholera. Furthermore, because of their roles in relation with household water supply and domestic chores, women are particularly at risk for such diseases. A direct correlation has been observed between women's status in society and the likelihood of their receiving health care. If climate change has been observed to "exacerbate current gender inequalities", due to the increase of pressure on societies, negative consequences on women's health can be expected.

Extreme weather effects such as flooding, landslides, and storms, cause death and injury, and may affect women and men differently, depending on the means at the disposal of each to ensure their own safety. It has been noted that in Bangladesh, cultural constraints on women's mobility hinder access to shelter and health care in the context of cyclones and floods.

Adaptation to climate change challenges traditional coping mechanisms. Traditional risk-sharing mechanisms, based on kin and social groups, may not be adequate, since whole regions are affected. If the natural resource base is degraded to the point of being insufficient to support the livelihoods of the current population, drastic measures are implemented, such as selling off of assets and migration. Across regions, women and/or men, and of different age groups migrate for different reasons. However, in the case of male migration, female-headed households have been identified as particularly vulnerable, since women must assume traditionally male responsibilities, without having equal or direct access to all resources, financial, technological, and social. Remittances, which can help mitigate the vulnerability of these households, depend on the economic opportunities offered to the migrant. In the context of decreased productivity that goes along with climate change, economic and social development is expected to slow down.

Global Action for Women towards Sustainable and Equitable Development Programme Area Basis for action

The international community has endorsed several plans of action and conventions for the full, equal and beneficial integration of women in all development activities; in particular the Nairobi Forward-looking Strategies for the Advancement of Women, which emphasize women's participation in national and international ecosystem management and control of environment degradation. Several conventions, including the Convention on the Elimination of All Forms of Discrimination against Women and conventions of ILO and UNESCO have also been adopted to end gender-based discrimination and ensure women access to land and other resources, education and safe and equal employment. Also relevant are the 1990 World Declaration on the Survival, Protection and

Development of Children and the Plan of Action for implementing the Declaration. Effective implementation of these programmes will depend on the active involvement of women in economic and political decision-making and will be critical to the successful implementation of Agenda 21.

Objectives

The following objectives are proposed for national Governments:

- To implement the Nairobi Forward-looking Strategies for the Advancement of Women, particularly with regard to women's participation in national ecosystem management and control of environment degradation;
- To increase the proportion of women decision makers, planners, technical advisers, managers and extension workers in environment and development fields;
- To consider developing and issuing by the year 2000 a strategy of changes necessary to eliminate constitutional, legal, administrative, cultural, behavioural, social and economic obstacles to women's full participation in sustainable development and in public life;
- To establish by the year 1995 mechanisms at the national, regional and international levels to assess the implementation and impact of development and environment policies and programmes on women and to ensure their contributions and benefits;
- To assess, review, revise and implement, where appropriate, curricula and other educational material, with a view to promoting the dissemination to both men and women of gender-relevant knowledge and valuation of women's roles through formal and non-formal education, as well as through training institutions, in collaboration with non-governmental organizations;

- To formulate and implement clear governmental policies and national guidelines, strategies and plans for the achievement of equality in all aspects of society, including the promotion of women's literacy, education, training, nutrition and health and their participation in key decision making positions and in management of the environment, particularly as it pertains to their access to resources, by facilitating better access to all forms of credit, particularly in the informal sector, taking measures towards ensuring women's access to property rights as well as agricultural inputs and implements;
- To implement, as a matter of urgency, in accordance with country-specific conditions, measures to ensure that women and men have the same right to decide freely and responsibly the number and spacing of their children and have access to information, education and means, as appropriate, to enable them to exercise this right in keeping with their freedom, dignity and personally held values;
- To consider adopting, strengthening and enforcing legislation prohibiting violence against women and to take all necessary administrative, social and educational measures to eliminate violence against women in all its forms.

Activities

Governments should take active steps to implement the following:

- Measures to review policies and establish plans to increase the proportion of women involved as decision makers, planners, managers, scientists and technical advisers in the design, development and implementation of policies and programmes for sustainable development;
- Measures to strengthen and empower women's bureaux, women's non-governmental organizations

and women's groups in enhancing capacity-building for sustainable development;

- Measures to eliminate illiteracy among females and to expand the enrolment of women and girls in educational institutions, to promote the goal of universal access to primary and secondary education for girl children and for women, and to increase educational and training opportunities for women and girls in sciences and technology, particularly at the post-secondary level;
- Programmes to promote the reduction of the heavy workload of women and girl children at home and outside through the establishment of more and affordable nurseries and kindergartens by Governments, local authorities, employers and other relevant organizations and the sharing of household tasks by men and women on an equal basis, and to promote the provision of environmentally sound technologies which have been designed, developed and improved in consultation with women, accessible and clean water, an efficient fuel supply and adequate sanitation facilities;
- Programmes to establish and strengthen preventive and curative health facilities, which include women-centred, women-managed, safe and effective reproductive health care and affordable, accessible, responsible planning of family size and services, as appropriate, in keeping with freedom, dignity and personally held values. Programmes should focus on providing comprehensive health care, including pre-natal care, education and information on health and responsible parenthood, and should provide the opportunity for all women to fully breastfeed at least during the first four months post-partum. Programmes should fully support women's productive and reproductive roles and well-being and should pay special attention to the need to provide equal and

improved health care for all children and to reduce the risk of maternal and child mortality and sickness;

- Programmes to support and strengthen equal employment opportunities and equitable remuneration for women in the formal and informal sectors with adequate economic, political and social support systems and services, including child care, particularly day-care facilities and parental leave, and equal access to credit, land and other natural resources;
- Programmes to establish rural banking systems with a view to facilitating and increasing rural women's access to credit and to agricultural inputs and implements;
- Programmes to develop consumer awareness and the active participation of women, emphasizing their crucial role in achieving changes necessary to reduce or eliminate unsustainable patterns of consumption and production, particularly in industrialized countries, in order to encourage investment in environmentally sound productive activities and induce environmentally and socially friendly industrial development;
- Programmes to eliminate persistent negative images, stereotypes, attitudes and prejudices against women through changes in socialization patterns, the media, advertising, and formal and non-formal education;
- Measures to review progress made in these areas, including the preparation of a review and appraisal report which includes recommendations to be submitted to the 1995 world conference on women.

Governments are urged to ratify all relevant conventions pertaining to women if they have not already done so. Those that have ratified conventions should enforce and establish legal, constitutional and administrative procedures to transform agreed rights into domestic legislation and should adopt measures to implement them in order to strengthen the legal capacity of women for full and equal participation in issues and decisions on sustainable development. States parties to the Convention on the Elimination of All Forms of Discrimination against Women should review and suggest amendments to it by the year 2000, with a view to strengthening those elements of the Convention related to environment and development, giving special attention to the issue of access and entitlements to natural resources, technology, creative banking facilities and low-cost housing, and the control of pollution and toxicity in the home and workplace. States parties should also clarify the extent of the Convention's scope with respect to the issues of environment and development and request the Committee on the Elimination of Discrimination against Women to develop guidelines regarding the nature of reporting such issues, required under particular articles of the Convention. A) Areas requiring urgent action.

Countries should take urgent measures to avert the ongoing rapid environmental and economic degradation in developing countries that generally affects the lives of women and children in rural areas suffering drought, desertification and deforestation, armed hostilities, natural disasters, toxic waste and the aftermath of the use of unsuitable agro-chemical products.

In order to reach these goals, women should be fully involved in decision-making and in the implementation of sustainable development activities. Research, data collection and dissemination of information Countries should develop gender-sensitive databases, information systems and participatory action-oriented research and policy analyses with the collaboration of academic institutions and local women researchers on the following: Knowledge and experience on the part of women of the management and conservation of natural resources for incorporation in the databases and information systems for sustainable development; The impact of structural adjustment programmes on women. In research done on structural adjustment programmes, special attention should be given to the differential impact of those programmes on women, especially in terms of cut-backs in social services, education and health and in the removal of subsidies on food and fuel; The impact on women of environmental degradation, particularly drought, desertification, toxic chemicals and armed hostilities; Analysis of the structural linkages between gender relations, environment and development; The integration of the value of unpaid work, including work that is currently designated "domestic", in resource accounting mechanisms in order better to represent the true value of the contribution of women to the economy, using revised guidelines for the United Nations System of National Accounts, to be issued in 1993; Measures to develop and include environmental, social and gender impact analyses as an essential step in the development and monitoring of programmes and policies; Programmes to create rural and urban training, research and resource centres in developing and developed countries that will serve to disseminate environmentally sound technologies to women.

International and Regional Cooperation and Coordination

The Secretary-General of the United Nations should review the adequacy of all United Nations institutions, including those with a special focus on the role of women, in meeting development and environment objectives, and make recommendations for strengthening their capacities. Institutions that require special attention in this area include the Division for the Advancement of Women (Centre for Social Development and Humanitarian Affairs, United Nations Office at Vienna), the United Nations Development Fund for Women (UNIFEM), the International Research and Training Institute for the Advancement of Women (INSTRAW) and the women's programmes of regional commissions. The review should consider how the environment and development programmes of each body of the United Nations system could be strengthened to implement Agenda 21 and how to incorporate the role of women in programmes and decisions related to sustainable development.

Each body of the United Nations system should review the number of women in senior policy-level and decision-making posts and, where appropriate, adopt programmes to increase that number, in accordance with Economic and Social Council resolution 199/17 on the improvement of the status of women in the Secretariat. UNIFEM should establish regular consultations with donors in collaboration with UNICEF, with a view to promoting operational programmes and projects on sustainable development that will strengthen the participation of women, especially low-income women, in sustainable development and in decision-making. UNDP should establish a women's focal point on development and environment in each of its resident representative offices to provide information and promote exchange of experience and information in these fields. Bodies of the United Nations system, governments and non-governmental organizations involved in the follow up to the Conference and the implementation of Agenda 21 should ensure that gender considerations are fully integrated into all the policies, programmes and activities. Means of implementation Financing and cost evaluation. The Conference secretariat has estimated the average total annual cost (1993-2000) of implementing the activities of this chapter to be about $40 million from the international community on grant or confessional terms. These are indicative and order-of-magnitude estimates only and have not been reviewed by Governments. Actual costs and financial terms, including any that are non-concessional, will depend upon, inter alia, the specific strategies and programmes Governments decide upon for implementation. Gender Mainstreaming in the Plan of Implementation of the World Summit on Sustainable Development–Extracts.

Good governance within each country and at the international level is essential for sustainable development. At the domestic level, sound environmental, social and economic policies, democratic institutions responsive to the needs of the people, the rule of law, anti-corruption measures, gender equality and an enabling environment for investment are the basis for sustainable development. As a result of globalization, external factors have become critical in determining the success or failure of developing countries in their national efforts. The gap between developed and developing countries points to the continued need for a dynamic and enabling international economic environment supportive of international cooperation, particularly in the areas of finance, technology transfer, debt and trade and full and effective participation of developing countries in global decision-making, if the momentum for global progress towards sustainable development is to be maintained and increased.

Poverty Eradication

Eradicating poverty is the greatest global challenge facing the world today and an indispensable requirement for sustainable development, particularly for developing countries. Although each country has the primary responsibility for its own sustainable development and poverty eradication and the role of national policies and development strategies cannot be overemphasized, concerted and concrete measures are required at all levels to enable developing countries to achieve their sustainable development goals as related to the internationally agreed poverty-related targets and goals, including those contained in Agenda 21, the relevant outcomes of other United Nations conferences and the United Nations Millennium Declaration. This would include actions at all levels to: (d) Promote women's equal access to and full participation in, on the basis of equality with men, decision-making at all levels, mainstreaming gender perspectives in all policies and strategies, eliminating all forms of violence and discrimination against women and improving the status, health and economic welfare of women and girls through full and equal access to economic opportunity, land, credit, education and health - care services;

Changing unsustainable patterns of consumption and production Call upon Governments, as well as relevant regional and international organizations and other relevant stakeholders, to implement, taking into account national and regional specificities and circumstances, the recommendations and conclusions of the Commission on Sustainable Development concerning energy sustainable development adopted at its ninth session, including the issues and options set out below, bearing in mind that in view of the different contributions to global environmental degradation, States have common but differentiated responsibilities. This

would include actions at all levels to: (m) Promote education to provide information for both men and women about available energy sources and technologies; IV. Protecting and managing the natural resource base of economic and social development 25. Launch a programme of actions, with financial and technical assistance, to achieve the Millennium development goal on safe drinking water. In this respect, we agree to halve, by the year 2015, the proportion of people who are unable to reach or to afford safe drinking water, as outlined in the Millennium Declaration, and the proportion of people without access to basic sanitation, which would include actions at all levels to: (a) Mobilize international and domestic financial resources at all levels, transfer technology, promote best practice and support capacity -building for water and sanitation infrastructure and services development, ensuring that such infrastructure and services meet the needs of the poor and are gender-sensitive; Facilitate access to public information and participation, including by women, at all levels in support of policy and decision -making related to water resources management and project implementation; Mountain ecosystems support particular livelihoods and include significant watershed resources, biological diversity and unique flora and fauna. Many are particularly fragile and vulnerable to the adverse effects of climate change and need specific protection.

Actions at all levels are required to: Develop and promote programmes, policies and approaches that integrate environmental, economic and social components of sustainable mountain development and strengthen international cooperation for its positive impacts on poverty eradication programmes, especially in developing countries; Implement programmes to address, where appropriate, deforestation, erosion, land degradation, loss of biodiversity, disruption of water flows and retreat of glaciers; Develop and implement, where appropriate, gender-sensitive policies and programmes, including public and private investments that help eliminate inequities facing mountain communities; Implement programmes to promote diversification and traditional mountain economies, sustainable livelihoods and small-scale production systems, including specific training programmes and better access to national and international markets, communications and transport planning, taking into account the particular sensitivity of mountains; Promote full participation and involvement of mountain communities in decisions that affect them and integrate indigenous knowledge, heritage and values in all development initiatives;

Health and Sustainable Development

Strengthen the capacity of health-care systems to deliver basic health services to all in an efficient, accessible and affordable manner aimed at preventing, controlling and treating diseases, and to reduce environmental health threats, in conformity with human rights and fundamental freedoms and consistent with national laws and cultural and religious values , and taking into account the reports of relevant United Nations conferences and summits and of special sessions of the General Assembly. This would include actions at all levels to: (i) Ensure equal access of women to health -care services, giving particular attention to maternal and emergency obstetric care; (j) Address effectively, for all individuals of appropriate age, the promotion of healthy living, including their reproductive and sexual health, consistent with the commitments and outcomes of recent United Nations conferences and summits, including the World Summit for Children, the United Nations Conference on Environment and Development, the International Conference on Population and Development, the World Summit for Social Development and the Fourth World Conference on Women, and their respective reviews and reports; (l) Transfer and disseminate, on mutually agreed terms, including through public-private multisector partnerships, with international financial support, technologies for safe water, sanitation and waste management for rural and urban areas in developing countries and countries with economies in transition, taking into account country-specific conditions and gender equality, including specific technology needs of women.

Gender Mainstreaming in the United Nations

Convention to Combat Desertification™Extracts Prologue: Stressing the important role played by women in regions affected by desertification and/or drought, particularly in rural areas of developing countries, and the importance of ensuring the full participation of both men and women at all levels in programmes to combat desertification and mitigate the effects of drought; PART II: General Provisions Article 5: Obligations of affected country Parties Promote awareness and facilitate the participation of local populations, particularly women and youth, with the support of nongovernmental organizations, in efforts to combat desertification and mitigate the effects of drought; PART III: Action Programmes, scientific and Technical Cooperation and Supporting Measures Section 1: Action programmes Article 10: National action programmes. National action programmes shall specify the respective roles of government, local communities and land users and the resources available and needed. They shall, inter alia: (f) provide for effective participation at the local, national and regional levels of nongovernmental organizations and local populations, both women and men, particularly resource users, including farmers and pastoralists and their representative organizations, in policy planning, decision-making, and implementation and review of national action programmes; and Section 3: Supporting measures Article 19: Capacity building, education and public awareness 1. The Parties recognize the significance of capacity building-that is to say, institution building, training and development of relevant local and national capacities—in efforts to combat desertification and mitigate the effects of drought. They shall promote, as appropriate, capacity building: (a) through the full participation at all levels of local people, particularly at the local level, especially women and youth, with the cooperation of non-governmental and local organizations; The Parties shall cooperate with each other and through competent intergovernmental organizations, as well as with non-governmental organizations, in undertaking and supporting public awareness and educational programmes in both affected and, where relevant, unaffected country Parties to promote understanding of the causes and effects of desertification and drought and of the importance of meeting the objective of this Convention. To that end, they shall: assess educational needs in affected areas, elaborate appropriate school curricula and expand, as needed, educational and adult literacy programmes and opportunities for all, in particular for girls and women, on the identification, conservation and sustainable use and management of the natural resources of affected areas.

14

Indigenous Knowledge and Biodiversity

Indigenous Peoples

Indigenous people number over 300 million. They live in about 75 of the world's 184 countries and are inhabitants of practically each main biome of the earth. Indigenous peoples, also called tribal, aboriginal or autochthonous peoples, national minorities or first peoples, are best defined by using several criteria. Indigenous peoples may have all or part of the following criteria: (a) are the descendants of the original inhabitants of a territory which has been overcome by conquest; (b) are "ecosystem peoples", such as shifting or permanent cultivators, herders, hunters and gatherers, fishers and/or handicraft makers, who adopt a multi-use strategy of appropriation of nature; (c) practice a small-scale, labor-intensive form of rural production which produce little surplus and has low energy needs; (d) do not have centralized political institutions, organize their life at the level of community, and make decisions on a consensus basis; (e) share a common language, religion, moral values, beliefs, clothing and other identifying characteristics as well as a relationship to a particular territory; (f) have a different world-view, consisting of a custodial and non materialist attitude to land and natural resources based on a symbolic interchange with the natural universe; (g) are subjugated by a dominant culture and society; and (h) consist of individuals who subjectively consider themselves to be indigenous.

It is possible to find indigenous peoples carrying out many different activities of use and management of planet's ecosystems: As forest-dwellers in the tropical lowlands or in the mountains, as pastoralists in savannas and other grasslands, or as nomadic or semi-nomadic hunthers and gatherers in forests, prairies and deserts. Fishing is, in addition, the principal economic activity and source of food for several million coastal and island dwellers, as well as many indigenous peoples inhabiting margins of rivers. Large numbers of indigenous peoples are, however, peasant producers and therefore can be indistinguishable from the non-indigenous peoples living nearby. In the Andean and Mesoamerican countries of Latin America, for instance, indigenous peoples farm like mestizo peasants. Similarly, in India distinctions between scheduled tribes and non-tribal peoples cannot be made solely on the basis of productive activities. In these and other many cases non-indigenous peasants and indigenous peoples produce the same crops with the same farming methods. Since in numberless countries many mestizo peasants are direct descendants of the indigenous peoples and retain most of their cultural traits, it has been pointed out that a broader definition of indigenous peoples might increase the real numbers. Thus, by considering other characteristics than language it is possible to enlarge the number of indigenous peoples in the contemporary world. Given the above, some authors like J. Burger think that the number of indigenous people can double the previously estimated. Thus, in the contemporary world there may be as many as 600 million indigenous peoples. However, such figures need qualification.

Based on percentage of total population identified as belonging to indigenous peoples, it is possible to recognize a group of selected nations with a strong presence of these peoples: Papua New Guinea (77%), Bolivia (70), Guatemala (47), Peru (40), Ecuador (38), Mynamar (33), Laos (30), Mexico (12) and New Zealand (12). On the other hand, the absolute number of people recognized as indigenous allow to identify nations with high indigenous population such as India (over 100 million) and China (between 60 and 80 million).

Biodiversity and Biomass Appropriation: The Role of Indigenous Peoples

Biodiversity conservation can not be separated from natural resources utilization. Human appropriation of nature inflows minerals, water, solar energy and principally living beings (biomass) from ecosystems. World statistics indicate that almost half the inhabitants of the planet are still people engaged in the appropriation of natural resources. This appropriation is carried out by a myriad of rural or primary producers through the management of terrestrial, marine and freshwater ecosystems. Forty five per cent of the total human population has been recorded by the Food and Agricultural Organization (FAO) as agricultural population (FAO, 1991). It can be estimated that between 60 and 80 percent of this agricultural population is represented by small-scale, solar-energized productive units based on a multi-use management of nature (Toledo, 1990). In fact, the statistical record shows that by 1990 around 1,200 million rural people were practicing agricultural activities on areas of 5 hectares or less. This figure coincides with the last available world census of agriculture by the FAO in 1970, where more than 80% of all reported holdings were smaller than 5 hectares. A similar pattern is found in world's fisheries where more than 90 percent are small-scale, artesanal operators, acting in a great variety of coastal habitats.

Most of these small-scale farmers and fishers develop its production activities not as socially isolated households but as familial nuclei belonging to specific village communities, many of which, in turn, correspond to cultures that can be considered as indigenous. Morever, within the core of these community based producers, those identified as indigenous people are who carry out the biomas's extraction affecting at the lowest level their local ecosystems. Called "ecosystem people" by some authors such as R.F. Dassman and M. Gadgil, these producers subsist by appropriating a diversity of biological resources from their immediate vicinity. Their quality of life is therefore intimately linked to the maintenance of certain levels of local biodiversity (Gadgil, 1993). As a consequence, they are productive actors in little transformed habitats of the world, including the main forest and sea wellers, slash and bum agriculturalists, some 25-30 million nomadic herders or pastoralists in East Africa, Sahel and Arabian peninsula, most of the 15-21 million world fishers, and all the half a million hunters and gatherers still recognized as citizens of the contemporary world.

In conclusion, indigenous peoples are the fraction of human appropriators of biomass determining the lowest ecological impacts. They generally live in what may be termed "frontier lands" or "refuge regions"; thus remote areas of great "wilderness" where the structure, not the components, of original ecosystems remains more or less untouched. In many cases these lands and waters are untamed, unknown, unowened and unclaimed.

Biodiversity and Indigenous People's Lands and Waters

Indigenous peoples occupy a substantial share of the world's little disturbed tropical and boreal forests, mountains, grasslands, tundra and desert, along with large areas of the world's coasts and near-shore waters (including mangroves and coral reefs) (Durning, 1993). The importance of indigenous territories to biodiversity conservation is therefore evident. In fact, indigenous peoples control, legally or not, immense areas of natural resources. Among the most remarkable examples are the cases of the Inuit people (formerly known as Eskimo) who govern a region covering one fifth of the territory of Canada (222 million hectares), the indigenous communities of Papua New Guinea whose lands represent 97 % of the national territory, and the tribes of Australia with nearly 90 million hectares (Figure 1). Although numbering only above 250,000, the indians of Brazil possess an area of over 100 million hectares, mainly in the Amazon basin, distributed in 565 territories (Figure 2 and Table V). Nearly 60% of the priority areas on central and southern Mexico recommended for protection are also inhabited by indigenous peoples (Figure 3), and half of the 30,000 rural communities are distributed in the ten most biologically rich states of the

Mexican territory. In summary, on the global scale it is estimated that the total area under indigenous control probably reach between 12 and 20 percent of the earth's land surface. The best example of notable overlaps between indigenous peoples and biological rich areas is the case of tropical humid forests, In fact, there is a clear correspondence between areas of remaining tropical forests and the presence of indigenous peoples in Latin America, the Congo Basin in Africa, and several countries of tropical Asia such as Philippines, Indonesia and New Guinea, The strong presence of indigenous peoples in Brazil, Indonesia and Zaire alone, which accounts for the 60 per cent of all the tropical forest of the world, is remarkable.

In Latin America, this geographical relationship has been strikingly verified for the Central American countries by a National Geographic Society map produced by a project headed by Mac Chapin in 1992. The same pattern can be found in the tropical humid areas of Mexico inhabited by 1.6 million of indigenous people, and for many regions of the Amazonia basin (see the case of Brazil in Figure 2). It has been estimated that in Amazonia above 1 million indigenous people of eight countries posses over 135 million hectares of tropical forests (Davis & Wall, 1994). Many temperate forests of the world also overlap with indigenous territories as for example in India (See Figure 4), Mynamar, Nepal, Guatemala, the Andean countries (Ecuador, Peru and Bolivia) and Canada. On the other hand, over two million of islanders of the South Pacific, most of whom are indigenous peoples, continue fishing and harvesting marine resources in high biodiversity areas (as coral reefs).

Biodiversity and Ethnoecology

Biodiversity is a very wide concept that refers to the variety of landscapes, ecosystems, species and genes, including their different functional processes. Therefore, maintenance and conservation of biodiversity demand efforts on these four levels. While the first level is oriented to preservation of assemblies of "ecosystems", the second ones focuses on protection of habitats in which the populations of species live.

At the species level, most knowledge on biodiversity concerns the large plants and animals such as flowering plants and vertebrates.

The extent of diversity of smaller plants and animals remains to be inventoried and protected. While most biological diversity is constituted by wild plants and animals, an important subset involves the diversity among of domesticated organisms. In this fourth level, the interest focuses on conservation of genetic variation of crops and domesticated animals. This section is dedicated to examine the potential role of indigenous peoples in biodiversity conservation from an ethnoecological perspective. Ethnoecology can be defined as an interdisciplinary approach exploring how nature is seen by human groups through a screen of beliefs and knowledge, and how humans in terms of their images use and/or manage natural resources. Thus, by focusing in the kosmos (the belief system or cosmovision), the corpus (the whole repertory of knowledge or cognitive systems) and the praxis (the set of practices), ethnoecology offers an integrative approach to the study of the process of human appropriation of nature (Toledo, 1992). This approach allows to recognize the value of the belief-knowledgepractice complex of indigenous peoples in relation to the conservation of biodiversity.

The Kosmos For indigenous peoples land and in general nature, has a sacred quality which is almost absent from Western thinking. Land is revered and respected and its inalienability is reflected in virtually every indigenous cosmovision. Indigenous people do not consider the land as merely an economic resource. Under indigenous cosmovisions, nature is the primary source of life that nourishes, supports and teaches. Nature is, therefore, not only a productive source but the center of the universe, the core of culture and the origin of ethnic identity. At the heart of this deep bond is the perception that all living and non-living things and natural and social worlds are intrinsically linked (reciprocity principle). Of particular interest is the research done by several authors (Reichel-Dolmatoff, E. Boege, Ph. Descola, C. van der Hammen) on the role played by the cosmology of several indigenous groups as a mechanism regulating the use and management of natural resources. In the indigenous cosmovision each act of appropriation of nature must be negotiated with all the existing things (living and non-living) through different mechanisms as agrarian rituals and shamanic acts (symbolic exchange). Humans are thus seen as a particular form of life participating in a wider community of living beings regulated by a single and totalizing set of rules of conduct.

The Corpus Indigenous societies house a repertory of ecological knowledge which generally is local, collective, diachronic and holistic. In fact, since indigenous peoples possess a very long history of resource-use practice, they have generated cognitive systems on their own circumscribed natural resources which are transmitted from generation to generation. The transmission of this knowledge is done through language, hence the corpus is generally an unwritten knowledge. Memory is, therefore, the most important intellectual resource among indigenous cultures. This body of knowledge is the expression of a certain personal wisdom and, at the same time, of a collective creation, it is to say, a historical and cultural synthesis turned into reality in the mind of a individual producer. For this reason, the corpus contained in a single producer's mind expresses a repertoire that is a synthesis of information from at least four sources: (a) the experience accumulated over historical time and transmitted from generation to generation by a certain cultural group; (b) the experiences socially shared by the members of a same time's generation or cohort; (c) the experience shared into the household or the domestic group to which the individual belongs; and (d) the personal experience , particular to each individual, achieved through the repetition of the annual cycles (natural and productive), enriched by the perceived variations and unpredictable conditions associated with them.

Thus, indigenous ecological knowledge is normally restricted to the immediate environments and is an intellectual construction resulting from a process of accumulation of experiences over both the historical time and the social space. These three main features of indigenous ecological knowledge being local, diachronic and collective are complemented with a fourth characteristic, namely holistic.

Indigenous knowledge is holistic because it is intrincately linked to the practical needs of use and management of local ecosystems. Although indigenous knowledge is based on observations on a rather restricted geographic scale, it must provide detailed information on the whole scenery represented by the concrete landscapes where natural resources are used and managed. As a consequence, indigenous minds not only possess detailed information about species of plants, animals, fungi and some microorganisms; they also recognize types of minerals, soils, waters, snows, landforms, vegetations and landscapes.

Similarly, indigenous knowledge is not restricted to the structural aspects of nature, which are related to the recognition and classification (ethnotaxonomies) of elements or components of nature, it also refers to dynamics (which refers to patterns and processes), relational (linked to relationships between or among natural elements or events) and utilitarian dimensions of natural resources. As a result, it is possible to integrate a cognitive matrix (Figure 5) which certifies the holistic character of indigenous knowledge and serves as a methodological framework to ethnoecological research (Toledo, 1992).

The Praxis Indigenous societies generally subsist by appropriating a diversity of biological resources from their immediate vicinity. Thus, subsistence of indigenous peoples is based more on ecological exchanges (with nature) than on economic exchanges (with markets). They are therefore forced to adopt survival mechanisms that guarantee an uninterrupted flow of goods, materials, and energy from ecosystems. In this context a predominant use-value economic rationality is adopted, which in practical terms is represented by a multi-use strategy that maximizes the variety of goods produced in order to provide basic household requirements throughout the year (for further details on this strategy see Toledo, 1990). This main feature accounts for the relatively high self-sufficiency of indigenous households and communities.

Indigenous households tend to carry out a non-specialized production based on the principle of diversity of resources and practices. This mode of subsistence results in the maximum utilization of all the available landscapes of the surrounding environments, the recycling of materials, energy and wastes, the diversification of the products obtained from ecosystems and, especially, the integration of different practices: agriculture, gathering, forest extraction, agroforestry, fishing, hunting, smallscale cattle-raising, and handicrafts. As a result, indigenous subsistence implies the generation of a myriad of products including food, domestic and work instruments, housing materials, medicines, fuelwoods, fibers, animal forage, and others. Under the multi-use strategy, indigenous producers manipulate the natural landscape in such a way that two main characteristics are maintained and favored: habitat patchiness and heterogeneity and biological as well as generical variation. In the spatial dimension, indigenous become a complex landscape mosaic in which agricultural fields, fallow areas, primary and secondary vegetation, household gardens, cattle-raising areas, and water bodies are all segments of the entire production system. This mosaic represents the field upon which indigenous producers, as multi-use strategists, play the game of subsistence through the manipulation of ecological components and processes (including forest succession, life cycles, and movement of materials).

It has been demostrated that some natural disturbances can increase biodiversity if they increase habitat heterogeneity, reduce the influence of competitively dominant species, or create opportunities for new species to invade the area On the other hand, number of species is commonly relatively small in highly disturbed biotic communities, because few populations are able to re-establish themselves before they are reduced by later disturbances. In contrast, a low rate of disturbance provides few opportunities for pioneer species and might allow competitively dominant species to usurp limiting resources. Therefore, biodiversity is often greater at intermediate levels of disturbances than either lower or higher rates.

The creation of landscape mosaics under the indigenous multi-use strategy in areas originally covered by only one natural community represents a human-originated mechanism which theoretically tends to maintain (and even increase) biodiversity. Several authors have already stressed the importance of the models of low intensity mosaic usage of the landscape by indigenous peoples and other small-landowner populations for biodiversity conservation. The same diversified arrangement found in indigenous landscapes tends to be reproduced at a micro-level, with multi-species, multi-story crops or agroforests favored over monocultures. As a consequence, animal and especially plant genetic resources tend to be maintained in indigenous agricultural fields, aquaculture systems, homegardens and agroforests (Gadgil, et al 1993). Polycultural systems managed by indigenous agriculturalists and agroforesters are relatively well known and the recent specialized literature is plenty of case studies illustrating such designs. Especially notable are the homegardens and agroforestry systems of the tropical and humid regions of the world, which operate as human-made refuge areas for many species of plants and animals, notably in areas strongly affected by deforestation.

At farm level, it is broadly recognized that crop populations are more diverse in indigenous farming systems than in agricultural areas dominated by agroindustriality. Therefore, indigenous peoples are recognized as key agents of on-farm preservation of plant genetic resources threatened by agricultural modernization (genetic erosion). The loss of biodiversity is also experienced in farming systems as indigenous cropping polycultural patterns are replaced by fossil-fueled monocrops. Indigenous agricultural systems and landscapes are then acknowledged as designs that preserve not only landraces of crop species, but semi domesticated and wild crops relatives and even non domesticated species.

Conserving Biodiversity by Empowering Indigenous Peoples

During the past three decades, as the loss of landscapes, habitats, species and genes, has become an issue of international concern, the protected areas of the world have increased notably both in size and number. However, as protected areas expanded, it became evident that the North originated model of uninhabited national parks could not be applied worldwide. Today, there are just nearly 10,000 nationally protected areas (parks and other reserves) in more than 160 countries, covering some 650 million of hectares, which represents over 5 percent of earth's land surface. Many of the areas that have been established as protected areas and many of those that are suitable for future addition to the protected area network are the homelands of indigenous peoples. In Latin America alone, over 80 per cent of protected areas are estimated to have indigenous people living within them. On the other hand, large tracts of the territories under indigenous control, estimated in between 12 and 20 per cent of the earth's surface, are in the scope of conservationists as future reserves. Morever, some authors like B. Nietschmann and J. Alcorn (1994) think the bulk of the world's biodiversity is embodied within the limits of the indigenous territories of the tropical countries. Given the above, as well as the evidences offered and discussed in the previous sections, the idea that biodiversity conservation is impossible without the participation of indigenous communities is increasingly gaining recognition in national and international conservation circles. For example, in its latest guidelines, IUCN's Commission on National Parks and Protected Areas (1994) consider that indigenously established "protected territories" can now be recognized as national parks, wilderness areas, protected landscapes and managed resource protected areas. On the other hand, the international conservation community is beginning to realize that sacred forests, mountains, lakes, rivers, and deserts can be considered protected areas, as well as managed reefs, lagoons, rivers and grasslands.

Protected areas based on consultation, co-management and even indigenous management, are likely to be increasingly important in coming years as the key role of indigenous cultures is being gradually recognized. It is important, however, not to over-idealize indigenous peoples and their resource management strategies and stewardhip skills. Conservationists have been frequently criticized for over-romanticize indigenous peoples, creating a late-twentieth-century version of "the noble savage". Acknowledgment of the positive links between indigenous peoples and biodiversity has been increasingly tempered by the recognition that under certain circumstances (high population densities, market pressures, unsuitable technologies, local disorganization) indigenous peoples can act as disruptive, not as conservationist, actors.

Biological diversity and sustainable development, are today two of the most powerful and central concepts in environmental protection. In recent years, special attention is being payed to the sustainable development of community-based peoples, as a key mechanism for the reinforcement of correct participation of local communities in biodiversity conservation. It is possible to define sustainable community development as an endogenous mechanism that allows a local society to lake (or retake) control of the processes that affect it. In other words, self-determination and local empowerment, conceived as a "taking of control", have to be the central objectives in all community development. Given the demostrated importance of indigenous peoples for biodiversity conservation, it is essential to recognize the necessity of empowering local communities. That is to maintain, reinforce or give control to the indigenous communities on their own territories and natural resources as well as sufficient access to information and technology. Important here are legally recognized and enforceable rights to lands and waters, which give the communities both an economic incentive and a legal basis for stewardship. In many countries, national recognition and policy support for existing, community-based property rights systems are crucial. In many Asian and African countries, returning a measure of control over public lands and resources to local communities is also fundamental to slowing biodiversity loss in threatened regions.

Similarly, it is very important to establish new resource-management partnerships between local communities and the state and other society institutions to maintain biodiversity. Local stewardship in conjunction with external governmental and non-governmental agencies and institutions is perhaps the best way to guarantee effective protection of landscapes, habitats, species and genes worldwide, and specially in tropical countries.

Local Vegetation Management

Over thousands of years local people have developed a variety of vegetation management practices that continue to exist in tropical Asia (Pandey, 1998), South America (Atran *et al.*, 1999; Gomez-Pompa and Kaus, 1999), Africa (Getz *et al.*, 1999; Infield, 2001), and other parts of the world (Brosius, 1997; Berkes, 1999). People also follow ethics that often help them regulate interactions with their natural environment (Callicott, 2001). Such systems are often integrated with traditional rainwater harvesting that promotes landscape heterogeneity through augmented growth of trees and other vegetation, which in turn support a variety of fauna (Pandey, 2002).

In India these systems can be classified in several ways:

- Religious traditions: temple forests, monastery forests, sanctified and deified trees
- Traditional tribal traditions: sacred forests, sacred groves and sacred trees
- Royal traditions: royal hunting preserves, elephant forests, royal gardens etc.
- Livelihood traditions: forests and groves serving as cultural and social space and source of livelihood products and services

The traditions are also reflected in a variety of practices regarding the use and management of trees, forests and water. These include:

- Collection and management of wood and non-wood forest products
- Traditional ethics, norms and practices for restraint use of forests, water and other natural resources
- Traditional practices on protection, production and regeneration of forests.
- Cultivation of useful trees in cultural landscapes and agroforestry systems
- Creation and maintenance of traditional water harvesting systems such as tanks along with plantation of the tree groves in the proximity

These systems support biodiversity, which is although less than natural ecosystems but it helps reduce the harvest pressure. For instance, there are 15 types of resource management practices that result in biodiversity conservation and contribute to landscape heterogeneity in arid ecosystems of Rajasthan. Environmental ethics of Bisnoi community suggest compassion to wildlife, and forbid felling of *Prosopis cineraria* trees found in the region. Bisnoi teachings proclaim: "If one has to lose head (life) for saving a tree, know that the bargain is inexpensive" (Pandey, 2002).

In India, local practices of vegetation management perhaps emanate from the basic ecological concepts of local communities reflected in "ecosystem-like concepts in traditional societies" (Berkes *et al.* 1998). Two key characteristics of these systems are that the unit of nature is often defined in terms of a geographical boundary; and abiotic components, plants, animals, and humans within this unit are considered to be interlinked. Many local knowledge systems are similar in temperament to the emerging scientific view of ecosystems as unpredictable and uncontrollable, and of ecosystem processes as nonlinear, multiequilibrium, and full of surprises (Berkes *et al.* 1998).

Biodiversity in Sacred Cliffs

Cliffs are completely forgotten cultural landscape elements that support a variety of species of plants and animals in India.

As humans have special fascinations to such areas often cliffs across the country are considered sacred. Cliffs elsewhere have been found to support undisturbed ancient woodland, dominated by tiny, slow-growing and widely spaced trees. Vertical cliffs often support populations of widely spaced trees that are exceptionally old, deformed and slow growing. Some of the most ancient and least-disturbed wooded habitats on Earth are found on cliffs, even if such sites are close to intensive agricultural and industrial development. The age of the trees on cliffs may indicate the age and growth rates of the entire plant communities on the cliffs. Cliffs across the world may support ancient, slow-growing, open woodland communities that have escaped major human disturbance, even when they are situated close to agricultural and industrial activity, which has destroyed or altered most other natural habitats (Larson *et al.*, 1999, 2000a & b; Peterken, 1996). Examples of such habitat in India abound. Cliffs in Udaipur and Kota districts of Rajasthan were surveyed (7 cliff with ancient vegetation). Cliffs were found to have more than 25 species of trees, several species of shrubs and herbs. Areas close to Bhopal have more than 50 cliffs in central India in a radius of about 100 kms. All the 7 cliffs surveyed in Rajasthan are sacred. They are often part of the sacred corridors along the riverbank escarpment with several meters of precipitous fall. Attempts have been made to regenerate the Gaipernath Cliff with the traditional species occurring in the area (*Lannea coromandelica, Boswellia serrata, Sterculia urens* etc. about 25 species). The result was very poor initially. But local ethnoforestry techniques of tucking the branch cuttings of coppicing species in whatever little crevices area may have were successful. Also, depositing the seeds same species that occur in crevices with the ball of moist earth has been found promising.

Farm Biodiversity

Throughout the Indian farms and field one finds strips of vegetation containing several species of plants and small animals. These strips are beneficial in several ways. Such strips on tropical lands have been found to accelerate natural successional processes by attracting seed-dispersing animals and increasing the seed rain of forest plants. Effects of these strips resemble the windbreaks on seed deposition patterns (Harvey, 2000). Isolated trees provide seed in the area for natural regeneration. The strips enhance seed rain, and connectivity. Because such strips trap large number of seeds of several species they help in further tree growth. Compared to open fields, farm boundaries with vegetation receive seed in greater densities and species-richness than open farms and pastures. All forms of seed dispersal help in the process but animal-dispersed (birds, bats, mammals etc.) seeds often occur in greater densities and species numbers. Presence of isolated trees and shrubs or remnant trees helps. Farm boundaries maintained throughout the country are often self regenerating and require only management as these barriers considerably increase the deposition of tree and shrub seeds within the cultural landscape. Indeed considerable biodiversity is found within these strips. This is a practice that needs to be maintained as it has several socio-economic benefits as well.

Value of traditional agroecosystems in supporting the plant and animal diversity (Kunte *et al.* 1998) is immense. Tree diversity in farms and agroecosystems is often the product of interaction of local and formal knowledge. A recent study by Shastri et al. (2002) provides interesting insights on the tree-growing practices and associated biodiversity in Karnataka. Shastri *et al.* (2002) found trees belonging to 93 species in a sampled area of 1.7 ha of Sirsimakki agro-ecosystem. Additional 44 species were noted on non-agricultural lands in the village ecosystem, which included *soppina betta*, minor forest and reserve forest. The overall agroecosystem had 556 trees/ha, while the non-agroecosystem had only 354 trees/ha. The overall, tree density of 418.8 per ha was present in the village. There were 144 species in the village ecosystem with 2238 individuals in the sampled area of 5.34 ha. The total number of species in non-agro ecosystem was 104 with 1286 individuals. Home-gardens are notable with 93 tree species in just about 1.7 ha. The number of tree species varies between 20 and 40 in home-gardens, indicating that home-gardens in Karnataka villages are highly biodiverse in comparison to those in Mexico and Brazil (Shastri *et al.* 2002).

Farms themselves have domesticated biodiversity essential for survival and subsistence. One such example is by Kimata *et al.* (2000) form South India on the cultivation and process of domestication of *Brachiaria ramosa* cultivated in pure stands. Its grains are used in nine traditional food preparations in South India. Another crop *Setaria glauca* is cultivated in mixed stands along with little millet (*Panicum sumatrense*). In Orissa state and in Southern India the grains are used to make at least six traditional supplementary foods. The weedy forms of these species were found by the researchers growing with upland rice and some millets in diverse agro-ecological niches. The domestication process is supposed to have gone through three phases: first growing in association with weed and with upland rice and other millets; a secondary crop mixed with *kodo* millet; and finally as an independent crop.

Cultivation of Medicinal Plants

There are numerous examples of medicinal plant cultivation by local people in India. Socio-culturally valued species find place in home gardens and courtyards. For example, Around the Nanda Devi Biosphere Reserve in the western Himalaya, the Bhotiya community, whose livelihood is depends on local natural resources, practices seasonal and altitudinal migration and stay inside the buffer zone for only 6 months (May-October). A survey in 5 villages in Pithoragarh District, found that Bhotiya people cultivate medicinal plants on their agriculture fields. Of a total of 71 families, 90% cultivated medicinal plants on 78% of the total reported cultivated area (15.29 ha). Around 12 species of medicinal plants were under cultivation. Survey also found that a family earned about Rs.2423 +/- 376.95 per season from the sale of medicinal plants in 1996 (Rs.38 = US$1 in 1996). Thus, supporting medicinal plant cultivation at high altitudes in the Himalayas may help to generate additional support to people as well as conserve the species in the wild (Silori and Badola, 2000, see also, Maikhuri *et al.* 1998). Another study (Satyal *et al.* 2002) on traditional knowledge of Kumaun Higher Himalaya found that Bhotia tribes use 34 species of medicinal plants native to the region. Among these, *Angelica glauca* and *Allium stracheyi* are narrow range endemic and *Allium stracheyi*, *Picrorhiza kurrooa* and *Nardostachys grandiflora* have been recorded in the Red Data Book of Indian Plants. Interestingly, the annual production of medicinal plants has been found to be comparable with the annual production of traditional crops. Thus, cultivation, and harvesting can help in livelihood security and *in situ* conservation of these species.

Similarly, juang and Munda tribes of the Keonjhar district of eastern India use 215 plants, belonging to 150 genera and 82 families (Mahapatra and Panda 2002). This suggests a wealth of traditional knowledge on biodiversity and herbal health care in tribes of eastern India. Tribes in the region are dependent on forests for other species as species of mushrooms, wild berries, tubers, and flowers that are included in their diet including cooking oil. Understanding of traditional knowledge on biodiversity of the region will be most helpful in planning for sustainable forest management.

Traditional Ethos

Similarly, in spite of the modernization, traditional ecological ethos continue to survive in many other local societies, although often in reduced forms. Investigations into the traditional resource use norms and associated cultural institutions prevailing in rural Bengal societies (Deb and Malhotra, 2001) demonstrate that a large number of elements of local biodiversity, regardless of their use value, are protected by the local cultural practices. Some of these may not have known conservation effect, yet may symbolically reflect, a collective appreciation of the intrinsic or existence value of life forms, and the love and respect for nature. Traditional conservation ethics are still capable of protecting much of the country's decimating biodiversity, as long as the local communities have even a stake in the management of natural resources.

Traditional ethos is reflected in a variety of practices including sacred groves and sacred landscapes. They are fairly well described.

One example from northeast India is particularly notable (*et al.* 1998). The tribal communities of Meghalaya-Khasis, Garos, and Jaintias-have a tradition of environmental conservation based on various religious beliefs. As elsewhere in India, particular patches of forests are designated as sacred groves under customary law and are protected from any product extraction by the community. Such forests are very rich in biological diversity and harbor many endangered plant species including rare herbs and medicinal plants. Tiwari *et al.* (1998) identified 79 sacred groves and their floristic survey revealed that these sacred groves are home to at least 514 species representing 340 genera and 131 families. The status of sacred groves was ascertained through canopy cover estimate. About 1.3% of total sacred grove area was undisturbed, 42.1% had relatively dense forest, 26.3% had sparse canopy cover, and 30.3% had open forest. Notably, the species diversity indices were higher for the sacred grove than for the disturbed forest.

Another notable example is from peninsular India. Study (Ramanujam and Kadamban 2001) on two sacred groves, Oorani and Olagapuram, situated on the north-west of Pondicherry found a total of 169 angiosperms from both sites. The Oorani grove (3.2 ha) had 74 flowering plant species distributed in 71 genera and 41 families; 30 of them are woody species, 8 are lianas and 4 are parasites. The Olagapuram grove (2.8 ha) was more species-rich with 136 species in 121 genera of 58 families; woody species were fewer (21) while 9 lianas and 3 parasites occurred. Associated local knowledge, cultural and religious rituals of local people sustain such diversity.

Another tradition worth mention is use of plants in mural painting. Such paintings are found, for example, in the Ajantan mural art. The practice spanned a whole millennium from the second century B.C. to the eighth century A.D. The tradition continued up to the nineteenth century under the support of different dynasties in India, but declined by the end of that century. Nayar *et al.* (1999) note that the art is kept alive by a few artists in Kerala who practice even today the methods and techniques of mural paintings similar to those practiced by the Ajantan mural painters. Various plant species provided materials for mural painting. Such knowledge can be very helpful in providing livelihood security to practitioners.

Traditional water harvesting structures too are also habitat for a variety of species. Even if pond size is small, as is the case in about 60% (out of 1.5 million total tanks) in India (Pandey, 2001) it may still be useful habitat for many species in rural ecosystems. Indeed, the island biogeography theory - valid in numerous cases–suggesting that larger areas support more species did not stand in case of 80 ponds in Switzerland (Oertli *et al.*, 2002).

Theoretical predictions and empirical support suggests that although intentional[4] conservation may be rare among small-scale societies as Smith and Wishnie (2000) have pointed out, but practices that actually result in what we today call 'sustainable use and management' of resources and habitats by local people is widespread globally that contribute to in biodiversity conservation and enhancement through creation of habitat mosaics (Smith and Wishnie, 2000).

Formal conservation efforts in India have relied heavily on the recently declared official protected areas in various categories for biodiversity conservation. However, ancient and widespread human practice to set aside areas for the preservation of natural values in India can be seen in several examples of sacred groves, royal hunting forests, and sacred gardens (Gadgil 1982, Pandey, 1991; Gadgil *et al.*, 1993; Kanowski *et al.*, 1999; Chandrashekara and Sankar, 1998). Several of these areas became national parks and wildlife sanctuaries in India and elsewhere (Pandey, 2001). It must be noted here that much of the India's biodiversity lies outside the officially declared protected areas. Indeed, biodiversity occurs in landscape continuum (Figure 1; Table 1 & 2). Other areas protect ecosystem services such as the delivery of clean water or the supply of timber, or mitigate the expected adverse effects of over-clearing (Grove, 1992). Others protect recreational and scenic values and some have been planned to foster international cooperation (Hanks, 1997). Many of these areas meet the World Conservation Union's definition of a strictly protected area (IUCN categories I-IV) (IUCN, 1994).

In view of accelerating biological and cultural landscape degradation, a better understanding of interactions between landscapes and the cultural forces driving them is essential for their sustainable management. We need environmental and cultural revolution, aiming at the reconciliation of human society with nature (Naveh, 1995).

Traditional Knowledge, Water, and Biodiversity

Simple local technology and an ethic that exhorts "capture rain where it rains" have given rise to 1.5 million traditional village tanks, ponds and earthen embankments that harvest substantial rainwater in 660,000 villages in India (Pandey, 2001a), and encourage growth of vegetation in commons and agroecosystems. If India were to simply build these tanks today it would take at least US $ 125 billion (Pandey, 2002).

Humans have virtually appropriated fresh water. Humanity now uses 26 percent of total terrestrial evapotranspiration and 54 percent of runoff that is geographically and temporally accessible. New dam construction could increase accessible runoff by about 10 percent over the next 30 years, whereas population is projected to increase by more than 45 percent during that period (Postel *et al.*, 1996).

Over thousands of years societies have developed a diversity of local water harvesting and management regimes that still continue to survive, for example, in South Asia, Africa, and other parts of the world (Agarwal and Narain, 1997). Such systems are often integrated with agroforestry (Wagachchi and Wiersum, 1997) and ethnoforestry practices (Pandey, 1998). Recently it has been suggested that market mechanisms for sustainable water management such as taxing users to pay commensurate costs of supply and distribution and of integrated watershed management and charging polluters for effluent treatment can solve the problem

(Johnson *et al.*, 2001). Such measures are essential although, but they are insufficient and would need to draw on the local knowledge on rainwater harvesting across different cultures (Pandey, 2001).

Rainwater harvesting in South Asia is different from other parts of the world in that it has a continued history of practice for at least over 5000 years. Similarly, Balinese water temple networks as complex adaptive systems are also very useful systems (Falvo 2000). Although hydraulic earthworks are known to have occurred in ancient landscapes in many regions, they are no longer an operational systems among the masses in the same proportion as in South Asia. For instance, remains of earthworks and water storage adaptations are found in Mayan lowlands in South America (Mann, 2000). Such systems had been used for prehistoric agriculture in Mayan lowlands (Turner, 1974; Coe, 1979), and for fish culture in Bolivian Amazon (Erickson, 2000).

Rainwater harvesting have been found to be scientific and useful for rainfed areas (Li *et al.*, 2000). For instance, a validation comes from the Negev. Ancient stone mounds and water conduits are found on hillslopes over large areas of the Negev desert. Field and laboratory studies suggest that ancient farmers were very efficient in harvesting water. A comparison of the volume of stones in the mounds to the volume of surface stones from the surrounding areas indicates that the ancient farmers removed only stones that had rested on the soil surface and left the embedded stones untouched. According to results of simulated rainfall experiments, this selective removal increased the volume of runoff generated over one square meter by almost 250% for small rainfall events compared to natural untreated soil surfaces (Lavee *et al.*, 1997).

One of the principle tree genus growing in association with tanks and ponds in India is *Ficus* which is culturally valued throughout the country. It is a keystone genus and supports a variety of other species. Records of frugivory from over 75 countries for 260 *Ficus* species (approximately 30% of described species) suggest that in addition to a small number of reptiles and fishes, 1274 bird and mammal species in 523 genera and 92 families are known to eat figs (Shanahan *et al.* 2001).

Conservation Principles in Ancient Texts

Natural Resource Management has been in the traditions of the Indian society, expressing itself variously in the management and utilization practices. This evolved through the continued historical interaction of communities and their environment, giving rise to practices and cultural landscapes such as sacred forests and groves, sacred corridors and a variety of ethnoforestry practices. This has also resulted in conservation practices that combined water, soil and trees. Nature-society interaction also brought about the socio-cultural beliefs as an institutional framework to manage the resultant practices arising out of application of traditional knowledge. The attitude of respect towards earth as mother is widespread among the Indian society.

Local knowledge has proved useful for forest restoration and protected area management in Rajasthan – one of the driest regions of India with scanty rainfall. Cultural landscapes in rural and urban areas and agroecosystems, created by the application of scientific and local knowledge, also support a variety trees, birds and other species, and provide opportunity of integration of nature and society (Taylor, 2002).

Ancient texts make explicit references as to how forests and other natural resources are to be treated. Sustainability in different forms has been an issue of development of thought since ancient times. For example, robust principles were designed in order to comprehend whether or not the intricate web of nature is sustaining itself. These principles roughly correspond with modern understanding of conservation, utilization, and regeneration.

Conservation Principles: Atharva Veda (12.1.11) hymn, believed to have been composed sometime at around 800 BC, somewhere amidst deep forests reads: "O Earth! Pleasant be thy hills, snow-clad mountains and forests; O numerous coloured, firm and protected Earth! On this earth I stand, undefeated, unslain, unhurt." Implicit here are the following principles:

- It must be ensured that earth remains forested.
- It must be understood that humans can sustain only if the earth is protected.
- To ensure that humans remain 'unslain' and 'unhurt', the ecosystem integrity must be maintained.
- Even if vaguely, it also makes reference to ecology, economy and society concurrently.

Utilization and Regeneration Principles: Another hymn from Atharva Veda (12.1.35) reads: "Whatever I dig out from you, O Earth! May that have quick regeneration again; may we not damage thy vital habitat and heart". Implicit here are the following principles:

- Human beings can use the resources from the earth for their sustenance,
- Resource use pattern must also help in resource regeneration,
- In the process of harvest no damage should be done to the earth,
- Humans are forewarned not against the use of nature for survival, but against the overuse and abuse.

Although not in modern terminology, the three segment of sustainability-ecology, economy and society seem to get addressed simultaneously. Similarly, water management and associated tree growing has been the subject of ancient text. Tanks have been the most important source of irrigation in India. Some tanks may date as far back as the *Rig Vedic* period, around 1500 BC. The *Rig Veda* refers to lotus ponds (*5.78.7*), ponds that give life to frogs (*7.103.2*) and ponds of varying depths for bathing (*10.71.7*). Reference to the tanks is also found in the *Arthashastra* of Kautilya[5] written around 300 BC (Rangarajan 1987: 231-233). The *Arthashastra* refers to the ownership and management of the village tanks in the following verses:

Waterworks such as reservoirs, embankments and tanks can be privately owned and the owner shall be free to sell or mortgage them (*3.9.33*)[6]. The ownership of the tanks shall lapse, if they had not been in use for a period of five years, excepting in case of distress (*3.9.32*). Anyone leasing, hiring, sharing or accepting a waterworks as a pledge, with a right to use them, shall keep them in good condition (*3.9.36*). Owners may give water to others in return for a share of the produce grown in the fields, parks or gardens (3.9.35). In the absence of owners, either charitable individuals or the people in village acting together shall maintain waterworks (*3.10.3*). No one will sell or mortgage, directly or indirectly, a bund or embankment built and long used as a charitable public undertaking except when it is in ruins or has been abandoned (*3.10.1,2*).

The earliest scholar to have commented on the relationship of tanks and trees is Varahamihira who described the detailed technical instructions for the tank constructions in his famous work *Brahatsamhita* (550 AD): Without the shade of the trees on their sides, water reservoirs do not look charming; therefore, one ought to plant the gardens on the banks of the water (*55.1*)[7] Commenting on the species to be planted on the embankments of the tank, after its construction, Varahamihira writes: The shoreline (banks) of the tanks should be shaded (planted) with the mixed stands of Arjun (*Terminalia arjuna*), Vata (*Ficus benghalensis*), Aam (*Mangifera indica*), Pipal (*Ficus religiosa*), Nichul (*Nauclea orientalis*), Jambu (*Syzygium cuminii*), Vet (*Calamus?*), Neep (*Mitragyna parvifolia*), Kurvak (?), Tal (*Borassus flabellifer*), Ashok (*Saraca asoka*), Madhuk (*Madhuca indica*), and Bakul (*Mimusops elengi*) (*54.119*).

For example, there is a considerable overlap in the formal and scientific forestry policy and practice, which provides hope that traditional knowledge systems can contribute to the management of natural resources. It would be pertinent to quote Gadgil and Guha (1992: 51) in this context: "Indeed one could argue that scientific prescriptions in industrial societies show little evidence of progress over the simple rule-of-thumb prescriptions for sustainable resource use and the conservation of diversity which characterized gatherer and peasant societies. Equally, the legal and codified procedures which are supposed to ensure the enforcement of scientific prescriptions work little better than earlier procedures based on religion or social convention".

Incorporating Traditional Knowledge in Practice

Any attempt, endeavouring to integrate traditional knowledge for biodiversity conservation and sustainability of natural resources should be based on the principle that traditional knowledge often cannot be dissociated from its cultural and institutional setting. Regarding the cultural and institutional the following suggestions may be useful:

1. Each programme aiming at the promotion of traditional knowledge should be based on the recognition that natural resource rights and tenurial security of local communities forms the fundamental basis of respecting traditional knowledge.
2. More attention is needed on protection of intellectual property rights of traditional people.
3. Innovative projects may need to be developed that aim at the enhancement of the capacity of local communities to use, express and develop their traditional knowledge on the basis of their own cultural and institutional norms.

There is an urgent need for the integration of Traditional and formal sciences. Following considerations may be useful in this regard:

1. Development of methods for mutual learning between local people and the formal scientists.
2. State forest policies and sustainable forest management processes need to give full attention to ethforestry and local institutional arrangements to incorporate traditional knowledge in forest management and development projects.
3. Traditional knowledge and traditions can contribute to the preparation of village microplans, which are prepared for eco-development, joint forest management and rural development. The plans should be

based on both geographic and traditional community boundaries rather than only on administrative boundaries.

4. Revival of the traditional water management systems that have served the society for hundreds of years but are currently threatened
5. There is a clear need to integrate traditional and formal sciences for participatory monitoring, and taking feedback to achieve adaptive strategies for management of natural resources.

In spite of the value of traditional knowledge for biodiversity conservation and natural resource management there still is a need to further the cause. The following consideration may be useful in this respect:

1. Encouraging the documentation of indigenous knowledge and its use in natural resource management. Such documentation should be carried out in participation with the communities that hold the knowledge. Due attention should be given to document the emic perspectives regarding IK rather than only the perspectives of professional outsiders. The documentation should not only consist of descriptions of knowledge systems and its use, but also information on the threats to its survival. People's biodiversity registers are a case in point (Gadgil 1994 & 1996, Gadgil *et al.* 2000). The program of People's Biodiversity Registers promotes folk ecological knowledge and wisdom by devising a formal means for their maintenance, and by creating new contexts for their continued practice. PBRs document traditional ecological knowledge and practices on use of natural resources, with the help of local educational institutions, teachers, students and NGOs working in collaboration with local, institutions. Such a process and the resulting documents, could serve a significant role in "promoting more sustainable, flexible, participatory systems of management and in ensuring

a better flow of benefits from economic use of the living resources to the local communities" (Gadgil *et al.* 2000).

2. Facilitating the translation of available and new documents describing Indic traditions such as ancient texts on medicinal plants, into local languages and dissemination of these documents amongst local people. Such a translation is indeed required because texts are often available in languages (e.g. Sanskrit) not understood by many in contemporary India. On the other hand, translation of local knowledge into formal scientific terminology will provide space to external researchers, policy makers, and practitioners to comprehend and support people's knowledge systems and initiatives.
3. Facilitating the exchange of information amongst practitioners of local knowledge.
4. Developing clear and concise educational material on traditional knowledge systems to be used in communication programmes to impart information regarding the merits and threats to indigenous knowledge systems to both policy makers and the general public.

Scientific institutions have an important role to play in supporting the knowledge systems. As has been pointed out earlier, it is now recognised that a dichotomy between local and formal systems of knowledge is not real, and that any knowledge is based on a set of basic values and beliefs and paradigms. Therefore, there is a definite need to further develop systematic insight into the nature and scope of traditional knowledge.

15

Biodiversity Education

Biodiversity Education Methods

Life on Earth is extraordinarily diverse.

From microscopic bacteria to mammoth whales, from firmly rooted redwoods to highly mobile humans, from marshy wetlands to the high chaparral, life on Earth is diverse, at the level of species as well as at the level of ecosystems. In Biodiversity Counts: Students will look closely at organisms, searching for answers that relate to their class site. They will investigate such questions as: What is out there?

- How many species of living things are there?
- How are they related?
- What is the life history of an organism?
- W h e re does it live?
- How does it obtain nutrients?
- What does it do?
- What is its role in the habitat?
- How does it reproduce?
- How does it defend itself?
- How are plants and arthropods adapted to the habitats at our site?

Species are the entities we observe, describe, and discuss when we study biodiversity Species are genetically distinct entities with discrete histories. They retain their genetic distinction because, under most circumstances in nature, a member of a particular species can breed only with another member of the same species. The genetic code that distinguishes a species usually determines a set of unique characteristics that enables us visually to tell one species from another. Each species is somehow distinct in form and has a unique evolutionary history that is not shared with any other species. Although 1.75 million species have been described so far, scientists estimate that as many as 10 million distinct species exist on Earth today.

In Biodiversity Counts: Students will emulate the work of systematic biologists and conservation biologists to identify plants and arthropods at their site. Living things are bound together in intricate, interconnected webs. The natural world is a vast network of complex communities, connected through shared environments. The interconnected webs serve to sustain biodiversity. For example, certain flowering plants depend on bees for pollination. The bees depend on the flowers for their food i.e. nectar or pollen. After pollination, the plants produce fruits, receptacles for the seeds of the next generation. Birds are attracted by the bright color and sweet aroma, and they eat the fruits. The seeds pass through the birds' digestive system and are dispersed in their droppings, and new plants grow, often some distance away from the parent plants. Beetles may eat their way into the same fruit. Small rodents gnaw into the fruits searching for beetles. The rodents themselves may be prey for birds, reptiles, and larger mammals. If one component is removed – if the bees are killed by insecticides, for example reverberations are felt throughout the network. Without bees to transport pollen, plants cannot produce seeds to ensure a new generation. Without seeds and their fruity packaging, there is no food for other wildlife. In Biodiversity Counts: Students will have opportunities to observe the complex ecological relationships among the different living things at their site.

Human activity on the planet has had a profound effect on the diversity of species. Biological diversity is threatened by an array of human factors, including overpopulation, loss and fragmentation of habitat, introduction of nonnative, or exotic species, overexploitation of commercially valuable species, pollution and contamination due to pesticides and acid rain, and changes of global dimensions such as warming and the depletion of the ozone layer. The result is that species loss seems to be occurring at a higher rate than ever before as a result of human activity.

In Biodiversity Counts: Students will examine and assess the impact of human activity on their local site. The causes of biodiversity loss are complex, and the consequences are equally complex, extensive, and long term. In order to make sound political, economic, and ecological decisions, we will need to know more about what species exist, where and how species live, how they are related, and how they interact both with each other and with their environments. There is hope, and there are positive steps we can take. To begin, we must stabilize population growth, reduce our consumption and waste, develop environmentally clean technologies, and limit the harvesting of our natural resources. In Biodiversity Counts: Students can learn more about the body of knowledge that exists about biodiversity. It is hoped that their hands-on experience at the site will increase their appreciation and their feeling of stewardship for the natural world, first for their home site and then for the world as a whole.

Going Outdoors

Depending on location, students may or may not be accustomed to going outdoors for part of their academic day. In most cases, a trip outside generates a lot of excitement, and while it is important that students enjoy the experience, it is also important in this initial foray to set a purposeful tone, to establish some routines, and to develop safety rules. It is also a good idea to enlist the help of other adults to supervise students outdoors. You may be able to find support staff in the school, or among parents or college students who are willing to help you manage field trips. This first outdoor task is less directed than future tasks. There is, however, a stated objective: Students are to observe, record, and then interpret their impressions of living things at the site. The purpose of the activity is to provide the class with a vehicle for contrasting this type of impressionistic interpretation with the more factual and objective types to come. The activity also points out the need for accurate record keeping in science, since memory is not always a reliable tool. Field trip lessons and field trip activities are to be identified: Each field trip has specific tasks and focus questions for students to help guide them in their field work. The tasks and focus questions are organized in matrices within the lessons and as single sheets that you can copy and take out into the field.

Enlisting the Help of Local Specialists

We encourage you to enlist the help of local specialists who are knowledgeable about plants and arthropods. They may well become some of your most valuable resources. They can help you select an appropriate site for the project, guide students in identification, and provide background information on the local ecosystem. Try making contacts at the local university, botanical garden, museum, nature center, or garden club.

Selecting the Site

Selecting an appropriate study site for the project is a crucial first step. If you have already recruited local naturalists in your area to work with you on the project, they may also be able to assist you in site selection. There are a number of criteria to consider when selecting the study site for our class:

Size of the Area

A site measuring somewhere between the size of a basketball court and a football field is ideal. Students need room to spread out and to walk around without stepping in one another's plots. The area should be large enough so that students can find a variety of interesting life-forms yet still be one that the teacher can manage easily. That is, students stay in sight, and you can circulate and communicate easily with the group. Urban environments present special problems because space is often limited, and what exists tends to be highly maintained and restricted and therefore less rich in biodiversity. If there are no school grounds, or if they are not suitable, consider getting permission to use a vacant lot, a neighborhood park, a community garden, or even a nearby graveyard.

Safety and Permission

Safety is of paramount concern, as is respect for private and public property. Consider the risks of the prospective site and how you would get proper permission to use it.

Convenience and Accessibility

An ideal site is within easy walking distance. It also is available at various times of the year and preferably over several years. It is better to use a site that is slightly less stimulating but easily accessible than to select a great site that is difficult to visit.

Biotic Richness

Because the study focuses primarily on plants and arthropods, a site with a variety of vegetation works best. This might include trees, bushes, weeds, grasses, herbs, and wildflowers, as well as dead plant materials such as fallen limbs and leaf litter. Stones, boards, trash, and other materials on the ground may also provide microhabitats for invertebrates. Water of any kind adds to the possibilities for finding diverse species. Look for areas with puddles, ponds, fountains, or sprinklers.

Field Journals

Record keeping is essential to the success of Biodiversity Counts. Make sure that each student has a journal in which to record data in many ways: text, measurements, and drawings. Every field activity in the project requires that students make observations, collect data, and record their findings.

They then use their records in a number of different ways:

- To compare data, students may use their own data to make comparisons with findings made by classmates.
- To create graphic displays of data, students increase their understanding and analytical abilities through multiple representations of data. As a means of consolidating and synthesizing information, students use their own data to develop several different types of graphs such as histograms, line graphs, pie charts. In their final presentations they can use the graphs and charts to communicate their information to a larger audience. All of these activities underscore and validate the reasons for keeping good records. By engaging in these activities, students emulate the work of real scientists make observations, record data, communicate and compare their findings with colleagues, consolidate and organize their data, and formally present their research to others.

Observational Rendering

Many people find drawing a stumbling block, and are reluctant to include drawings in their journals. It is important to make the point with students that, in science, drawings serve a special purpose. Although scientific illustrations can be quite beautiful, their purpose is not primarily to please the eye, but to record and communicate information.

In order to draw what they see, students are forced to make close and focused observations. Observational renderings then, are an aid to observation itself, and help the observer to key in on details. The renderings are also a memory aid, graphic reminders of what they found, saw, and did. Your students will find they see more when they draw than when they photograph.

Teacher Reading: Helpful Hints for Field Sketching

Seeing and observing are two different things. When we look at something, we usually are seeing just to identify. We might look at a person, a tree, or a building, and we may later recall the identity of the person we were looking at, or tell if a tree was in bloom or even if the building was made of brick or concrete. However, if asked whether the person had narrow-set eyes, or whether he or she held one shoulder lower than the other, or what exactly was the color of his or her hair well, our ability to recall becomes less exact. Was the building or tree or person! taller than it was wide? How would you describe the color of the leaves?

Were they yellow-green or blue-green? You get the picture, don't you? Observation is a discipline, and drawing is, in a sense, a way of training ourselves to observe. Many people assume that they don't have talent, that they lack that ability to draw. They think that drawing skills are somehow magically bestowed on "artists" alone. As with many assumptions, this one is incorrect. Drawing is a skill that one can learn. As with many other skills, some people will exhibit more aptitude than others, but everyone can improve his or her abilities with some time and effort.

The kind of drawing we are discussing here is observational rendering—trying to capture on paper in two dimensions some aspect of what you are observing. The first thing to do is determine what aspect(s) of your subject you want to record. For example, do you want to know the relative height and width of your subject? Are you recording color? Do you want to show the volume of the subject? How much detail are you interested in? Some of these questions are answered by two factors: the medium you are using and the amount of time you want to spend on drawing. Obviously, if you are using a graphite pencil or charcoal on paper, you are not recording color. And if you intend to spend only a few minutes drawing, you will not be recording a lot of detail.

Hint 1 Proportion

One of the most difficult aspects of drawing is perhaps the most easily resolved. When you draw you are usually creating an image that is smaller in size than the object you are rendering. You do not have to know the actual measurements, only the relative height and width. For example if you are drawing a sleeping cat, you need to know how wide the animal is in relation to how tall it appears to you from your perspective. That is, from where you are observing the animal, you will see a particular and unique set of relationships between height and width. If you move to a new observing position, you will see a different set of relationships. So how do you determine what those relationships are? Re m e m b e r, you do not have to know the actual size of the object you are observing. What you must do is first establish one element. Let's start with the width of our sleeping cat. Simply mark off on your drawing how wide you want your cat to appear on the page. This can be accomplished by making two light marks to establish these outer boundaries. Next you must establish a relationship between the tallest part of the sleeping cat as you view it from your observing point and your now established width. A simple way to accomplish this task is to hold your pencil or even a stick or any straight object at arm's length and sight down your arm to the subject. Turn the stick sideways and align one end with the right or left side of the object and move your thumb to mark the end of the other side on the length of the stick. You now have created your own optical measurement from your viewing location. You can use this measurement to establish the relative sizes of anything you see from your particular and unique viewing position. For example, you can turn your stick vertically, while still holding it at arm's length with your thumb still holding the mark, and compare the width of the cat with its height, as seen from your vantage point. You can now use that relative size to mark off the upper boundaries of the size of the cat on your page you can use this method to establish all of the proportions of what you see on your drawing, such as the distance between the cat's eyes, or the size of the space between the sleeping cat and the rock on the ground next to it. And so on. You will find that with practice you can create very naturalistically proportioned drawings.

Hint Perspective

Things appear smaller the farther away from you they are. This phenomenon is called perspective. For example, if you look at a tree that is a few feet away from you and, using the method for establishing relative size from hint 1, compare the size of a tree as observed from a fixed viewing point to a similarly size tree that is farther away, you will see that it appears smaller. The actual relationship of size of distant objects and their placement on the page can be established the same way you drew the proportions of the cat.

If you are trying to create a sense of volume or fullness of what you are drawing, you need to establish the source of light. Light falling across a form creates a sense of volume or fullness in a three-dimensional object. Volume is established by the relative lightness or darkness of areas of a volumetric surface as is viewed. For example, if you look at light falling on a tree trunk, one side probably will appear darker than the other. The side of the trunk that is in the direction of the light source will appear lighter, and the side of the trunk farther away from the source of light will appear darker. If you squint your eyes when looking at a subject, it helps to establish these relative values of lightness and darkness, because we are not confused by the details. And speaking of details, when you draw the rest of the tree, don't try to draw it a leaf at a time. Again squint your eyes and look at the volume created by the whole bunch of branches and leaves and draw that volume rather than each individual leaf. You can always add detail later if you want to or have the time.

As with drawing the leaves of tree, look for the general shapes. Start with the basic proportions and then add detail. If you start with the details first, you will have only a record or inventory of objects. You will not have a convincing drawing. Drawing is a discipline that can be learned. The more you do it, the better you get. Like playing ball or learning to play the piano, the more you practice, the better you get. This is the most important hint. Have fun. You will be surprised at how much "talent" you never knew you had.

Lesson 1 What is Biodiversity?

Class sessions MATERIALS Chart paper and markers Student journals optional: An artistic interpretation of a site or of a living thing such as a poem, painting, photo, song, or dance. PREPARATION. Prepare two charts for recording students' ideas. Label one Rules for Field Trips and the other Explanations of Biodiversity. Display both of these throughout the unit, so that students can add to them. OPTIONAL: Select an artistic interpretation of some natural object(s) to share with the class. You may want to involve other faculty members, such as the English, art, history, language, and math teachers, in the site interpretation activity.

Student Reading Selections

Introduce students to the idea that they will take part in the Biodiversity Counts project. To open the discussion on biodiversity, you might ask: Ask students to use all this information to come to a simple definition of biodiversity. Have them record what they understand of the concept in their journals. Ask students to discuss the following points: Give some examples of the variety of biologically diverse forms of life that exist on Earth. Name some of the diverse types of habitats in which life exists. How do humans depend on other life-forms? Give some specific examples. What are some of the reasons that increasing numbers of species are becoming extinct? What are some of the solutions to the problem of biodiversity loss?

Possible Responses

The common thread is that they are all living organisms. In spite of their remarkable diversity, they all share the characteristics that distinguish living from nonliving things. Living things grow, respond to stimuli, adapt to changes in their environment, reproduce, and carry on respiration. Students may point out differences in size, shape, color, mobility, body covering, habitat, food, and behavior, among other things.

Questions what do all of the following have in common: dogs, mosquitoes, daffodils, seventh graders, plankton, eagles, squid, rice, frogs, elephants, amoebas, and rats? List some characteristics of living things. How do organisms differ? Then work together to construct a collaborative explanation of what students understand about biodiversity based on new information. Record this explanation and save it until the end of the section. At that time you will develop another explanation and compare the two.

Assessment: Baselines

The individual explanations of biodiversity students re corded in their journals will give you a good sense of each student's baseline knowledge on the topic. You may want to refer to them at the end of the unit as evidence of student progress. The collaborative explanation may seem simple or naive at this point, but that is to be expected. Let students know that you will return to this explanation often to revise and expand it as they work through the unit. To bolster students' recording skills and to encourage them to draw what they observe, ask them to read the essays "Drawing as a Way of Looking at the Natural World,"

"Keeping a Field Journal 1," and "Keeping

- How do they like to begin making a drawing?
- Explain how you can draw proportions accurately.

Brian Boom has a system for record keeping in the field and for translating information to a specimen label. Explain his system in your own words. Could you adopt some of his techniques for your field work, or do you have other ideas for a recording system of your own?

Tell students that, over the coming months, they are going to study an outdoor site in great detail, just as scientists do. They will look primarily at arthropods, plants, or both. Explain that they will engage in all the activities that scientists do: observing, collecting, classifying, recording, analyzing data, and communicating their findings with other students. The first visit is to get to know their site. They are free to use any means they like to record their impressions. Ask for their ideas on possible ways to record and express what they experience. Ideas may include drawing, writing a short story or poem, composing a song or a dance, or taking photographs. Ask them to bring in any special equipment they might need for the activity. Optional: Share with the class an artistic interpretation of nature such as a painting, photograph, poem, or musical selection. Before you go outside for the first time, develop a list of safety rules with the class.

Lesson Experiencing the Outdoor Site

Materials

A variety of materials for recording student impressions of the outdoor site. These may include art supplies, audiovisual equipment, cameras, and musical instruments. PREPARATION

- Discuss the decisions involved in your selection of a site for the outdoor activity.
- Have students bring in materials for recording their impressions.
  - In the classroom, ask students to predict what they think they'll see when they visit their site: How many arthropods or plants? How many different kinds?
  - Explain to students that they are free to observe, explore, and enjoy the area in as many ways and using as many senses—but never taste—as they safely can.
  - Take them to the area that you have selected for the activity. Remind them of their safety rules. Be sure to point out the boundaries of the area and ask students to stay within them. Set a time limit and ask that students report back to you at a specified time. Review the senses that are relatively safe to use—sight, hearing, and smell. They may work individually or in small groups. Then send them off to begin recording their impressions of the living things at the site.

- As students work, circulate among them to observe their approaches to the task. You might use some of these questions to help them keep on task:

  Point out to students that they will be sharing information with other students. Ask them to begin thinking about which of their observations they would want to share with others.
- Before you leave the site, have students check that it is in good order and that they have left nothing behind.

Focus Questions

What living things have you observed so far? Have you looked at both plants and animals? How many different senses have you used to make observations? How are you recording your observations? How would you report them to someone who has not been to the site?

Tasks To observe living things at the site To use more than one sense to make observations To record observations The task and focus questions on the following page can be copied and distributed to students.

Tasks and Focus Questions
Field Trip Experiencing the Outdoor Site

Focus Questions

What living things have you observed so far?
Have you looked at both plants and animals?
How many different senses have you used to make observations?
How are you recording your observations?
How would you report them to someone who has not been to the site?

To observe living things at the site To use more than one sense to make observations To record observations LESSON 3 REPORTING BACK ON SITE EXPERIENCES TIME 2 or 3 class sessions MATERIALS Student products generated during the field trip For the Optional Mapping Activity: A blank map of the outdoor site on overhead transparency film, markers, and an overhead projector; or a large sheet of craft paper and markers PREPARATION For the Optional Mapping Activity: On a sheet of transparency film or on a large sheet of paper, draw an outline of the outdoor area students visited.

1. Allow students time to organize their observations and to prepare brief presentations for three to five minutes. Then let them share their impressions of the living things at the outdoor site. Presentations will vary, so make the point that there are many different ways to look at a place and many ways to record observations of an area. Each type is good for a different purpose, and each type yields different information.
2. Hold a discussion on the difference between evidence and impressions.

   Use some of these questions to guide the discussion: If your task in the field has been to collect facts and evidence about the variety of life-forms you observe, how would your presentations have been different? Students should mention that scientific observations would require collecting and recording data about the site, taking precise measurements, and gathering information that can be replicated.

ASSESSMENT: After discussing the differences between impressionistic observe actions and scientific observations, ask students to list the criteria they think are important for making good scientific observations. Do you think your presentations included every living thing at the site? What else might be there? What might be there that you didn't see?

Optional Mapping Activity

Activity Creating a Map of the Outdoor Site

1. When students have finished their presentations, show a blank map of the area. Ask students to come up to the map

and add details based solely on the information they included in their presentations. They may not have included much detailed or factual information in their renderings and so may have little to add to the map. Assure them that this is acceptable for this activity. Make it clear that nothing they present is either right or wrong, that all suggestions are valid.

Recognize, too, that students may argue or debate some points. Encourage them to do so. Then ask what evidence they need to collect to settle the debate. 2. After they have filled in the map as best they can, ask: Would someone else be able to recognize the area from this map? If not, why not? What is missing? How could you gather more factual data about the site? Note: Label the map Number One and save it to use again.

Assessment: Digging Deeper

In Lesson 1, students recorded their first explanations of what they understand about biodiversity. These explanations are useful as measures of their baseline knowledge. Then students began to acquire new information through the video, their first experiences in the field, and discussions. Hold a brief discussion as closure.

Teaching Tip

One way to track the development of students' understanding of biodiversity is to keep a running record of their explanations. You may keep them in a list form, or construct a concept web or mind map to which students can add information as new ideas develop. Webs and maps have the advantage of laying out students' thinking in a visual form at. What is Biodiversity?

For Students

Drawing as a Way of Looking at the Natural World Keeping a Field Journal 1 Keeping a Field Journal.

Target Question

"If each individual talks about what he or she likes about what was observed and why he or she thinks the rest of us should care about it, the result could be a very rich and personal tour of the ecosystem." – What is biodiversity?

How To–Site Selection (Teacher Preparation)

Prerequisites: What is Biodiversity?

In the opening discussion, students begin to learn about the term biodiversity at its most basic level. Students continue to develop their understanding of the concept throughout the unit. On their first field trip, students observe a natural space and report on what they see. They also begin to see the need for keeping written records, since memory is not always a good enough tool. As the curriculum develops, students will learn how to make scientific observations of their site and become more systematic in their recording methods. Both teachers and students will discuss their initial site experiences and observations and begin to use the resources available on the Web site.

Background Information for the Teacher

Biodiversity is the spectacular variety of life on Earth and the essential interdependence among all living things. Initially, students will find it difficult to verbalize even the most rudimentary understanding of that. In the course of the unit, however, they will have many opportunities to develop an understanding of the term. By the conclusion of the unit, students will have had opportunities to do firsthand work in the field. These ex p experiences will help each student develop a more sophisticated definition of biodiversity, acquire important scientific skills, and develop a sense of stewardship for both the local and global environment. Biodiversity Counts is based on five key concepts that provide the context for the unit. These key concepts are the foundation for all the activities.

Brief Summary of the Five Key Concepts

1. Life on Earth is extraordinarily diverse. From microscopic bacteria to mammoth whales, from firmly rooted redwoods to highly mobile humans, from marshy wetlands to the high

chaparral, life on Earth is diverse, at the level of species as well as at the level of ecosystems. In Biodiversity Counts: Students will look closely at organisms, searching for answers that relate to their class site. They will investigate such questions as: What is out there? How many species of living things are there? How are they related? What is the life history of an organism? W h e re does it live? How does it obtain nutrients? What does it do? What is its role in the habitat? How does it reproduce? How does it defend itself? How are plants and arthropods adapted to the habitats at our site? Species are the entities we observe, describe, and discuss when we study biodiversity.

Species are genetically distinct entities with discrete histories.

They retain their genetic distinction because, under most circumstances in nature, a member of a particular species can breed only with another member of the same species. The genetic code that distinguishes a species usually determines a set of unique characteristics that enables us visually to tell one species from another. Each species is somehow distinct in form and has a unique evolutionary history that is not shared with any other species. Although 1.75 million species have been described so far, scientists estimate that as many as 10 million distinct species exist on Earth today. In Biodiversity Counts: Students will emulate the work of systematic biologists and conservation biologists to identify plants and arthropods at their site. Living things are bound together in intricate, interconnected webs. The natural world is a vast network of complex communities, connected through shared environments. The interconnected webs serve to sustain biodiversity. For example, certain flowering plants depend on bees for pollination. The bees depend on the flowers for their food nectar or pollen. After pollination, the plants produce fruits, receptacles for the seeds of the next generation. Birds are attracted by the bright color and sweet aroma, and they eat the fruits. The seeds pass through the birds' digestive system and are dispersed in their droppings, and new plants grow, often some distance away from the parent plants. Beetles may eat their way into the same fruit. Small rodents gnaw into the fruits searching for beetles. The rodents themselves may be prey for birds, reptiles, and larger mammals. If one component is removed – if the bees are killed by insecticides, for example reverberations are felt throughout the network. Without bees to transport pollen, plants cannot produce seeds to ensure a new generation. Without seeds and their fruity packaging, there is no food for other wildlife.

Teacher Reading

Seeing and observing are two different things. When we look at something, we usually are seeing just to identify. We might look at a person, a tree, or a building, and we may later recall the identity of the person we were looking at, or tell if a tree was in bloom or even if the building was made of brick or concrete. However, if asked whether the person had narrow-set eyes, or whether he or she held one shoulder lower than the other, or what exactly was the color of his or her hair – well, our ability to recall becomes less exact. Was the building or tree – or person! – taller than it was wide? How would you describe the color of the leaves? Were they yellow-green or blue-green? You get the picture, don't you? Observation is a discipline, and drawing is, in a sense, a way of training ourselves to observe. Many people assume that they don't have talent, that they lack that ability to draw. They think that drawing skills are somehow magically bestowed on "artists" alone. As with many assumptions, this one is incorrect. Drawing is a skill that one can learn. As with many other skills, some people will exhibit more aptitude than others, but everyone can improve his or her abilities with some time and effort.

The kind of drawing we are discussing here is observational rendering – trying to capture on paper in two dimensions some aspect of what you are observing. The first thing to do is determine what aspect(s) of your subject you want to record. For example, do you want to know the relative height and width of your subject? Are you recording color? Do you want to show the volume of the subject? How much detail are you interested in? Some of these questions are answered by two factors: the medium you are using and the amount of time you want to spend on drawing. Obviously, if you are using a graphite pencil or charcoal on paper, you are not recording color. And if you intend to spend only a few minutes drawing, you will not be recording a lot of detail.

One of the most difficult aspects of drawing is perhaps the most easily resolved. When you draw you are usually creating an image that is smaller in size than the object you are rendering. You do not have to know the actual measurements, only the relative height and width. For example if you are drawing a sleeping cat, you need to know how wide the animal is in relation to how tall it appears to you from your perspective. That is, from where you are observing the animal, you will particular and unique set of relationships between height and width. If you move to a new observing position, you will see a different set of relationships. So how do you determine what those relationships are? Re m e m b e r, you do not have to know the actual size of the object you are observing. What you must do is first establish one element.

Background Information for the Teacher

Why begin with plants? Plants hold a primary place in the world of living organisms because of their ability to manufacture their own food from simple inorganic materials (carbon dioxide and water). This process, called photosynthesis, is powered by energy that plants capture from the sun. Thus plants feed not only themselves, but are food for herbivorous animals and indirectly for the carnivores that feed on the herbivores. They also produce virtually all of the oxygen in the atmosphere, since they give off oxygen as a b y-product of photosynthesis. In many ways, plants determine the numbers and kinds of other organism that can live and thrive in an area. In addition to food, they supply protection from the elements, nesting sites, and homes for small creatures. They are host to a wide range of mammals, birds, fish, worms, arthropods, fungi, and bacteria. Finally, plants are enormously important to human welfare. They are a source of food, clothing, housing materials, medicines, and dyes. They replenish the air we breathe, cool the atmosphere that surrounds us, hold our soils in place, and please our senses. They contribute not only to our physical but also to our emotional well-being.

Classification: Getting to the Species Level

Biological scientists around the globe are working to discover, describe, and classify species. In order to communicate clearly with one another, they all use the same hierarchical classification system and the same naming system - binomial nomenclature.

Classification System

Animals Plants Kingdom Phylum Division Class Order Family Genus Species As you move through the categories from kingdom to species, each category contains fewer organisms, and the organisms in each category have more features in common. The fundamental unit of all biology is the species; thus, every individual can be described as belonging to a distinct species. Scientific research on biodiversity focuses primarily on species and it is the unit that scientists use to discuss biodiversity. Identification to the species level is a difficult process and may require special equipment.

Binomial Nomenclature

In binomial nomenclature, an organism is given two Latin or Latinized names. Latin is the recognized language of science. It has the advantage of being a language that belongs to no single country or culture and one that is no longer changing. The first name of the organism is its genus, always capitalized. The second name is its species, always written in lower case. Both are usually italicized but are sometimes underlined or written in bold type. Most conventional keys and identification guides use this system. For example, the white willow belongs to the genus Salix, along with more than 200 other species such as the weeping willow, the crack willow, and the black willow. Its full name, Salix alba, distinguishes it from the other willows in the genus by adding the species name alba, or white. Not all species names are so conveniently descriptive. It is a challenge to identify plants all the way to the species level. But it is important that students engage in the process. They will become more familiar with the language of science, and they will be working the same way that scientists do to describe and identify organisms.

Using Dichotomous Keys

A dichotomous key from the Greek dicho meaning "two" and to mous meaning "to cut" is a system for identifying organisms based on a series of choices between two alternative descriptions. It is an extremely useful tool for identification. Unfortunately, keys do not exist down to the species level for all the plants your students may discover, and there is much work yet to be done in that field. You will find some dichotomous keys to conifer genera. They may be used on-line or printed out and placed in plastic sheet protectors for students to carry into the field. Keys are fun to use—and quite simple, once you get the idea. It's like solving a mystery from the clues. No two species are exactly alike, so by following a series of clues and eliminating suspects along the way, you can arrive at a positive identification most of the time.

Please see How to Use a Dichotomous Key on the Web for more information. Most standard biology textbooks also have information on dichotomous keys. Morphospecies A morphospecies may be described as a group of organisms that, based on external form and structure, appears to be distinct from other groups of organisms.

The group of individuals is recognized solely on the basis of similarities in external characteristics. When faced with a bewildering array of organisms, it is useful to begin by sorting them into groups based on their observable characteristics. At the most basic level, species may be defined as a unique entity. This entity, or population of individuals, is genetically distinct, and its genetic distinctions are reflected by form and structure or morphology. In fact, new species are described because they demonstrate observable and inheritable differences in morphology or shape. Thus, the amateur or student scientist may legitimately begin the process of classification by sorting organisms into categories based on what they look like. Using observable attributes such as shape, color, size, texture, numbers of structures, and the absence or presence of structures, students can make a good start on the road to identification. Later, as more data are collected, the original identification may be either verified or disproved.

In the process of sorting, students will have to observe the organism in great detail in order to discover how it is similar to or diff e rent from the others—a useful exercise in itself. Even if the observation does not lead to identification of the species, it may well lead to identification at a higher taxonomic level, such as order or family. This is a real accomplishment.

Teaching Tip: Identification is a Challenge

It is not always possible to reach the species level of identification—for students or scientists. Recognize that it will be a challenging task and that students may often be frustrated in their attempts. While identification to the species level is important, it is not the main objective of the study. Help students get into the habit of consulting the on-line resource sections often. Some areas that you and your students will find useful are the external links to botanical gardens, education programs, environmental organizations, entomological societies, herbaria, natural history museums and science centers, national parks, and zoological parks.

Measuring Environmental Factors

Much of what scientists do is look for patterns in nature. They may look for patterns of plant distribution, patterns of evolutionary relationships, or correlations between one variable and another to try to understand the factors that determine biodiversity. Scientists have observed that the occurrence of many plants is determined by environmental factors. There is a correlation between the presence of a particular plant, for example, and the environmental conditions that favor its survival. These conditions include temperature range, light levels, soil composition, and average rainfall.

By recording these environmental factors over a long period of time, we could expect to see patterns that show which factors are important to a species. Activity sheet per team Reference materials such as field guides and keys to identification Some Clues to Describing and Understanding Organisms. 3. Ask students what tools and equipment they will need to take measurements of the environmental factors at their site. Ideally, they should plan to use thermometers, a soil test kit, a light meter, and a rain gauge. 4. Distribute copies of Some Clues to Describing and Understanding Organisms to your students and go over it to help prepare for the field activities.

Questions What factors determine the number and kinds of plants you will find at the site? How do plants provide for the needs of other living things?

How can you tell one plant from another? What features could you look for? Possible Responses Factors include: Rainfall, humidity, temperature, soil composition, and hours of sunlight, activities of people and other animals. They provide food, shelter, homes, nesting places. By leaf or needle size and shape, arrangement of plant parts on the stem, smell, texture, vein patterns, flower and bud types, bark, seeds, cones fruits, or overall growth patterns.

Field Trip Close Observations Materials

Understanding Organisms Camera and drawing materials Field guides and keys to identification Tell students that they are to make close observations of at least one plant each. If possible, they should try to identify the plant. Have them use the "Some Clues to Describing and Understanding Organisms" to guide their observations. Ask them to record their observations by drawing and writing in their journals. Circulate among the teams as they work and use some of the questions on the next page to prompt them to make more detailed observations. While still in the field, encourage students to use field guides and keys to try to identify their plants. Later they may use additional on-line resources as well. In many cases, students will be able to identify only fertile specimens.

Continue the process until students have recorded all the plants in their plot and identified them to the best of their ability. Identification is an ongoing process, and although students may not be able to identify a specimen right away, they may be able to do so at a later date. Which of your senses have you used to observe? Have you touched the plant, smelled it, looked at it with a hand lens? What patterns have you noticed on the plant? How are the leaves or needles arranged along the stem? What patterns do the veins make? Do all the blossoms have the same shape and the same number of petals? Have you measured the plant? What parts have you measured? What is the plant doing right now? Is it producing buds, seeds, fruits, leaves, or blossoms? Do you think the plant may have an ecological relationship to another plant, or to an animal? Why do you think so? Where have you recorded your observations? Have you included written descriptions, measurements, and drawings?

What environmental factors are you measuring? What tools are you using to take the measurements? To prompt students to use more than one sense to make observations To seek patterns To encourage measurement•To focus on plant processes To look for relationships To ensure record keeping To measure and record environmental factors The task and focus questions on the following page can be copied and distributed to students. Which of your senses have you used to observe? Have you touched the plant, smelled it, looked at it with a hand lens? What patterns have you noticed on the plant? How are the leaves or needles arranged along the stem? What patterns do the veins make? Do all the blossoms have the same shape and the same number of petals? Have you measured the plant? What parts have you measured? What is the plant doing right now? Is it producing buds, seeds, fruits, leaves, or blossoms? Do you think the plant may have an ecological relationship to another plant, or to an animal? Why do you think so? Where have you recorded your observations? Have you included written descriptions, measurements, and drawings? What environmental factors are you measuring? What tools are you using to take the measurements? To prompt students to use more than one sense to make observations To seek patterns To encourage Measurement To focus on plant processes To look for relationships To ensure record keeping To measure and record environmental factors. Before the trip, remind students of their own list of rules for responsible collecting, go over safety rules, and check that you have all the equipment needed. Students could also review the on-line reading selections that refer to collecting. You may find it easier to take a plant press into the field, rather than have students carry specimens back to class. The advantage is that you can supervise what goes into the press, make sure that specimens are properly identified, and limit the amount the class collects.

At the site, encourage student teams to work together to decide what to collect. They should select the most complete specimens in their plot. Continue the process until students have identified as many of the plants in their plot as possible. When they are unable to identify a plant, they should keep a record of the plant's characteristics and assign it a working name, such as mystery species. Remind students to keep a numbered list in their journals of specimens they collect and to label each plant specimen as they go. The label should tell the location where the plant was found, the date, and the name of the collector. Have students use field guides and keys to try to identify the plant while still in the field.

Glossary

Ametabolous	–	A group of insects which undergo simple metamorphosis, they moult as they grow but do not change form.
Antenna	–	Jointed feeler found on the head (p.antennae).
Apex	–	Tip or summit (pl.apices).
Apical	–	At the tip (end) of any organ or cell structure.
Aquatic	–	Living in or close to water
Biological species	–	Population of individuals that can interbreed, that is, a true species
Biology	–	The study of living organisms.
Caudal	–	Towards the tail end.
Cercius	–	Appendage at the end of the abdomen (pl.cerci)
Classification	–	The arrangement of living organisms by similarities and differences, closely reflecting related evolutionary groups whenever possible.
Cline	–	The gradual change in characteristics from one area to another
Clubbed	–	Having the end of antennae enlarged
Compound eye	–	Each eye made up of a number of individual units called omatidia
Conspicuous	–	Obvious

Costal cell	–	The area of an insect's wing between the costa and subscosta; anterior vein or margin
Couplets	–	A set of two opposing characteristics in an identification key
Coax	–	Leg joint nearest the body
Crochets	–	Used in locomotion of larvae (hooks)
Cylindrical	–	Cylinder - like
Distal	–	End of any cell structure or organ
Ecology	–	The study of interactions between organisms and their environment
Ectognathous	–	Mouthparts always protruding
Ectoparasite	–	An external parasite
Elytron	–	Forewing of beetles, which is a hard protective covering for the hindwing and body, not used in flight (pl.elytra)
Endopterygota	–	A group of insects which undergo complete metamorphosis, (larvae to pupae to adult) and develop wings internally.
Exopterygota	–	A group of insects which undergo gradual metamorphosis (nymph to adult) and develop wings in stages outside the body.
Extant	–	Still living
Extinct	–	No longer living
Exudation	–	A cell or organism discharging material
Facet	–	Surface of an ommatidium (a unit making up a compound eye).
Femur	–	From the body it is the third segment of an insect or spider's leg.
Filaments	–	Generally any thread-like structure

Filioform	–	Thread-like antennae
Flabellate	–	Shaped like a fan
Forceps	–	The cerci of some insects that are clasper-shaped
Forewings	–	The first pair of wings
Galea	–	External lobe of the maxillary palp which is soft
Hypognathous condition	–	Position of the mouth-parts relative to the head
Labium	–	Lower lip.
Labrum	–	Anterior lip.
Lacinia	–	External lobe of the maxillary palp which is hard and toothed
Larva	–	Immature stage, for example, caterpillar, grub.
Mandible	–	Lower jaw
Maxilla	–	Upper jaw
Maxillary palps	–	Appendage of the insect's maxilla
Membranous	–	Thin, dry and transparent (of wing)
Mesothorax	–	The middle segment of an insect's thorax
Metamorphosis	–	Change in form, for example, caterpillar to moth or butterfly
Metathorax	–	Posterior section of the thorax
Moniliform	–	Appearance being like beaded necklace
Morphology	–	Form and structure
Morpho species	–	Species defined by similarities in morphological characters only
Natural species	–	Population of individuals that can interbreed, that is, a true species
Nomenclature	–	The process of naming organisms within a pre-existing classification

Nymphs – Juvenile form without wings in insects with incomplete metamor phosis

Ocellus – Simple eye (pl.ocelli)

Ommatidia – Light receptive organ that appears as a separate facet (outer lens)

Ovipositor – A structure in insects that is specialized for the depositing of eggs

Phylogenetic – The elucidation of patterns of relationships and interpretation of their origins

Pincers – Claspers, for example, cerci of Dermaptera (earwigs) used in defense.

Plumose – Appearance being like features.

Predacious – Predator; an organism that catches and eats other organisms

Pretarsus – Bearing claws

Proboscis – Trunk-like projection of head used for feeding

Proleg – Unjointed abdominal appendage, found in larvae of lepidoptera and some other arthropods

Prathorax – First segment of an insects thorax

Pterothorax – Wing-bearing area of the thorax (mesothorax and metathorax)

Pupa – Resign stage in an insect's life cycle

Repose – At rest

Sedentary – Not free-moving, attached

Serrate – With edges notched like the teeth of a saw

Spiracle – Aperture in the sides of the thoracic and abdominal segments of insects, through which the tracheal respiratory system connects to the outside.

Springing apparatus – An appendage that allows insect to spring, thereby projecting themselves into the air usually in a forward direction, for example, springtails (Collembola)

Style – Bristle-like process on the clasper (cercus) of male insect.

Systematics – The scientific study of the diversity of and relationships between different organisms

Tarsus – Fifth segment of leg usually made up of a number of smaller segments (pl.tarsi)

Taxon – Defined units in the classifications of living organisms (pl.taxa)

Taxonomic species – Species defined by similarities in morphological characters only

Taxonomy – The study of the rules, principles and practice of classifying. The theory and practice of describing, naming and classifying

Terrestrial – Living or found on land

Tibia – Fourth segment of an insect leg

Transverse suture – A line of junction, lying across or between a section, that is, two parts

Transverse veinlets – Cross veins (of wing)

Trochanter – Second segment of an insect's leg

Venation – The arrangement of the veins of an insect's wings

Ventral – The underside of an animal

Vestigial – Having smaller and simpler structure when compared with an ancestor

References

Arnason, R. (1998), Ecological Fisheries Management Using Individual Transferable Share Quotas. *Ecological Applications.* 8(1):S151–S159._. 1998b. Minimum Information Management in Fisheries. *Canadian Journal of Economics* 23:630–53.

Beverton, R., J. Cooke, J. Csirke, R. Doyle, G. Hempel, S. Holt, A. MacCall, D. Policansky, J. Roughgarden, J. Shepherd, M. Sissenwine, and P. Wiebe (1984), Dynamics of Single Species—Group Report. In *Exploitation of Marine Communities Dahlem Konferenzen,* R. May, ed. New York: Springer Verlag.

Blindheim, J., and H.R. Skjoldal (1993), Effects of Climatic Changes on the Biomass Yield of the Barents Sea, Norwegian Sea, and West Greenland Large Marine Ecosystems. In *Large Marine Ecosystems: Stress, Mitigation, and Sustainability,* K. Sherman, L.M. Alexander, and B.D. Gold, eds. Washington, DC: AAAS Press.

Dickson, R.R., J. Meincke, S. Malmberg, and A.J. Lee (1988), The "Great Salinity Anomaly" in the Northern North Atlantic 1968–1982. *Prog. Oceanog.* 20:103–51.

Dragesund, O., J. Hamre, and O. Ulltang (1980), Biology and Population Dynamics of the Norwegian Spring-spawning Herring. *Rapp. P.-v. Reun. Cons. Int. Explor. Mer* 177:43–71.

Flaaten, O. (1991), Bioeconomics of Sustainable Harvest of Competing Species. *Journal of Environmental Economics and Management* 20:163–80.

Glantz, M., and J. Thompson, eds. (1981), *Resource Management and Environmental Uncertainty.* New York: John Wiley & Sons.

Glantz, M.H. (1979), Science, Politics and Economics of the Peruvian Anchoveta Fishery. *Marine Policy.* _. 1980. El Niño: Lessons for Coastal Fisheries in Africa? *Oceanus* 23(January):9–17.

Hannesson, R. (1983), Optimal Harvesting of Ecologically Interdependent Fish Species, *Journal of Environmental Economics and Management* 10(4):329–45.

Hargis, W.J., and D.S. Haven (1988), Rehabilitation of the Troubled Oyster Industry of the Lower Chesapeake Bay. *Journal of Shellfish Research* 7:271–79.

Holling, C.S., D.W. Schindler, B.W. Walker, and J. Roughgarten (1995), Biodiversity in the Functioning of Ecosystems: An Ecological Synthesis. In *Biodiversity Loss: Economic and Ecological Issues*, C. Perrings, K. Maler, C. Folk, C. Holling, and B-O. Jansson, eds. Cambridge: Cambridge University Press.

Hovgaard, H., and E. Buch (1990), Fluctuation in the Cod Biomass of the West Greenland Sea Ecosystem in Relation to Climate. *Large Marine Ecosystems: Patterns, Processes and Yields*, K. Sherman, L.M. Alexander, and B.D. Gold, eds. Washington, DC: AAAS Press.

Jakobsson, J. (1992), Recent Variability in the Fisheries of the North Atlantic. *ICES-Marine Science Symposia* 195:291–315.

Johnston, R., and J.G. Sutinen. 1996. Uncertain Biomass Shift and Collapse: Implications for Harvest Policy in the Fishery. *Land Economics* 72(4):500–18

Jones, J.B. (1992), Environmental Impact of Trawling on the Seabed: A Review. *New Zealand Journal of Marine and Freshwater Research*. 26:59–67.

Lauck, T., C.W. Clark, M. Mangel, and G.R. Munro (1998), Implementing the Precautionary Principle in Fisheries Management Through Marine Reserves. *Ecological Applications*. 8(1):S160–S164.

Myers, R.A., and G. Mertz (1998), The Limits of Exploitation: A Precautionary Approach. *Ecological Applications*. 8(1):S165–S169.

National Research Council (1995), Committee on Biological Diversity in Marine Systems.

Understanding Marine Biodiversity: A Research Agenda for the Nation. Washington, DC: National Academy Press.

OECD (1996), *Saving Biological Diversity. Economic Incentives*. OECD, Paris.

Patterson, K.R., J. Zuzunaga, and G. Cárdenas. Size of the South American Sardine (*Sardinops sagax*) Population in the Northern Part of the Peru Upwelling Ecosystem af *Thalassorama 83* ter Collapse of Anchoveta (Engraulis ringens) Stocks. *Canadian Journal of Fisheries and Aquatic Sciences* 49:1762–69.

Pauly, D., V. Christensen, J. Dalsgaard, R. Froese, and F. Torres, Jr. (1998), Fishing Down Marine Food Webs. *Science* 279:861–63.

Perring, C., and D. Pearce (1994), Threshold Effects and Incentives for the Conservation of Biodiversity. *Environmental and Resource Economics* 4(1):13–28.

Perrings, C., and H. Opschoor (1994), The Loss of Biological Diversity: Some Policy Implications. *Environmental and Resource Economics* 4(1):1–12

Schrank, W. (1997), The Newfoundland Fishery: Past, Present and Future. In *Subsidies and Depletion of World Fisheries: Case Studies.* World Wildlife Fund, U.S., Washington, DC 1997.

Sherman, K. (1994), Sustainability, Biomass Yields, and Health of Coastal Ecosystems: An Ecological Perspective. *Marine Ecology Progress Series* 112:277–301.

Skjoldal, H.R., H. Gjosaeter, and H. Loeng (1992), The Barents Sea Ecosystem in the 1980s: Ocean Climate, Plankton, and Capelin Growth. *ICES-Marine Science Symposia* 195:278–90.

Steele, J. (1984), Kinds of Variability and Uncertainty Affecting Fisheries. In *Exploitation of Marine Communities Dahlem Konferenzen,* R. May, ed. New York: Springer Verlag.

Thompson, J. (1981), Climate, Upwelling, and Biological Productivity: Some Primary Relationships.

In *Resource Management and Environmental Uncertainty,* Glantz, M., and J. Thompson, eds. New York: John Wiley & Sons.

Watling, L., and R. Langton (1994), Fishing, Habitat Disruption, and Biodiversity Loss. *EOS Transaction and American Geophysical Union* 75(3):210

Weitzman, M. (1992), On Diversity. *Quarterly Journal of Economics* 107(2):363–405._. 1993. What to Preserve? An Application of Diversity Theory to Crane Conservation. *Quarterly Journal of Economics* 108(1):157–83.

Zaitsev, Y., and V. Mamev (1997), *Marine Biological Diversity in the Black Sea: A Study of Change and Decline*. United Nations Publications, New York, New York.

Constanza R, d'Arge R, de Groot R, Fabes S, Grano M, Hannon B, Limburg K, Naeem S, O'Neil R V, Pareulo J, Raskin R, Sulton P and van der Belt M. (1997), The Value of the World's Ecosystem Services and Natural Capital. Nature 387: 254-260.

CBD Secretariat (2007), Gincana 3: Biodiversity and Climate Change. CBD Secretariat, Montreal, Canada.

Millennium Ecosystem Assessment (2005), Ecosystems and human well-being: Synthesis, World Resources Institute, Washington, USA.

CBD Secretariat (2007), Biodiversity and Climate Change. CBD Secretariat, Montreal, Canada .

IUCN (2006), The IUCN Red List of threatened species. IUCN Cambridge, UK.

Thomas, C.D. et al (2004), Extinction Risk from Climate Change. Nature 427:145-148.

CBD Secretariat (2003), Interlinkages between Biological Diversity and Climate Change: Advice on the Integration of Biodiversity Considerations into the Implementation of the United Nations Framework Convention on Climate Change and its Kyoto Protocol.

CBD Secretariat Technical Series No. 10. CBD Secretariat, Montreal, Canada. UNEP/CBD/SBSTA/9/8 2002. CBD Secretariat, Montreal, Canada. Page, S. E. et al (2002), The Amount of Carbon released from Peat and Forest Fires in Indonesia during 1997. Nature 420:61-65.

Sciare, J. et al (2000), Inter-annual Variability of Atmospheric Dimethylsulfide in the Southern Indian Ocean. Journal of Geophysical Research 105:26,369-26,377.

Intergovernmental Panel on Climate Change (2001), Climate Change 2001: Impacts, Adaptation, and Vulnerability. IPCC Secretariat, Bonn.

Hassol, S.J. (2004), Impacts of a Warming Arctic. Arctic Climate Impact Assessment (ACIA). Cambridge University Press, UK.

NASA, Goddard Space Flight Center (2006), Warming Climate May Put Chill on Arctic Polar Bear Population, accessed online at http://www.nasa.gov/centers/goddard/news/topstory/2006/polar_bears.html

Ad hoc Technical Expert Group on Biological Diversity and Climate Change (2003), CBD Technical Series No.10, Secretariat of the Convention on Biological Diversity.

Honchin, D. (1994), *Providing for Reasonable Use: Jurisdictional Issues.* Unpublished paper. Kenchington, R. A. (1990) *Managing marine environments.* Taylor & Francis, New York. Whitehouse, J. F. (1993) *A review of the Great Barrier Reef Marine Park Authority.* Department of Environment, Sport and Territories, Canberra.

ANZECC (1995), *National Strategy for the Conservation of Australia's Biological Diversity.* Department of Environment, Sport and Territories, Canberra.

George, R.J., McFarlane, D.J., and Speed, R. (1995), 'The Consequences of a Changing Hydrologic Environment for Native Vegetation in Southwestern Australia'. *Nature conservation 4: the role of networks.* Saunders, D. A., Craig, J. L., Mattiske, E. M. (eds) Surrey Beatty and Sons, Chipping Norton, NSW. Pp 9-22.

Hopper, S.D., van Leeuwen, S., Brown, A.P., Patrick, S. (1990), *Western Australia's Endangered Flora and Other Plants under Consideration for Declaration.* Department of Conservation and Land Management, Perth, Western Australia.

Kitchener, D.J., Chapman, A., Dell, J., Muir, B.G., Palmer, M. (1980), 'Conservation Value for Mammals of Reserves in the Western Australian Wheatbelt-Some Implications for Conservation'. *Biological Conservation* 18, 179-207.

Lambeck, R. J., Wallace, J. F. (1993), *Assessment of the Conservation Value of Remnant Vegetation in the Central Wheatbelt of Western Australia using Landsat TM Imagery.* Report for Australian National Parks and Wildlife Service "Save the Bush" program.

Lambert, J., Elix, J. (1996), 'Bioregional Planning for Biodiversity Conservation'. In; *Approaches to Bioregional Planning, Part 2, Background Papers. Biodiversity Series Paper No. 10,* Biodiversity Unit, Department of the Environment, Sport and Territories, Canberra.

Leigh, J., Boden, R. and Briggs, J. (1984), *Extinct and Endangered Plants of Australia.* MacMillan, Melbourne, Victoria.

Pittock, A.B. (1988), 'Actual and Anticipated Changes in Australia's Climate'. *Greenhouse. Planning for climate change.* Pearman, G. I. (ed) CSIRO, Australia. pp. 35-51.

Saunders, D.A., Curry, P.J. (1990), 'The Impact of Agricultural and Pastoral Industries on Birds in the Southern Half of Western Australia: Past, Present and Future'. *Proceedings of the Ecological Society of Australia* 16: 303-321.

Saunders, D.A., Ingram, J.A. (1995), *Birds of Southwestern Australia: An Atlas of Changes in Distribution and Abundance of the Wheatbelt Avifauna.* Surrey Beatty and Sons, Chipping Norton, NSW.

Commonwealth of Australia (1992), *National Strategy for Ecologically Sustainable Development.* AGPS, Canberra.

ANZECC (1995), *National Strategy for the Conservation of Australia's Biological Diversity.* Department of Environment, Sport and Territories, Canberra.

Commonwealth of Australia (1995), *National Forest Conservation Reserves Commonwealth proposed criteria a position paper July 1995.* Canberra.

Commonwealth of Australia (1995), *Regional Forest Agreements – The Commonwealth Position.* Canberra.

Department of the Environment, Sport and Territories (1995), *Guidelines for the Assessment of Biological Diversity.* AGPS, Canberra.

Joint ANZECC-MCFFA NFPS National Forest Policy Statement Implementation Sub-committee (1995), *Broad Criteria for the Establishment of a Comprehensive, Adequate and Representative Forest Reserve System in Australia draft report July 1995.* Canberra.

Joint ANZECC-MCFFA NFPS National Forest Policy Statement Implementation Sub-committee (1995), *The Development of Consistent Nationwide Baseline Environmental Standards for Native Forests Draft Report July 1995.* Canberra.

Abbott, I., Van Heurck, P. and Wong, L. (1984), Responses to long-term Fire Exclusion: Physical, Chemical and Faunal Features of Litter and Soil in a Western Australian Forest. *Aust. For.* 47:237-242.

Andersen, A.N. (1985), *Seed Predation by Insects in Sclerophyllous Vegetation at Wilson's Promontory, Victoria, with particular reference to ants*. Unpubl. PhD Thesis, School of Botany, University of Melbourne. 241 pp.

Andersen, A.N. (1988), Immediate and Longerterm Effects of Fire on Seed Predation by Ants in Sclerophyllous Vegetation in South-eastern Australia. *Aust. J. Ecol. 13*:285-293.

Andersen, A.N. (1990), The Use of Ant Communities to Evaluate Change in Australian Terrestrial Ecosystems: A Review and a Recipe. *Proc. Ecol. Soc. Aust.* 16, 347- 357.

Andersen, A.N. and Ashton, D.H. (1985), Rates of Seed Removal by Ants at Heath and Woodland Sites in Southeastern Australia. *Aust.J.Ecol.* 10:381-390.

Andersen, A.N. and McKaige, M.E. (1987), Ant Communities at Rotamah Island Victoria, with particular reference to disturbance and *Rhytidoponera tasmaniensis*. *Proc. Roy. Soc. Vict 99*: 141-146.

Beattie, A.J. (ed.) (1995), *Biodiversity, Australia's Living Wealth*. Research Unit for Biodiversity and Bioresources, School of Biological Sciences, Macquarie University, Australia. 131pp.

Beattie, A.J. and Oliver, I. (1994), Taxonomic Minimilisation. *Trends in Ecology and Evolution* 9:488-490.

Benson, D.H. (1985), Maturation Periods for Fire Sensitive Shrub Species in Hawkesbury Sandstone Vegetation. *Cunninghamia 1*:339- 349.

Birk, E.M. and Bridges, R.G. (1989), Recurrent Fires and Fuel Accumulation in Even-aged Blackbutt (*Eucalyptus pilularis*) forests. *For. Ecol. and Manage.* 29:59-79.

Bornemisszа, G.F. (1969), The re-invasion of Burnt Woodland Areas by Insects and Mites. *Proc. Ecol. Soc. Aust. 4*:138.

Bradstock, R.A. and Myerscough, P.J. (1981), Fire Effects on Seed release and the Emergence and Establishment of Seedlings in *Banksia Ericifolia* L.f. *Aust. J. Bot. 29*:521-532.

Brian, M.V., Mountford, M.D., Abbot, A. and Vincent, S. (1976), The Changes in Ant Species Distribution during Ten Years Postfire Regeneration of a Heath. *J. Anim. Ecol.45*:115-133.

Burbidge, A.H., Leicester, K., McDavitt, S. and Majer, J.D. (1992), Ants as Indicators of Disturbance at Yanchep National Park, Western Australia. *J.Royal Soc.West.Aust.* 75:89-95.

Campbell, A.J. (1973), *The Effects of Prescribed Burning on Surface Active Invertebrate Fauna in Pine and Eucalypt Forest Within the Australian Capital Territory*. BSc. (Hons) Thesis. ANU. Canberra, ACT. Australia.

Campbell, A.J. and Tanton, M.T. (1981), Effect of Fire on the Invertebrate Fauna of Soil and Litter of a Eucalypt Forest. In: *Fire and the Australian Biota.* (Eds Gill, A.M., Groves, R.H. and Noble, I.R.). Australian Academy of Science, Canberra.

Cary, G.J. and Morrison, D.A. (1995), Effects of Fire Frequency on Plant Species Composition of Sandstone Communities in the Sydney Region: Combinations of Inter-fire Intervals. *Aust.J.Ecol. 20*:418-426.

Catling, PC. (1991), Ecological Effects of Prescribed Burning Practices on the Mammals of Southeastern Australia. In: *Conservation of Australia's Forest Fauna.* (ed Lunney, D.), pp. 353-63.

Royal Zoological Society of NSW, Mosman. Caughley, G. and Gunn, A. (1996), *Conservation Biology in Theory and Practice.* Blackwell Science, Massachusetts USA. 459 pp.

Christensen, N. (1985), Shrubland Fire Regimes and their Evolutionary Consequences. pp.86-99 in S.T.A. Pickett and P.S.White (Eds) *The Ecology of Natural Disturbance and Patch Dynamics.* Academic Press, Orlando, Florida.

Cornaby, B.W. (1977), Saprophagous Organisms and Problems in Applied Resource Partitioning. pp. 96-100 in Mattson, W.J. (ed) (1977) *The Role of Arthropods in Forest Ecosystems.* Springer-Verlag, New York.

Coy, R. (1996), The Impact of Fire on Soil Invertebrates in *E.regnans* forest at Powelltown, Victoria. In: *Fire and Biodiversity. The Effects and Effectiveness of Fire Management.* pp. 183-198. Biodiversity Series, Paper No. 8. Biodiversity Unit, Department of the Environment Sport and Territories. Canberra, Australia. CSIRO (1991) *The Insects of Australia.* Second Edition. Melbourne University Press. Dindal, D.L. and Metz, L.J. (1977) Community structure of Collembola affected by fire frequency. pp.88-95 in *The Role of Arthropods in Forest Ecosystems.* Mattson, W.J. (ed) Springer-Verlag, New York.

Dolva, G. (1993), *The Effect of Fire on the Ecology and Life-history of the Mood Cricket* Nambungia Balyarta *(Nemobiinae:Gryllidae).* Unpublished Masters Thesis, University of Western Australia.

Duncan, K.W. (1969), The Ecology of Two Species of Terrestrial Amphipoda (Crustaceae: Family Tallitridae) living in waste grassland. *Pedobiologia* 9:323-341.

Elmes, G.W. and Wardlaw, J.C. (1982), A Population Study of the Ants *Myrmica sabuletti* and *Myrmica scabrinodis*, living at two sites in the south of England. II. Effect of above-nest vegetation. *J.Anim.Ecol.* 51:665-680.

Engelmann, M.D. (1961), The Role of Soil Arthropods in the Energetics of an Old Field Community. *Ecol. Monogr. 31*:221-238.

ESDWG-Forest Use (1991), *Ecologically Sustainable Development Working Groups. Final Report - Forest Use.* Australian Government Publishing Service, Canberra. 227 pp.

Forestry Commission of NSW (1992), 1992-1995 Corporate Plan. For.Com.NSW, Sydney. 26 pp.

Fox, B.J. (1979), An Objective Method of Measuring the Vegetation Structure of Animal Habitats. *Aust.Wildl.Res.* 6:297-303.

Friend, G. (1996), Fire Ecology of Invertebrates- Implications for Nature Conservation, Fire Management and Future Research. In *Fire and Biodiversity. The Effects and Effectiveness of Fire Management*. pp.155-162.

Friend, J.A. and Richardson, A.M.M. (1986), Biology of Terrestrial Amphipods. *Ann. Rev. Entomol.* 31: 25-48.

Gill, A.M. (1981), Adaptive Responses of Australian Vascular Plant Species to Fires. In: *Fire and the Australian Biota* (eds. A.M. Gill, R.H. Groves and I.R. Noble), pp.243-272.

Aust. Acad. Sci., Canberra. Goldstein, E.L. (1975), Island Biogeography of Ants. *Evolution* 29:750-762.

Greenslade, P. and Greenslade, P.J.M. (1987), Ecological Strategies in Collembola: A New Approach to the use of Terrestrial Invertebrates in Environmental Assessment. In: B.R. Striganova (Ed.), *Soil Fauna and Soil fertility*, pp.245-252.

Nauka, Moscow Greenslade, P.J.M. (1984), Invertebrates and Environmental Assessment. *Environment and Planning* 3:13-15.

Greenslade, P.J.M. (1985), Pterygote Insects and the Soil: Their Diversity, their Effects on Soils and the Problem of Species Identification. *Quaest. Ent. 21*:571-585. 263

Bushfire and forest invertebrates Greenslade, P.J.M. and Mott, J.J. (1979), Ants of Native and sown Pastures in the Katherine Area, Northern Territory, Australia. (Hymenoptera:Formicidae). pp.153-156

Crosby, T.K. and Pottinger, R.P. (eds) proceedings of *The 2nd Australasian Conference on Grassland Invertebrate Ecology*. Palmerston North, New Zealand. 22-26th May, 1978. Wellington Govt. Printer. Greenslade, P.J.M. and Thompson, C.H. (1981) Ant distribution, vegetation and soil relationships in the Coolola-Noosa River area, Queensland. In: *Vegetation Classification in Australia* (eds A.N. Gillson & D.J. Anderson), pp. 192-207. CSIRO and Australian National University Press, Canberra.

Harvey, M.S. and Yen, A.L. (1989), *Worms to Wasps. An Illustrated Guide to Australia's Terrestrial Invertebrates*. Oxford University Press. 203 pp.

Hobbs, R.J. (1992), Function of Biodiversity in Mediterranean Ecosystems in Australia: Definitions and Background. In: *Biodiversity of Mediterranean Ecosystems in Australia* (ed. R.J. Hobbs), pp. 1-25.

Surrey Beatty & Sons. Holldobler, B. and Wilson, E.O. (1990), *The Ants*. Springer-Verlag, Berlin. 732 pp.

Huhta, V. (1971), Succession in the Spider communities of the Forest Floor after Clearcutting and Prescribed Burning. *Ann. Zool. Fennici. 8*:483-542.

Huhta, V.E., Karpinnen, E., Nurminen, M. and Valpus, A. (1967), Effects of Silvicultural Practices upon Arthropod Annelid and Nematode Populations in Coniferous Forest Soil. *Anneles Zoologie Fennici* 4:87-135.

Hurlbert, S.H. (1984), Pseudoreplication and the Design of Ecological Field Experiments. *Ecol. Monogr. 54*:187-211.

IUCN (1980), International Union for the Conservation of Nature, United Nations Environment Program (UNEP) and World Wildlife Fund (WWF) – World Conservation Strategy - Living Resource Conservation for Sustainable Development. IUCN, Gland, Switzerland.

Krebs, C.J. (1985), *Ecology. The Experimental Analysis of Distribution and Abundance*. Third Edition. Harper & Row, New York. 800pp.

Kremen, C. (1992), Assessing the Indicator Properties of Species Assemblages for Natural Areas Monitoring. *Ecological Applications* 2:203-217.

Lambert, M.J. (1982), *Methods for Chemical Analysis.* Forestry Commission of NSW. Sydney, Australia. 3rd Edition. Lawrence, J.F. and Britton, E.B. (1994) *Australian Beetles.* Melbourne University Press. 192pp.

Leonard, B.V. (1972), *The Effect of Fire upon Selected Small Mammals and Leaf Litter Fauna in Sclerophyll Forest in Southern Australia.* Unpubl. MSc. Thesis. Monash University, Victoria, Australia. Lowman, M.D. (1982) Seasonal variation in insect abundance among three Australian rain forests, with particular reference to phytophagous types. *Aust. J. Ecol. 7*:353-361.

MacArthur, R.H. and MacArthur, J.W. (1961), On Bird Species Diversity. *Ecology* 42:594-598. MacFadyen, A. (1962) Soil arthropod sampling. *Adv. Ecol. Res. 1*:1-34.

MacQuillan, P.B. (1985), The Identification of Root-feeding Cockchafer larvae (Coleoptera: Scarabaeidae) found in pastures in Tasmania. *Aust.J.Zool.* 33: 509-46.

Majer, J.D. (1980), Report on a Study of Invertebrates in relation to the Kojonup Nature Reserve fire management plan. *Bull. Dept. Biol. West. Aust. Inst. Tech.* 2:1-22.

Nedlands. Majer, J.D. (1985), Recolonisation by Ants of Rehabilitated Mineral Sand Mines on North Stradbroke Island, Queensland, with particular reference to seed removal. *Aust. J. Ecol. 10*:31-48.

Majer, J.D. (1992), Ant Recolonisation of Rehabilitated Bauxite Mines of Pocos de Caldas, Brazil. *J.Trop.Ecol.* 8:97-108.

Majer, J.D., Satori, M., Stone, R. and Perriman, W.S. (1982), Recolonisation by Ants and Other Invertebrates in Rehabilitated Mineral Sand Mines near Eneabba, Western Australia. *Reclaim. Reveg. Res. 1*:63-81.

Majer, J.D., Day, J.E., Kabay, E.D. and Perriman, W.S. (1984), Recolonisation by Ants in Bauxite Mines Rehabilitated by a Number of Different Methods. *J. Appl. Ecol. 21*:355-375. 264

Australia's Biodiveristy- Responses to Fire Margules, C.R., Nicholls, A.O. and Pressey, R.L. (1988) Selecting networks of reserves to maximise biological diversity. *Biological Conservation* 43:63-76.

Mattson, W.J. (ed) (1977) *The Role of Arthropods in Forest Ecosystems.* Springer-Verlag, New York. Mawson, P.R. (1986), A Comparative Study of Arachnid Communities in Rehabilitated Bauxite Mines. *W.A.I.T. School of Biology Bulletin No.14.*

McIver, J.D., Parsons, G.L. and Moldenke, A.R. (1992), Litter Spider Succession after Clearcutting in a Western Coniferous Forest. *Canad.J.For.Res.* 22:984-992.

Merrett, P. (1976), Changes in the Ground-living Spider Fauna after Heathland Fires in Dorset. *Bull. Br. Arachnol. Soc.* 3:214-221.

Metz, L.J. and Dindal, D.L. (1975) Collembola Populations and Prescribed Burning. *Environ. Entomol.* 4:583-587.

Metz, L.J. and Farrier, M.H. (1973), Prescribed Burning and Populations of Soil Mesofauna. *Environ. Entomol.* 2:433-440.

Morris, M.G. (1975), Preliminary Observations on the Effects of Burning on the Hemiptera (Homoptera and Auchenorhyncha) of limestone grassland. *Biol. Conserv.* 7:312-319.

Moulton, T.P. (1982), *The Effect of Prescribed Burning and Simulated Burning on Soil and Litter Arthropods in Open Forest at Cordeaux, N.S.W. Australia.* Unpubl. PhD. Thesis. Macquarie University.

Neumann, F.G. (1992), Effects of Low Intensity Prescribed Fire on Invertebrates in Mixed Eucalypt Forest- Wombat State Forest. In Tolhurst, K. And Flinn, D. (eds.) *Ecological Impacts of Fuel Reduction Burning in Dry Sclerophyll Forest: First Progress Report.* Conservation and Environment, Victoria, Australia.

Neumann, F.G. and Tolhurst, K. (1991), Effects of Fuel Reduction Burning on Epigeal Arthropods and Earthworms in Dry Sclerophyll Eucalypt Forest of West-central Victoria. *Aust. J. Ecol.* 16, 315-330.

New, T.R. (1984), *Insect Conservation: An Australian Perspective.* Dr. W.Junk Publishers, Netherlands. New, T.R. (1995) *Introduction to Invertebrate Conservation Biology.* Oxford University Press, New York. 194 pp.

NFPS (1992), *National Forest Policy Statement. A New Focus for Australia's Forests.* Commonwealth of Australia. Canberra. Nichols, O.G. and Bunn, S. (1980) Termite utilisation of rehabilitated bauxite mined areas. Alcoa of Australia Limited. *Environ. Res. Bull.No.9.*

Nieuwenhuis, A. (1987), The Effect of Fire Frequency on the Sclerophyll Vegetation of the West Head, New South Wales. *Aust.J.Ecol.* 12:373-385.

Noble, I.R. and Slatyer, R.O. (1981), Concepts and Models of Succession in Vascular Plant Communities Subject to Recurrent Fire. In *Fire and the Australian Biota* (Gill, A.M., Groves, R.H. and Noble, I.R. eds.), pp.311-315.

Aust.Acad.Sci., Canberra. NSESD (1992), National Strategy for Ecologically Sustainable Development. Commonwealth of Australia. 128 pp.

Oliver, I. (1995), *Rapid Biodiversity Assessment and its Application to Invertebrate Conservation in Production Forests*. Unpubl. PhD Thesis. School of Biological Sciences, Macquarie University, Australia. Oliver, I. and Beattie, A. J. (1993) A possible method for the rapid assessment of biodiversity. *Cons.Biol.* 7:562-568.

Oliver, I. and Beattie, A. J. (1996a), Invertebrate Morphospecies as Surrogates for Species: A Case Study. *Cons. Biol.* 10:99–109.

Oliver, I. and Beattie, A. J. (1996b), Designing a Cost-effective Invertebrate Survey: A Test of Methods for Rapid Biodiversity Assessment. *Ecological Applications* 6:594-607.

Palmer, M.W. (1993), Putting things in Even Better Order: The Advantages of Canonical Correspondence Analysis. *Ecology* 74:2215- 2230.

Pickett, S.T.A. and White, P.S. (Eds) (1985), *The Ecology of Natural Disturbance and Patch Dynamics*. Academic Press, Orlando, Florida.

Punttila, P., Haila, Y., Payunen, T and Tukia H. (1991), Colonisation of Clearcut Forests by Ants in the Southern Finnish Taiga: A Quantitative Survey. *Oikos* 61, 250-262.

Ridley, D. (1993), Fire Fuel Management in Eden State Forest District. pp. 91-101 in Ross, J. (Ed) *The Burning Question: Fire Management in NSW*. Proceedings of a Conference held at the Department of Continuing Education, University of New England, Armidale. 265

Bushfire and Forest Invertebrates Samways, M.J. (1994), *Insect Conservation Biology*. Chapman & Hall, London. 358 pp.

Seastedt, T.R. (1984), Below-ground Macroarthropods of Annually Burned and Unburned Tallgrass Prairie. *Am. Midl. Nat. 111*:405-408.

Pringett, J.A. (1976), The Effect of prescribed Burning on the Soil Fauna and Litter Decomposition in Western Australian forests. *Aust.J.Ecol.* 1:77-82.

Pringett, J.A. (1979), The Effects of a Single Hot Summer Fire on Soil Fauna and on Litter Decomposition in Jarrah (*Eucalyptus marginata*) forest in Western Australia. *Aust. J. Ecol.* 4:279-91.

FNSW (1995), *Forest Management Zoning in State Forests (Draft)*. State Forests of New South Wales. Sydney, Australia. Taylor, R. W. (1983) Systematics and ecology: reflections from the interface. pp.50-79

Highley and R.W. Taylor (eds) *Australian Systematic Entomology: A Bicentenary Perspective*. CSIRO, Melbourne. Ter Braak, C. J. F. (1986) Canonical Correspondence Analysis: a new eigenvector technique for multivariate direct gradient analysis. *Ecology* 67:1167-1179.

Turner, J. (1993), *Research and Ecologically Sustainable Forest Management. A Discussion Paper*. Research Division, State Forests of New South Wales. 26 pp.

Van der Drift, J. (1958), The Role of the Soil Fauna in the Decomposition of Forest Litter. pp.357- 360 in *Proc. 15th Intern. Congr. Zool.*

Van Loon, A.P. (1970), Further Investigations in the Effects of Prescribed Burning on Young Even-aged Blackbutt. Taree Forestry Office Research Report.

WCED (1987), World Commission on Environment and Development. *Our Common Future*. Oxford University Press, Oxford, UK. 400pp.

Whelan, R.J., Langedyk, W. and Pashby, A.S. (1980), The Effects of Wildfire on Arthropod Populations in a *Jarrah-Banksia* woodland. *W.A. Naturalist* 14:214-220.

Whelan, R.J. (1995), *The Ecology of Fire*. Cambridge University Press. 346 pp.

Williams, J.E. and Gill, A.M. (1995), *The Impact of Fire Regimes on Native Forests in Eastern New South Wales*. Environmental Heritage Monograph Series No. 2. NSW National Parks and Wildlife Service. 68 pp.

Woodward, F.I. (1993), How many Species are Required for a Functioning Ecosystem? In *Ecosystem Function of Biodiversity*. Eds. E.D.
Schultz and H.A. Mooney. Springer-Verlag, Heidelberg.

Yeatman, E.M. and Greenslade, P.J.M. (1980), Ants as Indicators of Habitat in Three Conservation Parks in South Australia. *S.Aust.Nat.* 55:20-26, 30.

Yen, A.L. (1987), A Preliminary Assessment of the Correlation between plant, vertebrate and Coleoptera communities in the Victorianmallee. In : Majer, J.D. (ed.), *The Role of Invertebrates in Conservation and Biological Survey* pp.73-88.

W.A. Dept. of Conservation and Land Management Report. York, A. (1989), *The Response of Ant Communities to Fire-induced Habitat Disturbance*. Unpublished PhD. Thesis, School of Biological Sciences, University of New South Wales.

York, A. (1993). Towards Ecologically Sustainable Forest Use. In *Rapid Biodiversity Assessment* (ed Beattie, A.J.), pp. 51-55. Proceedings of the Biodiversity Workshop Macquarie University-1993. Research Unit for Biodiversity and Bioresources, School of Biological Sciences, Macquarie University. 266

Australia's Biodiveristy - Responses to Fire York, A. (1994), The Long-term Effects of Fire on Forest Ant Communities: Management Implications for the Conservation of Biodiversity. *Memoirs of the Queensland Museum* 36, 231-239.

York, A. (1996), Long-term Effects of Fuel Reduction Burning on Invertebrates in a Dry sclerophyll Forest. In *Fire and Biodiversity. The Effects and Effectiveness of Fire Management.* pp. 163-181. Biodiversity Series, Paper No. 8. Biodiversity Unit, Department of the Environment Sport and Territories.

Canberra, Australia. Zborowski, P. and Storey, R. (1995), *A Field Guide to Insects in Australia.* Reed Books, Sydney. 207pp.

Zedler, P.H., Gautier, C.R. and McMaster, G.S. (1983) Vegetation change in response to extreme events: the effect of a short interval between fires in California chaparral.

Burne, R.V. (1991/1992), Lilliput's Castles: Stromatolites of Hamelin Pool, *Landscope.* 7(2):34-41.

Oran, G.F. and Hopper, S.D. (1987), Conservation of the Genetic Resources of Rare and Widespread Eucalypts in Remnant Vegetation. In: Saunders, D.A. *et al* (eds.). *Nature Conservation: The Role of Remnants of Native Vegetation.* Surrey Beatty & Sons Pty Ltd and others, Norton, NSW.

Orld Conservation Monitoring Centre (WCMC) (comp.). (1992). *Global Biodiversity: Status of the Earth's Living Resources.* In collaboration with the Nat. History Museum, London; in assoc. with IUCN, UNEP, WWF & WRI, Chapman & Hall, London. pp. 1-3.

Orld Conservation Monitoring Centre (WCMC) (comp.). (1992), *Global Biodiversity Status of the Earth's Living Resources.* pg. xiii.

Magurran, A.E. (1988), *Ecological Diversity and Its Measurement.* Croom-Helm, London.

Melbourne, B. (1993), Insect Biodiversity and Its Assessment in Terrestrial Ecosystems: A Review of the Literature. Unpublished paper. Division of Botany and Zoology, The Australian National University.

Faith, D. (1992), Conservation Evaluation and Phylogenetic Diversity. *Biological Conservation* 61:1-10.

World Conservation Monitoring Centre (WCMC) (comp.). (1992), *Global Biodiversity Status of the Earth's Living Resources.* pp.11-12.

World Conservation Monitoring Centre (WCMC) (comp.). (1992), *Global Biodiversity Status of the Earth's Living Resources.* pp.xiii,17.

Nielsen, E.S. and Halliday, B. (1993), CSIRO. Personal Communication.

Based on Margulis, L. and Schwartz, K.V. (1988), *Five Kingdoms–An Illustrated Guide to the Phyla of Life on Earth. Second Edition.* W.H. Freeman and Company, New York NY.

Grassle, J.F., Lasserre, P.A., McIntyre, A.D., and Ray, G.C. (1991), Marine Biodiversity and Ecosystem Function. *Biol. Int.* Special issue 23: i-iv, 1-19. Cited in Ray, G.C. and Grassle, J.F. (1991). Marine Biological Diversity. *BioScience* 41 (7): 453-457.

Polunin, N.U.C. (1983), Marine "genetic" Resources and the Potential Role of Protected Areas in Conserving this. *Environmental Conservation* 10(1):31-41.

Read, Sturgess and Associates (1982), Evaluation of Economic Values of Wood and Water for the Thomson Catchment. Report prepared for Melbourne Water.

Attiwill, P.M. and Leeper, G.W. (1987), *Forest Soils and Nutrient Cycles,* Melbourne University Press, Melbourne.]

Based on World Conservation Monitoring Centre (WCMC) (comp.). (1992), *Global Biodiversity Status of the Earth's Living Resources.* pp.47-54.

Margulis, L. and Schwartz, K.V. (1988), *Five Kingdoms – An Illustrated Guide to the Phyla of Life on Earth. Second Edition.* W.H. Freeman and Company, New York NY.

Margulis, L. and Fester, R. (eds.) (1991), *Symbiosis as a Source of Evolutionary Innovation,* Massachusetts Institute of Technology Press, Cambridge, Mass.

Read, D.J. (1991), Mycorrhizas in Ecosystems–Nature's Response to the "Law of the Minimum". In: Hawksworth, D.L. (ed.), *Frontiers in Mycology.* CAB International, Wallingford. pp. 101-130.

Anderson, R.A. (1992), The Diversity of Eukaryotic Algae. *Biodiversity and Conservation.* 1(4):267-292.

Round, F.E. (1981), *The Ecology of Algae.* Cambridge University Press, New York.

Lal, R. (1991), Soil Conservation and Biodiversity. In: Hawksworth, D.L. (ed.), *The Biodiversity of Microorganisms and Invertebrates: its role in sustainable agriculture.* CAB International, Wallingford. pp. 89-104.

McKinnell, F.H., Hopkins, E.R. and Fox, J.E.D. (eds). (1991), *Forest Management in Australia.* Surrey Beatty and Sons Pty Ltd, Chipping Norton, NSW.

Plotkin, M.J. (1988), The Outlook for New Agricultural and Industrial Products from the Tropics. In: E.O. Wilson (ed) *Biodiversity.* National Academy Press, Washington DC. pp.106-116.

Reid, W.V. and Miller, K.R. (1989), *Keeping Options Alive: The Scientific Basis for Conserving Biological Diversity.* World Resources Institute, Washington DC.

House, A.P.N. and Harwood, C.E. (eds) (1992), *Australian Dry-Zone Acacias for Human Food. Proceedings of a workshop held at Glen Helen NT, 7-10 August 1991.* CSIRO, Canberra

Ralph, W. (1993), From Bush Tucker to Beans. *Ecos* 75 (Autumn): 14-18

Brand, J.C. and Cherikoff, V. (1985), Australian Aboriginal Bush Foods: The Nutritional Composition of Plants from Arid and Semi-arid Areas. *Australian Aboriginal Studies* 2:38-46.

Isaacs, J. (1987), *Bush Food.* Weldons Pty Ltd., Sydney. 30. Latz, P.K. (1982). *Bushfires and bushtucker: Aborigines and plants in Central Australia.* Unpublished MA thesis, University of New England, Armidale, NSW.

Reid, E.J. and Betts, T.J. (1979), Records of Western Australian Plants used by Aboriginals as Medicinal Agents. *Planta Medica* 36: 164-173.

Kalotas, A.C. (1993), Recording and Applying Aboriginal Botanical Knowledge in Western Australia: Some recent examples and future prospects. In: Williams, N.M. and Baines, G. (eds). *Traditional Ecological Knowledge: Wisdom for Sustainable Development.* Based on the Traditional Ecological Knowledge Workshop. CRES, ANU, 18-29 April 1988. CRES ANU, Canberra.

Low, T. (1987), Cures from the Canopy. *Australian Natural History* 22(4):185-188.

Meier, L. and Figgis, P. (eds). (1985), Rainforests of Australia. Weldon Pty Ltd, Sydney.

Beattie, A. (1993), Macquarie University. Personal Communication.

Tyler, M.J. (1989), *Australian Frogs.* Viking O'Neil, Melbourne.

Olsen, H.F. (1977), Great Sandy Strait fisheries habitat reserves. In: P. Lauer (ed) Occasional Papers in Anthropology No.8, 223-230. University of Qld, Australia.

Ivanovici A.M. (ed). (1983), *Inventory of Declared Marine and Estuarine Protected Areas in Australian Waters.* ANPWS, Canberra.

Iltis, H.H., (1988), Serendipity in the Exploration of Biodiversity: What Good are Weedy Tomatoes? In: E.O. Wilson (ed) *Biodiversity.*

Beattie, A. (1991), Biodiversity and Bioresources–The Forgotten McNeely, J.A. (1988). *Economics and Biological Diversity.* IUCN, Gland, Switzerland.

Ecologically Sustainable Development (ESD) Working Group on Tourism (1991), *Ecologically Sustainable Development Working Groups: Final Report–Tourism.* AGPS, Canberra.

DASETT (1991), Environmental Tourism. Draft Internal Document, Tourism Policy Branch. DASETT, Canberra.

Coombs, H.C., Brandl, M.M. and Snowdon, W.E. (1983), *A Certain Heritage.* CRES Monograph No.9. ANU, Canberra.

CSIRO (1990), Australia's Environment and Its Natural Resources: an outlook. CSIRO, Canberra. pg.14.

CSIRO (1990), Australia's Environment and Its Natural Resources: an outlook. CSIRO, Canberra. pg.2

Murray-Darling Basin Ministerial Council (MDBMC). (1990). *Natural Resources Management Strategy*, MDBMC, Canberra.

Hawke, R.J.L. (1989), *Our Country, Our Future*, Statement on the Environment by the Hon RJL Hawke, AC. Prime Minister of Australia. AGPS, Canberra.